# Computational Nanotechnology Using Finite Difference Time Domain

Edited by
Sarhan M. Musa

# Computational Nanotechnology Using Finite Difference Time Domain

CRC Press
Taylor & Francis Group
Boca Raton  London  New York

CRC Press is an imprint of the
Taylor & Francis Group, an **informa** business

CRC Press
Taylor & Francis Group
6000 Broken Sound Parkway NW, Suite 300
Boca Raton, FL 33487-2742

First issued in paperback 2017

Version Date: 20130626

ISBN 13: 978-1-4665-8361-0 (hbk)
ISBN 13: 978-1-138-07346-3 (pbk)

---

**Library of Congress Cataloging-in-Publication Data**

---

Computational nanotechnology using finite difference time domain / Sarhan M. Musa, editor.
    pages cm
    Summary: "Written in a manner that is easily digestible to beginners and useful to seasoned professionals, this book introduces the key concepts of computational Finite Difference Time Domain (FDTD) method used in nanotechnology. It covers future applications of nanotechnology in technical industry as well as new developments and interdisciplinary research in engineering, science, and medicine. It includes an overview of computational nanotechnologies using FDTD method and describes the technologies with an emphasis on how they work and their key benefits. "-- Provided by publisher.
    Includes bibliographical references and index.
    ISBN 978-1-4665-8361-0 (hardback)
    1. Nanophotonics--Data processing. 2. Finite differences--Data processing. 3. Time-domain analysis. I. Musa, Sarhan M., editor of compilation. II. Title: Computational nanotechnology using FDTD.

TA1530.C67 2013
620'.501515353--dc23
                                         2013023423

---

**Visit the Taylor & Francis Web site at**
**http://www.taylorandfrancis.com**

**and the CRC Press Web site at**
**http://www.crcpress.com**

*Dedicated to my late father, Mahmoud, and my late sisters Majida and Rasmia.*

# Contents

# Contents

# *Preface*

Today, nanotechnology and nanoscience are emerging as leading fields in the technological revolution of the new millennium. Research and development in nanotechnology and nanoscience are fast-growing industries worldwide. Nanotechnology involves interdisciplinary research such as engineering, physics, chemistry, biology, and medicine. It entails engaging with matter that is smaller than 100 nanometers and that has applications in the real world. Therefore, the ability to manipulate the atoms at nanoscale helps scientists create tiny devices with diameters of only a few atoms for many applications such as consumer supplies, electronics and computers (in the development of molecular-size transistors for microcomputer memory), information and biotechnology, aerospace, energy, optics, the environment, improvement of drug delivery in identifying cancer cells in medicine, energy production, and so on. Using nanotechnology brings vast benefits to industry such as reduced labor requirements and lower cost. This book will be used as a tool for engineers, scientists, and biologists involved in the computational technique using the finite-difference time-domain (FDTD) method in nanotechnology. The FDTD method and its advances support the improvement of computational nanotechnology applications. The FDTD method is an essential tool in modeling inhomogeneous, anisotropic, and dispersive media with random, multilayered, and periodic fundamental (or device) nanostructures due to its features of extreme flexibility and easy implementation. For example, there is interest in studying nanophotonics applications and design optimization using the FDTD method. Furthermore, FDTD studies of electromagnetic field enhancement in silver nano cylinders arranged in triangle geometry have been reported. In fact, researchers have demonstrated a technique of the FDTD method for modeling and simulation of nanointerconnects for nanophotonics and for studying optical properties of nanoshell dimers.

The challenges of the FDTD method continue to attract researchers around the world and have led to many new discoveries concerning the guided modes in nanoplasmonics waveguides. For example, the numerical simulation of plasmonic waveguide formed by seven elements shows that rows of silver nanoscale cylinders can guide the propagation of light due to the coupling of surface plasmons.

There is no doubt that the FDTD method shines light onto a number of different disciplines, including engineering, physics, chemistry, biology, medicine, and aerospace. Therefore, the study of the computational FDTD method at the nanometer scale is encouraged by the rapid progress in nanotechnology and nanoscience.

This book comprises seven chapters and three appendices.

Chapter 1 presents the FDTD method for photonics and the applications of FDTD in nanophotonics. Various examples are provided to cover a wide range of applications in the nanophotonic field in order to show the usefulness of the FDTD method in the area of nano-optics. The author proposes an efficient approach for obtaining second-order accurate sensitivities using the adjoint variable method applied on FDTD. This technique is utilized to obtain different responses. It has also been applied to calculate the sensitivity of the power reflectivity using no additional simulations that use a self-adjoint version of this approach. The sensitivity approach is accurate and efficient and proves an efficient method for design optimization of various photonics and nanophotonics structures.

Chapter 2 provides an overview of the recent developments, basic theories, core techniques, real-world applications, and future directions for the FDTD method. The authors review the unified theoretical frameworks of the FDTD method and its advances (Runge Kutta-FDTD, symplectic FDTD, alternative direction implicit-FDTD, high-order FDTD, multiresolution-TD, pseudospectral-TD, etc.) for solving Maxwell's equations systematically. In addition, they briefly describe core techniques of the FDTD method (involving the basic update equations, material averaging technique, perfectly matched layer, source excitation, near-to-far-field transformation, periodic boundary condition, and treatment of dispersive media) and particularly focus on those for nano-optics applications. Indeed, the authors demonstrate the powerful capabilities of the FDTD method to model versatile physical problems in the nano-optics field. The chapter discusses the case studies on plasmonic solar cells, nanoantennas, spontaneous emissions, and metamaterials with detailed physical understandings. Then, the authors present the numerical analyses and implementations of the FDTD method to simulate the Schrödinger equation. Also, they show several simple examples on the numerical solution of quantum physics problems with the FDTD method.

Chapter 3 presents the modeling of optical metamaterials using the FDTD method. The structures considered in the chapter are the optical cloaking structures for reducing scattering cross sections, the hyperlens structures for simultaneous imaging and magnification of subwavelength source distributions, and nanoplasmonic waveguides for enhancing light transmissions.

Chapter 4 presents a parallel FDTD model incorporating a four-level atomic system. The authors include detailed implication and efficiency of the method. Also, they use a one-dimensional gain slab to check the validity of the method. Indeed, the authors simulate three-dimensional fishnet and SRR structures with gain material underneath to demonstrate the strong interaction between structures and gain materials. Their model can be used as a guide for real pump-probe experiments in metamaterials and provide an insight into the dynamic interaction between nanostructure and gain materials. A significant advantage of their method is the complexity of the FDTD, which scales only linearly with time and space. However, the FDTD method is very demanding in terms of memory and speed of the available computer hardware when applied to practical 3D problems such as analysis of transmission spectra of photonic crystals waveguide in layered structures with 2D patterning. Thus, one can turn to the parallel of FDTD method.

Chapter 5 uses the three-dimensional FDTD method and Maxwell stress tensor to set up a simulation model and calculate the trapping force. By using VSWFs the transmission fields are obtained with linearly polarized and radially polarized beams in the vicinity of the focus point. The authors simulate and compare the trapping capacities on nanoscale-diameter nanowires based on both the linearly polarized beam and the radially polarized beam with each other. When the radially polarized beam is adopted for lower refractive index nanowires, the multiple trapping equilibrium positions beyond the focal plane exist. With the increase of the refractive indices of nanowires, the axial and radial forces of the radially polarized beam both increase greatly. They compare the radially polarized beam with the linearly polarized beam, demonstrating higher trapping efficiency on the higher refractive index nanowire. The authors show that the radially polarized beam is suitable for trapping those higher refractive index nanowires.

Chapter 6 addresses a detailed FDTD analysis of light absorption enhancement dependence on the groove shape of nanogratings etched into the surfaces of metal-semiconductor-metal photodetectors (MSM-PDs) structure. The author analyzes the light absorption enhancement dependence on the nanogratings groove shape into the surface of the MSM-PDs. The FDTD simulation results show that it is possible to obtain about 50 times better light absorption enhancement prediction for about 850 nm light due to improved optical signal propagation through the nanogratings in comparison to the conventional MSM-PD designs employing only the subwavelength apertures. The author also presents the nanograting groove shapes obtained typically in experiments using focused ion beam lithography and discusses the dependency of the light absorption enhancement on the geometric parameters of the nanogratings inscribed into the MSM-PDs. In addition, the author analyzes the nanograting phase-shift and groove profiles obtained using focused ion beam milling and atomic force microscopy and discusses the dependency of light absorption enhancement on the nanogratings phase-shift and groove profiles inscribed into MSM-PDs. The FDTD simulation results also show that the nanograting phase-shift red-shifts the wavelength at which the light absorption enhancement is maximum and that the combined effects of the nanograting groove shape and phase-shift degrade the light absorption enhancement by up to 50%.

Chapter 7 presents the applications of the FDTD method in medicine and the biomedical field.

Finally, the book concludes with appendices. Appendix A provides common material and physical constants, with the considerations that the material constant values varied from one published source to another due to many varieties of most materials and conductivity is sensitive to temperature, impurities, moisture content, and the dependence of relative permittivity and permeability on temperature, humidity, and the like. Appendix B provides equations for photon energy, frequency, and wavelength and electromagnetic spectrum including the approximation of common optical wavelength ranges of light. In addition, it provides a figure for wavelengths of commercially available lasers. Appendix C provides common symbols and useful mathematical formulas.

**Sarhan M. Musa**

# Acknowledgments

My sincere appreciation and gratitude go out to all the book's contributors. Thank you to Brian Gaskin and James Gaskin for their wonderful hearts and for being great American neighbors. It is my pleasure to acknowledge the outstanding help and support of the team at Taylor & Francis Group/CRC Press in preparing this book, especially from Nora Konopka, Michele Smith, Kari Budyk, and Charlotte Byrnes.

I wish to thank Dr. Kendall Harris, my college dean, for his constant support. Finally, the book would never have seen the light of day if not for the constant support, love, and patience of my family.

# *Editor*

**Sarhan M. Musa**, PhD, is currently an associate professor in the Department of Engineering Technology at Prairie View A&M University, Texas. He has been director of Prairie View Networking Academy, Texas, since 2004. Dr. Musa has published more than 100 papers in peer-reviewed journals and conferences. He is a frequent invited speaker on computational nanotechnology, has consulted for multiple organizations nationally and internationally, and has written and edited several books. Dr. Musa is a senior member of the Institute of  Electrical and Electronics Engineers (IEEE) and is also a LTD Sprint and a Boeing Welliver Fellow.

# Contributors

**Narottam Das**
Department of Electrical and
    Computer Engineering
Curtin University
Perth, Australia
and
School of Engineering
Edith Cowan University
Perth, Australia

**Zhixiang Huang**
The Key Laboratory of Intelligent
    Computing and Signal
    Processing
Ministry of Education
Anhui University
Hefei, China

**Jing Li**
Department of Precision Machinery
    and Precision Instrumentation
University of Science and
    Technology of China
Hefei, China

**Wei E. I. Sha**
Department of Electrical and
    Electronic Engineering
The University of Hong Kong
Pokfulam Road, Hong Kong

**Mohammed A. Swillam**
School of Science and Engineering
American University in Cairo
New Cairo, Egypt

**Viroj Wiwanitkit**
(Retired)
Hainan Medical University
Haikou, Hainan, China
and
University of Niš
Niš, Serbia
and
Joseph Ayo Babalola University
Ikeji-Arakeji, Osun, Nigeria

**Bo Wu**
The Key Laboratory of Intelligent
    Computing and Signal
    Processing
Ministry of Education
Anhui University
Hefei, China

**Xianliang Wu**
The Key Laboratory of Intelligent
    Computing and Signal
    Processing
Ministry of Education
Anhui University
Hefei, China

**Xiaoyan Y. Z. Xiong**
Department of Electrical and
    Electronic Engineering
The University of Hong Kong
Pokfulam Road, Hong Kong

**Xiaoping Wu**
Department of Modern
    Mechanics
University of Science and
    Technology of China
Hefei, China

**Yan Zhao**
International School of Engineering
Chulalongkon University
Pathumwan, Bangkok, Thailand

**Chunli Zhu**
Department of Precision Machinery
    and Precision Instrumentation
University of Science and
    Technology of China
Hefei, China

# 1

# Finite-Difference Time-Domain Method in Photonics and Nanophotonics

**Mohamed A. Swillam**

*School of Science and Engineering, American University in Cairo (AUC), New Cairo, Egypt*

## CONTENTS

## 1.1 Introduction

In this chapter, the solution of Maxwell's equations using the finite-difference time-domain (FDTD) method for photonics and nanophotonics is presented; dispersive and lossy as well as nonlinear media properties are included in the solution. The new trends and applications of this method in novel nanostructures are presented. These novel applications include solar cell, plasmonics devices, silicon photonics, and nanoparticles.

The newly presented theory for efficient sensitivity analysis using the FDTD method is also discussed in detail.

## 1.2 Finite Difference Time Domain (FDTD)

FDTD is a direct solution of the time-dependent differential form of Maxwell's equations. FDTD is mainly a discretization of Maxwell's equations in time domain using the central differences technique. This method was introduced by Yee [1] in 1966 for microwave applications. It was later utilized for modeling optical structures [2]. Here, the Cartesian discretization of the electric field and magnetic field based on Yee's algorithm is presented [2].

The electric field update equations [2] are given by

$$E_x\big|_{i,j,k}^{n+1} = C_a(i,j,k).E_x\big|_{i,j,k}^{n}$$

$$+ C_b(i,j,k).\left[ \frac{H_z\big|_{i,j,k}^{n+0.5} - H_z\big|_{i,j-1,k}^{n+0.5}}{\Delta y} - \frac{H_y\big|_{i,j,k}^{n+0.5} - H_y\big|_{i,j,k-1}^{n+0.5}}{\Delta z} - J_{ex}\big|_{i,j,k}^{n+0.5} \right] \quad (1.1)$$

$$E_y\big|_{i,j,k}^{n+1} = C_a(i,j,k).E_y\big|_{i,j,k}^{n}$$

$$+ C_b(i,j,k).\left[ \frac{H_x\big|_{i,j,k}^{n+0.5} - H_x\big|_{i,j,k-1}^{n+0.5}}{\Delta z} - \frac{H_z\big|_{i,j,k}^{n+0.5} - H_y\big|_{i-1,j,k}^{n+0.5}}{\Delta x} - J_{ey}\big|_{i,j,k}^{n+0.5} \right] \quad (1.2)$$

$$E_z\big|_{i,j,k}^{n+1} = C_a(i,j,k).E_z\big|_{i,j,k}^{n}$$

$$+ C_b(i,j,k).\left[ \frac{H_y\big|_{i,j,k}^{n+0.5} - H_y\big|_{i-1,j,k-1}^{n+0.5}}{\Delta x} - \frac{H_x\big|_{i,j,k}^{n+0.5} - H_x\big|_{i,j-1,k}^{n+0.5}}{\Delta y} - J_{ez}\big|_{i,j,k}^{n+0.5} \right] \quad (1.3)$$

where

$$C_a(i,j,k) = \frac{1 - \dfrac{\sigma(i,j,k)\Delta t}{2\varepsilon(i,j,k)}}{1 + \dfrac{\sigma(i,j,k)\Delta t}{2\varepsilon(i,j,k)}}, \quad \text{and} \quad C_b(i,j,k) = \frac{\dfrac{\Delta t}{\varepsilon(i,j,k)}}{1 + \dfrac{\sigma(i,j,k)\Delta t}{2\varepsilon(i,j,k)}} \quad (1.4)$$

where the indices $n$, $i$, $j$, and $k$ represent the discretization in time, $x$-direction, $y$-direction, and $z$-direction, respectively. In (1.1)–(1.3), $E$, $H$, and $J$ are the electric field, the magnetic field, and the electric current, respectively. σ and ε in (1.4) are the electric conductivity and the permittivity, respectively.

The update equations of the magnetic field are given by

$$
H_x\big|_{i,j,k}^{n+0.5} = D_a(i,j,k) \cdot H_x\big|_{i,j,k}^{n-.05} + D_b(i,j,k) \cdot \left[ \frac{E_z\big|_{i,j,k}^{n} - H_z\big|_{i,j+1,k}^{n}}{\Delta y} - \frac{E_y\big|_{i,j,k}^{n} - E_y\big|_{i,j,k+1}^{n}}{\Delta z} \right]
$$

(1.5)

$$
H_y\big|_{i,j,k}^{n+0.5} = D_a(i,j,k) \cdot H_x\big|_{i,j,k}^{n-0.5} + D_b(i,j,k) \cdot \left[ \frac{E_x\big|_{i,j,k}^{n} - E_x\big|_{i,j,k+1}^{n}}{\Delta z} - \frac{E_z\big|_{i,j,k}^{n} - E_z\big|_{i+1,j,k}^{n}}{\Delta x} \right]
$$

(1.6)

$$
H_z\big|_{i,j,k}^{n+0.5} = D_a(i,j,k) \cdot H_x\big|_{i,j,k}^{n-0.5} + D_b(i,j,k) \cdot \left[ \frac{E_y\big|_{i,j,k}^{n} - E_x\big|_{i,j+1,k}^{n}}{\Delta x} - \frac{E_x\big|_{i,j+1,k}^{n} - E_x\big|_{i,j,k}^{n}}{\Delta y} \right]
$$

(1.7)

where

$$
D_a(i,j,k) = \frac{1 - \dfrac{\sigma_m(i,j,k)\Delta t}{2\mu(i,j,k)}}{1 + \dfrac{\sigma_m(i,j,k)\Delta t}{2\mu(i,j,k)}}, \quad \text{and} \quad D_b(i,j,k) = \frac{\dfrac{\Delta t}{\mu(i,j,k)}}{1 + \dfrac{\sigma_m(i,j,k)\Delta t}{2\mu(i,j,k)}}
$$

(1.8)

where $\sigma_m$ and $\mu$ in (1.8) are the magnetic conductivity and the permeability, respectively.

As shown from the update equations, the electric field and the magnetic field components are located at alternative half space steps as shown in Figure 1.1. They are also updated at alternative half time step. This approach is called the leap-frog approach.

FDTD is a very rigorous technique and can solve a huge number of optical structures. However, this technique is slow for photonic structures. This is mainly due to the fact that most of the photonic structures are electrically large. FDTD also consumes huge computational resources. Thus most of the applications that can be handled on a personal computer are limited to the two-dimensional (2D) version of this approach. Complex three-dimensional (3D) structures are usually simulated using parallel computational techniques to provide the necessary processing power and memory requirements of this technique.

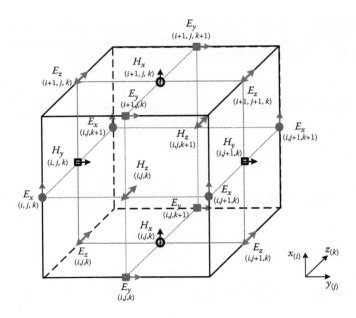

**FIGURE 1.1**
Yee's cell.

The stability of Yee's cell-based FDTD is mainly governed by the Courant relationship [2]. This relationship imposes a constraint on the time step of the stable algorithm to satisfy the following relationship

$$\Delta t \leq \frac{1}{C\sqrt{\dfrac{1}{(\Delta x)^2} + \dfrac{1}{(\Delta y)^2} + \dfrac{1}{(\Delta z)^2}}} \tag{1.9}$$

where $C$ is the velocity of the light in the modeled medium.

As shown from the above stability condition, the time step of the FDTD algorithm reduces with the reduction of the spatial grid size. As a result, for structures with fine details this algorithm takes longer to converge to a stable solution. This is one of the major drawbacks of the FDTD technique.

## 1.3 FDTD for Photonic Material

In general, photonic devices are considered open structures. An effective absorbing boundary condition (ABC) is thus needed in the implemented algorithm to avoid any reflection of the field from the edges of the finite numerical window. In general, the absorbing boundary should be a few cells only outside the computational domain. In the ideal case, the absorbing medium should be reflectionless to all impinging waves over their full frequency spectrum, highly absorbing, and effective in the near field of a source. In 1994, Berenger introduced a highly effective absorbing material called the perfectly matched layer (PML) [3]. It is based on matching the incident wave with arbitrary frequency and polarization at the boundary. Many versions and modifications were later introduced to Berenger's PML, such as stretched coordinate PML [4] and uniaxial PML [5]. These modifications have an effective impact on reducing the implementation efforts. They also make the update of the field components inside the absorbing medium similar to those inside the regular computational region.

In addition, most of the optical materials are dispersive (the refractive index has wavelength dependency). Thus, the typical algorithm of the FDTD has to be modified to account for the wavelength or frequency dependency of the refractive index.

This dispersion characteristic becomes crucial in the recent applications of nanophotonics, especially the ones that involve metals. The metals at optical frequencies are treated as dispersive materials. Noble metals such as gold, silver, aluminum, and copper have plasma frequencies that allow for creating surface plasmon polariton on the metal interface with dielectric or air [6]. FDTD is usually used to model such structure using modified update equations that take into account the time dependence of the dielectric constant. Techniques such as auxiliary differential equations (ADE) [2], piecewise linear recursive convolution method [2], and Z-transform technique are usually utilized for modeling the dispersive material using FDTD [7].

In the last few decades, FDTD proved to be an accurate tool for modeling electromagnetic devices in general and more specifically microwave structures. Even though FDTD is considered as an accurate approach for modeling photonic devices, its application was limited due to huge required resources. This is mainly due to the fact that the conventional photonic devices have very long electrical length (geometrical length per wavelength); for example, weakly guiding structures can be more than 1000 λ. As explained previously, FDTD has a maximum step to ensure stable performance. Thus simulating such long structure using FDTD is highly demanding. On the other hand, other efficient techniques such as beam propagation method [8] are considered as a fast modeling approach for electrically long structure. Thus, FDTD is considered as an inefficient method for modeling such structures.

In recent decades, with the emerging technology of nanophotonics, the structure size starts to shrink to be comparable to or smaller than the wavelength. With this small size and high field localization, FDTD is considered to be the most suitable tool for modeling these structures. The ability to model the surface plasmon wave of nanostructures and metamaterials is considered one of the main advantages of using FDTD in nanophotonics due to the limited availability of other accurate techniques. For such structures, the electric size is on the order of one to few multiples of the wavelength, which makes the computational resources for modeling these nanostructures using FDTD affordable using the recent advancement in technology.

FDTD provide an accurate technique for modeling nanophotonic devices. It has also been exploited in understanding various novel effects that have been recently developed in the area of nano-optics and photonics. Various commercial tools have been recently widely used to model such structures; among them, OPTI-FDTD [9] is considered one of the mature software tools and has been well known for a decade. Lumerical [10] is also fast growing with various applications and a friendly graphical user interface.

## 1.4 Recent Applications of FDTD in Nanophotonics

The power of FDTD has recently been demonstrated in accurate modeling of nanophotonic devices where other techniques fail either to obtain good accuracy or to provide the solution efficiently. In this section some of the novel applications are discussed and the ability of FDTD to accurately model the response is clearly demonstrated.

### 1.4.1 Application of FDTD in Plasmonic Devices

Plasmonic devices attracted attention in the last few decades due to their unique ability to guide the light in nanoscale confinement. This unique ability allows for a wide range of novel applications including on-chip devices [11] and biosensing [12]. Various plasmonic waveguide configurations have been proposed and studied. Among these structures, metal-insulator-metal (MIM) and insulator-metal-insulator (IMI) are widely used. FDTD has been utilized to understand and analyze the characteristics of these plasmonic waveguides. The guided light inside MIM waveguide is simulated using FDTD simulations as shown in Figure 1.2. The main component of the modal field for the MIM is the component normal to the metal-insulator interface (i.e., Ex in Figure 1.2). From this figure, it is clear that the field is confined inside the insulator slot. This slot can be a few nanometers only. This subwavelength confinement is a unique characteristic for such waveguides. This makes this waveguide a good candidate for high-density integration and various on chip sensing applications. On the other hand, the IMI waveguide has two modes as shown in Figure 1.3. The fundamental mode is an asymmetric mode with relatively high losses. The first-order mode is symmetric and has very low loss especially for small metal thickness. This mode is known for its ability to propagate over long distance and is called long-range surface plasmon (LRSPP) [13]. This mode has large spot size and can be optically excited by end fire coupling from optical fiber. It does not provide subwavelength confinement for the

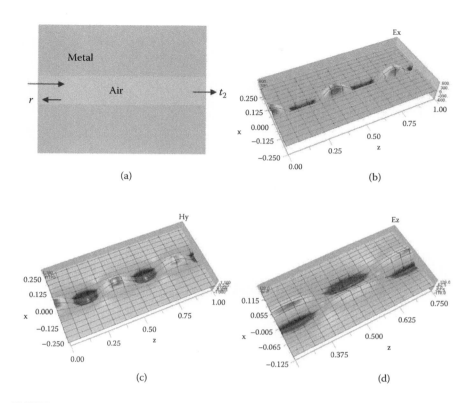

**FIGURE 1.2**
The propagation field inside a plasmonic slot for the different field components. (See color figure.)

modal field and requires symmetric material around the metal strip, which sometimes is challenging to attain practically.

FDTD has also been utilized to explore the properties and behavior of various plasmonic components based on the MIM configuration. For example, the ability of the MIM to bend the light over 90° has been demonstrated and understood through FDTD simulations [14]. This ability is unique for MIM plasmonic structures and is shown in Figure 1.4.

FDTD has also been utilized to understand the behavior of $T$ and $X$ junctions made using MIM waveguide [15, 16]. These junctions have the ability to couple a significant portion of the input field to the orthogonal arms, which is a very attractive characteristic that is unique for such MIM configuration. Thus, this platform can be exploited to propose highly compact structures.

Based on this configuration, various plasmonic devices have been recently proposed and modeled using the FDTD method. For example, a simple structure with feedback path can have various functionalities [15]. The structure is shown in Figure 1.6. By controlling the transmitted light through the

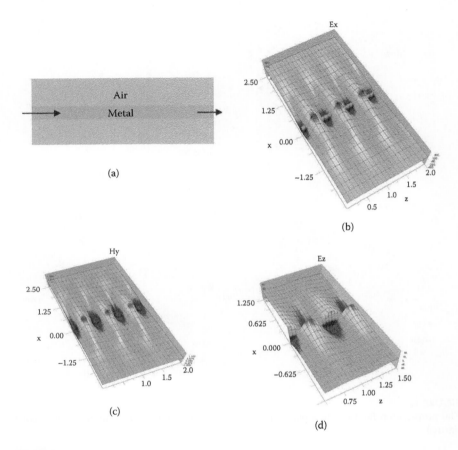

**FIGURE 1.3**
The propagation field inside an IMI for the symmetric mode (b,c,d) for different field components and for the asymmetric mode (e,f,g). (See color figure.) (*Continued*)

feedback U-shaped path various filter responses can be obtained. For example, a wide-band Lorentzian-like response is obtained as shown in Figure 1.7.

In this structure, the slot width of the input and access waveguide, $d_1$, is 30 nm, and $d_2 = d_3 = 60$ nm. The length and width of the feedback path $L_1$ and $L_2$ are given as 375 nm and 700 nm, respectively. The field intensity inside the structure is plotted using FDTD at the peak transmission wavelength of 950 nm and at the minimum transmission wavelength of 1250 nm as shown in Figure 1.8. From the figure, we can conclude that the optical field inside the feedback path interferes with the forward field. This interference can be constructive or destructive and hence determines the amount of the power at the output port.

Controlling the amount of the light coupled out from the feedback path can change the characteristics of the transmission response significantly. For

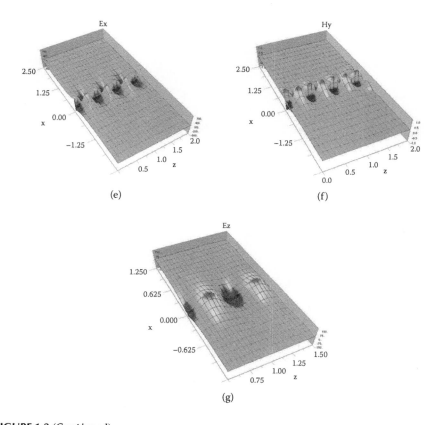

**FIGURE 1.3 (*Continued*)**
The propagation field inside an IMI for the symmetric mode (b,c,d) for different field components and for the asymmetric mode (e,f,g). (See color figure.)

example, in the aforementioned cases, most of the light is coupled out from this path over the band of interest. On the other hand, minimizing this coupling should create a resonance effect inside the feedback junction (the U shape). Thus, by adjusting this path length such that $L_1 + 2L_2 = \lambda/2n_{\text{eff}}$, the structure can operate as a resonant notch filter as shown in Figure 1.9. The control of the coupled light is mainly performed by engineering the ports of the *T*-junctions simply by controlling the width of the different waveguide sections, effectively changing their impendence and hence changing the transmission, through, and reflection properties. For example, for designing Resonator 1, shown in Figure 1.9, we utilize $d_1 = d_2 = 4d_3 = 200$ nm and $L_1 = 100$ nm, and $L_2 = 300$ nm. However, reducing the width of the slot increases the losses and degrades the performance of the resonators. To overcome this problem the design has been optimized. In this optimized design, namely Resonator 2, a narrow PSW is exploited at the *T*-junctions only ($d_3 = 40$ nm)

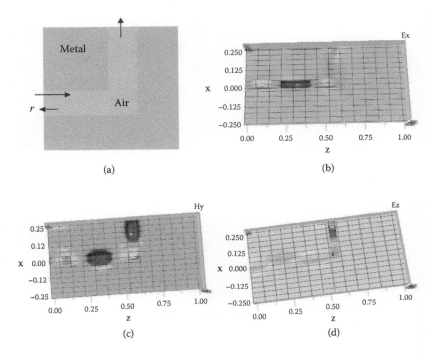

**FIGURE 1.4**
The field inside an L-shaped sharp bend. (See color figure.)

to obtain the required resonant effect. However, most of the propagation in the feedback path is done inside wider waveguide ($d_4 = d_1 = d_2$ 200 nm) as shown in the inset of Figure 1.9. The length of the junctions $L_1$, $L_2$, and $L_3$ are optimized to be 220 nm, 400 nm, and 380 nm, respectively. The transmission characteristics of this modified structure are also shown in Figure 1.9 with smaller loss and narrower resonance than Resonator 1.

Plasmonic slot waveguide is considered as the 3D practical realization of the MIM as shown in Figure 1.10. Due to the tight field confinement, coupling light to this waveguide is very challenging.

A novel coupling scheme between a silicon waveguide with 500 nm core to a plasmonic slot has been recently proposed and tested [17, 18]. In this technique, FDTD has been utilized to correctly simulate the coupling efficiency and the propagation loss inside the slot [18].

## 1.4.2 Applications of FDTD in Photonics Crystals

Another important class of structures is those based on the photonic band gap effect. For such structures, a periodic dielectric lattice is created and is known as photonic crystal [19–21]. This periodicity creates a band gap effect

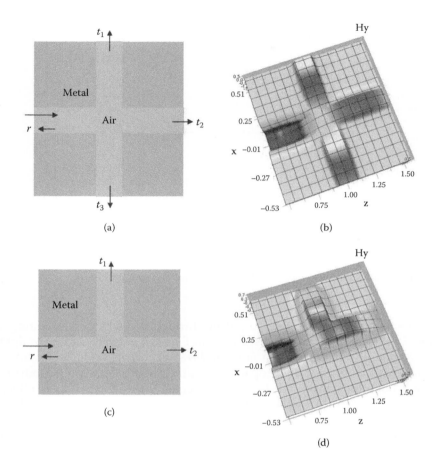

**FIGURE 1.5**
The field inside X and T junctions. (See color figure.)

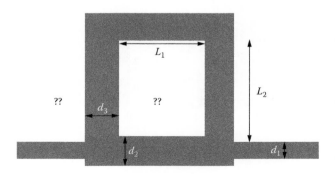

**FIGURE 1.6**
The schematics of the feedback structure with U-shaped stub.

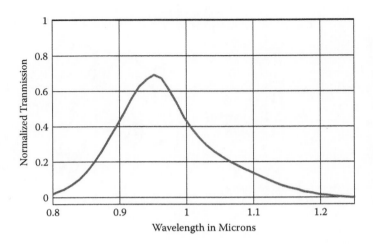

**FIGURE 1.7**
The transmission intensity of feedback structure with asymmetric junction.

that prevents the light from propagating through this structure. Creating a defect in this periodic lattice, as shown in Figure 1.12, may allow the light to propagate in the defect area only as shown in Figure 1.13. As shown in this figure, the light is only propagating in the area that has no periodicity and it is not allowed to enter the periodic region due to the band gap effect. It is also shown in Figure 1.13 that the light splits evenly to the two output arms. Photonic crystal has been utilized in various applications including interconnects, sensing [21], and solar energy [22]. Modeling of light propagation inside a photonic crystal is mainly done using FDTD as shown in

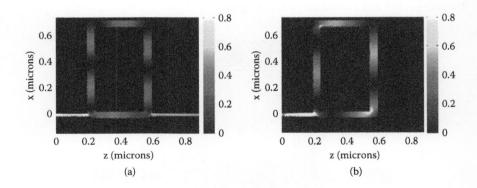

**FIGURE 1.8**
Field intensity of the asymmetric structure at wavelength (a) 950 nm and (b) 1250 nm. (See color figure.)

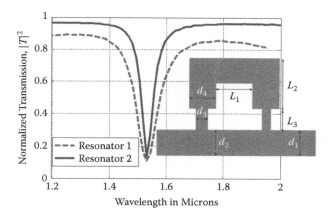

**FIGURE 1.9**
The transmission characteristics of two types of resonators using feedback plasmonics scheme. The inset is the schematics of Resonator 2.

Figure 1.13. For photonic crystal propagation problems, FDTD is considered the most suitable method.

### 1.4.3 Applications of FDTD in Nanoparticles and Nanostructures

In the last decade the ability to manipulate light in nanoscale has attracted attention. The ability to control light with the support of the surface plasmon field has been demonstrated in various applications including the extraordinary transmission (EOT) from hole array, subwavelength focusing, and directional excitation or beaming.

The EOT has been experimentally demonstrated through subwavelength holes at wavelengths that are 10 times the size of the hole [23]. This transmission is a few orders of magnitude more than the one estimated using the

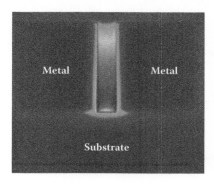

**FIGURE 1.10**
Schematic diagram of the plasmonic slot. (See color figure.)

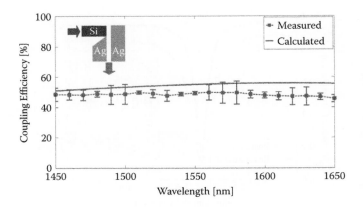

**FIGURE 1.11**
The coupling efficiency of hybrid orthogonal junction calculated using FDTD and experimentally measured.

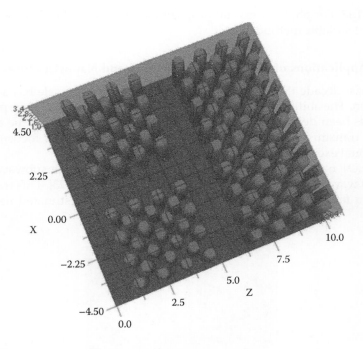

**FIGURE 1.12**
Schematic of photonic crystal with *T*-shaped defect.

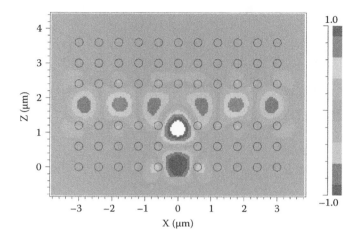

**FIGURE 1.13**
Field propagation inside the *T*-shaped photonic crystal. (See color figure.)

standard theory of diffraction from small holes [24]. It has been proved that the coupled surface plasmonic field is the main reason for such enhancement [23]. This hole array is very sensitive to the filling material and hence it has been exploited for biosensing applications [25].

FDTD has been utilized to model and characterize the enhanced transmission from the subwavelength hole array such as the one shown in Figure 1.14. In this configuration, air-filled subwavelength holes in gold film of thickness 200 nm are simulated using 3D FDTD. The size of the hole is 285 nm × 285 nm. The periodicity of the hole array is 425 nm. The normalized transmission is calculated using FDTD simulation and shown in Figure 1.15. The response shows the enhancement of the transmission over two wavelength bands.

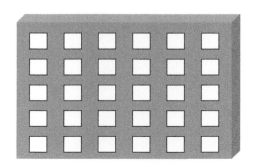

**FIGURE 1.14**
Schematic diagram of the hole array.

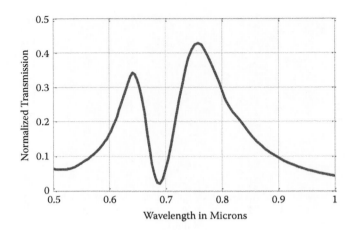

**FIGURE 1.15**
Extraordinary transmission from nanohole array.

The ability to enhance the optical transmission through a single sub-wavelength has also been recently demonstrated [26]. In this structure, a single hole or aperture is created with surrounded gratings as shown in Figure 1.16. The enhancement is mainly due to the excited surface plasmonic (SPP) field on the surface of the metal side metal. The SPP is coupled through the subwavelength hole to the surface and then diffracted by the grating to

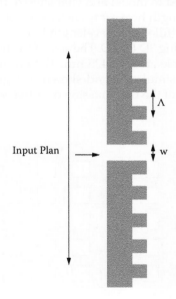

**FIGURE 1.16**
Schematic diagram of single slit with surrounding gratings.

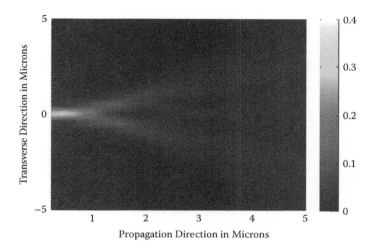

**FIGURE 1.17**
The field intensity for subwavelength focusing. (See color figure.)

constructively interfere, creating a subwavelength spot. This focusing effect is utilized in various applications including near- and far-field microscopes [27], small divergence angle laser sources [28], and light-harvesting silicon photodetectors [29]. FDTD has been utilized to demonstrate this effect as shown in Figure 1.17. Recently, to further demonstrate the essential rule played with the grating, an important experiment has been done using transient thermal grating that lasts for only a few picoseconds [30]. In this experiment, the FDTD has been utilized to accurately model the light behavior at different time intervals during the presence of the transient grating. This grating is created using the interference of two ultrafast laser beams with 150 fs pulse period. This interference effect creates a periodic change in the intensity over a thin-film planar metal film. This intensity change produces a local change in the permittivity of the metal film and hence creates a planar grating effect. The transient gratings last for a few picoseconds and hence the beaming effect from the single slit lasts for 3 ps.

Another important application for the FDTD is the ability to calculate the scattered light from nanoparticles, vertical nanowires, and nanorods. This is important to calculate the field localization and scattering from these nanostructures. For example, the field localization is calculated using FDTD for an aluminum nanorod of radius 30 nm. The simulation results are shown in Figure 1.18.

### 1.4.4 Applications of FDTD on Solar Cell

Energy harvesting is of prime importance to any sustainable society. Photovoltaics and thermal nanoantennas are considered as promising solutions for solar energy harvesting. The target in any of these systems is to

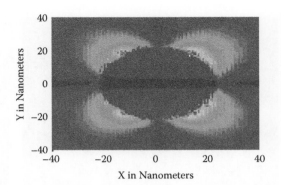

**FIGURE 1.18**
The electric field localization around Al nanorod of radius 30 nm. (See color figure.)

maximize the amount of light absorbed. The power absorbed in the complex design of a solar cell is mainly calculated using FDTD.

For example, one of the recent innovations in solar cell is to utilize plasmonic field enhancement by incorporating metal nanostructures with thin film material. These nanostructures can be fabricated using noble metals excited at their surface plasmon resonance.

These nanoparticles increase the scattering cross section and improve the optical localization of the light, which in turn increases the amount of absorbed light. It also increases the acceptance cone of incident angles of the absorbing layer. This in turn will have a huge impact on increasing the efficiency of the solar cell. The main mechanism of the light trapping is the excitation of the local surface plasmon polariton (LSPP) mode of the nanoparticle. Plasmonic solar cells (PSCs) have attracted attention in the last few years due to their promising performance [31]. Modeling such solar cells is complex and requires careful treatment and can be mainly done using FDTD simulation. One of the possible configurations can be done by depositing metal nanoparticles on the surface of the absorption layer as shown in Figure 1.19.

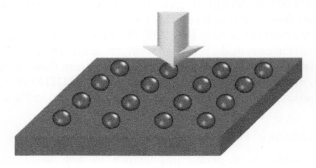

**FIGURE 1.19**
Schematic of plasmonic solar cell.

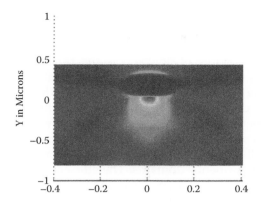

**FIGURE 1.20**
Enhancement of optical field due to the plasmonic nanoparticles. (See color figure.)

This arrangement can allow for field localization under the nanoparticles and inside the absorption layer as shown in Figure 1.20.

To further illustrate the powerfulness of the FDTD, a plasmonic solar cell as described in [32] is simulated using 3D FDTD simulation [9]. The amount of the absorbed power is calculated using FDTD and shown in Figure 1.21.

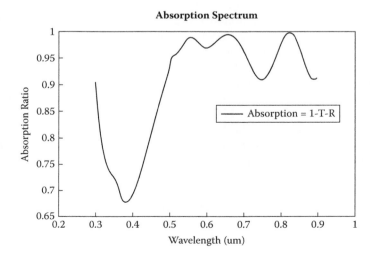

**FIGURE 1.21**
The absorbed power inside the solar cell.

## 1.5 Efficient Approach for Sensitivity Analysis Using FDTD

Calculating the sensitivities of a given objective function with respect to the design parameters of photonic structures is a desirable feature in any photonic solver. These sensitivities are required by both the designer and the manufacturer for design optimization, yield, and tolerance analyses. The adjoint variable method (AVM) is one of the most powerful methods for sensitivity analysis. It offers an efficient approach for design sensitivity analysis. The AVM is a well-established general technique for sensitivity analysis in both the time and frequency domains. The efficient application of this general approach to different numerical techniques is still the subject of ongoing research in both the microwaves and photonic communities. The AVM implementation differs with different numerical methods as the governing equations are different.

In the last few years, this approach has been efficiently applied to sensitivity analysis of complex microwave structures for different numerical methods using approximate approaches [33–41]. These methods include time-domain techniques such as the transmission line method (TLM) [35] and the FDTD method [36]. The AVM has also been applied to frequency domain techniques such as the method of moments (MoM) [33] and the frequency domain transmission line method (FD-TLM) [34].

Recently, the AVM has been applied to photonic devices using the finite-difference frequency-domain (FDFD) method [37]. In this work, the FDFD with nonuniform grid is used to extract the sensitivities of photonic crystals.

In general, the application of the AVM varies with the variation of the numerical techniques as mentioned earlier. More specifically, the time-domain and the frequency-domain approaches have conceptually different approaches to applying the AVM method. Even for the techniques that belong to the same domain, the application of the AVM may be different. This is mainly due to the fact that it depends on the system of equations that vary from one technique to the other.

As FDTD deals with a large class of photonic devices, it is of prime importance to exploit the AVM method for FDTD for photonic devices [36, 40]. In this section we briefly review the application of this approach with the FDTD method.

The AVM approach aims at estimating the sensitivities of a general objective function of the form [36]

$$F = \int_0^{T_{max}} \int_\Omega f(E, p) \, d\Omega \quad dt \tag{1.10}$$

where $T_{max}$ is the total simulation time, $E$ is the vector of the electric field temporal values in the computational domain, $p$ is the vector of the design parameters, and $\Omega$ is the observation domain where the objective function is evaluated. The sensitivity of the objective function with respect to the $n$th design parameter $p_n$ is given by

$$\frac{\partial F}{\partial p_n} = \frac{\partial^e F}{\partial p_n} + \int_0^{T_{max}} \int_\Omega \frac{\partial f}{\partial E} \cdot \frac{\partial E}{\partial p_n} \, d\Omega \, dt, \qquad n = 1,\dots,N \qquad (1.11)$$

where $\partial^e/\partial p_n$ represents the explicit dependence of $F$ on the design parameter $p_n$. The sensitivities (1.11) are usually estimated using finite differences at the response level. This requires $N$ extra FDTD simulations if forward differences are used. Additional $2N$ simulations are required if the more accurate central differences are utilized. This computational cost motivates research for a more efficient sensitivity estimation approach.

The algorithm presented in [36] aims at efficiently estimating the objective function gradient with the FDTD method. This approach utilizes the second-order vectorial wave equation for a lossy isotropic medium

$$\nabla \times \nabla \times E + \mu\varepsilon \frac{\partial^2 E}{\partial t^2} + \sigma\mu \frac{\partial E}{\partial t} = -\mu \frac{\partial J}{\partial t} \qquad (1.12)$$

where $J$, $\varepsilon$, $\sigma$, and $\mu$ are the electric current density, permittivity, conductivity, and permeability, respectively. This equation is then discretized into the matrix equation [36]

$$M\ddot{E} + N\dot{E} + KE = Q \qquad (1.13)$$

where $\dot{E}$ and $\ddot{E}$ are the first and second time derivatives of the electric field components, respectively. $M$, $N$, and $K$ are the symmetric system matrices, and $Q$ is the excitation vector.

If the $n$th design parameter is perturbed from $p_n$ to $p_{n+}\Delta p_n$, the corresponding perturbed system is given by

$$\tilde{M}_{p_n \Delta p_n} \ddot{E} + \tilde{N}_{p_n \Delta p_n} \dot{E} + \tilde{K}_{p_n \Delta p_n} E + \Delta_{p_n} R = 0 \qquad (1.14)$$

where

$$\tilde{A}_{p_n} = A + \Delta_{p_n} A, \qquad A = M, N, K, \text{ and, } Q,$$

$$\Delta_{p_n} R = \Delta_{p_n} M \ddot{E} + \Delta_{p_n} N \cdot \dot{E} + \Delta_{p_n} K E - \Delta_{p_n} Q \qquad (1.15)$$

The corresponding adjoint simulation is derived as follows: Multiplying (1.14) by an arbitrary variable $\lambda_{p_n}$, integrating twice, and imposing the terminal conditions $\lambda_{p_n}(T_{max}) = 0$ and $\dot{\lambda}_{p_n}(T_{max}) = 0$, we get

$$\int_0^{T_{max}} (\ddot{\lambda}_{p_n}^T \tilde{M}_{p_n} - \dot{\lambda}_{p_n}^T \tilde{N}_{p_n} + \lambda_{p_n}^T \tilde{K}_{p_n}) \Delta_{p_n} E \ dt = - \int_0^{T_{max}} \lambda_{p_n}^T \Delta_{p_n} R \ dt \tag{1.16}$$

By comparing (1.16) and (1.11), the expression for the AVM sensitivity is obtained as

$$\frac{\partial F}{\partial p_n} \approx \frac{\partial^e F}{\partial p_n} - \int_0^{T_{max}} \int_\Omega \lambda_{p_n}^T \cdot \frac{\Delta_{p_n} R}{\Delta p_n} d\Omega \cdot dt, \quad n = 1, \dots, N \tag{1.17}$$

where the adjoint vector $\lambda_{p_n}$ is obtained by solving

$$\tilde{M}_{p_n} \ddot{\lambda}_{p_n} - \tilde{N}_{p_n} \dot{\lambda}_{p_n} + \tilde{K}_{p_n} \lambda_{p_n} = \frac{\partial f}{\partial E}^T \tag{1.18}$$

The system (1.18) supplies the values of the adjoint variable $\lambda_{p_n}$ at all time steps. The expression (1.17) is then utilized to estimate the required sensitivities. To avoid carrying out $N$ such adjoint simulations, a one-to-one mapping is applied. We carry out only one adjoint simulation of the unperturbed system

$$M\ddot{\lambda} - N\dot{\lambda} + K\lambda = \frac{\partial f}{\partial E}^T \tag{1.19}$$

The second-order system of equations (1.10) is shown to be equivalent to the Maxwell's curl equations

$$\nabla \times \lambda^H = \frac{\partial \varepsilon \lambda}{\partial t} - \sigma \lambda + J^\lambda \tag{1.20}$$

$$\nabla \times \lambda = -\frac{\partial \mu \lambda^H}{\partial t} \tag{1.21}$$

with the terminal conditions $\lambda(T_{max}) = 0$, $\lambda^H(T_{max}) = 0$, where the adjoint excitation $J^\lambda$ is given by

$$\frac{\partial J^\lambda}{\partial t} = \frac{1}{\mu} \frac{\partial f}{\partial E}^T \tag{1.22}$$

In (1.20) and (1.21), $\lambda$ and $\lambda^H$ are the electric and magnetic fields of the adjoint problem, respectively. The update equations of the FDTD algorithm for the adjoint problem are the same as those of the original simulation if we solve for $(-\lambda, \lambda^H)$. Therefore, the same absorbing boundary conditions are used in both simulations, which simplify the implementation and allow for using the commercial tools.

### 1.5.1 AVM Approach for Dielectric Discontinuities

Since most of the photonic devices are dielectric-based structures, in this section we show how the AVM approach can be adapted to dielectric discontinuities. We illustrate the approach for the 2D case where the $TM_x$ mode is analyzed. In this case, the respective matrices have the form

$$M = -\mu_r B_{\varepsilon_r} \left( \frac{\Delta h}{c\Delta t} \right)^2 \tag{1.23}$$

$$N = -\frac{B_\sigma \mu_0 \mu_r}{\Delta t} \Delta h^2 \tag{1.24}$$

$$Q = \frac{\mu_0 \mu_r \Delta h^2}{\Delta t} D_t J_x \tag{1.25}$$

$$D_t J_x = J_x \left( t + \frac{\Delta t}{2} \right) - J_x \left( t - \frac{\Delta t}{2} \right) \tag{1.26}$$

Here, $B_{\varepsilon_r}$ and $B_\sigma \in \Re^{j_{max} k_{max}}$ where $j_{max}$ and $k_{max}$ are the number of cells in $y$- and $z$-directions, respectively. These diagonal matrices contain the values of $\varepsilon_r$ and $\sigma$ at each grid point. The double curl operator $K$ has three vector components and can be written as

$$(KE)_x = h_y^2 D_{yy} E_x + h_z^2 D_{zz} E_x - h_z h_x D_{zx} E_z - h_y h_x D_{yx} E_y, \tag{1.27}$$

$$(KE)_y = h_x^2 D_{xx} E_y + h_z^2 D_{zz} E_y - h_x h_y D_{xy} E_x - h_z h_y D_{zy} E_z, \tag{1.28}$$

$$(KE)_z = h_x^2 D_{xx} E_z + h_y^2 D_{yy} E_z - h_x h_z D_{xz} E_x - h_y h_z D_{yz} E_y \tag{1.29}$$

where

$$h_x = \frac{\Delta h}{\Delta x}, \quad h_y = \frac{\Delta h}{\Delta y}, \quad h_z = \frac{\Delta h}{\Delta z}, \quad \text{and} \quad \Delta h = \min(\Delta x, \Delta y, \Delta z).$$

The second-order difference with respect to variables κ, and $v(\kappa, v = x, y, z, t)$ are calculated using the central difference approach as

$$D_{\kappa\kappa}E = E(\kappa - \Delta\kappa) + 2E(\kappa) + E(\kappa + \Delta\kappa), \tag{1.30}$$

and

$$D_{\kappa v}E = E\left(\kappa + \frac{\Delta\kappa}{2}, v + \frac{\Delta v}{2}\right) + E\left(\kappa - \frac{\Delta\kappa}{2}, v - \frac{\Delta v}{2}\right)$$
$$- E\left(\kappa + \frac{\Delta\kappa}{2}, v - \frac{\Delta v}{2}\right) - E\left(\kappa - \frac{\Delta\kappa}{2}, v + \frac{\Delta v}{2}\right) \tag{1.31}$$

The expression (1.15) can thus be evaluated as

$$\Delta_{p_n}R = (\Delta_{p_n}M)D_{tt}E_x + (\Delta_{p_n}N)D_tE_x + (\Delta_{p_n}K)E_x - \Delta_{p_n}Q \tag{1.32}$$

Evaluating the sensitivity expression (1.17) requires storing the field information for the nodes corresponding to nonzero components of the vector $\Delta_{p_n}R$. Here, a virtual perturbation of the respective parameter is assumed in the forward direction only. The corresponding vector perturbation is denoted by $\Delta_{p_n}^+R$.

To illustrate this approach, consider the dielectric slab of width $L$, permittivity $\varepsilon_{r2}$, and conductivity $\sigma_2$ inserted between two perfect electric walls (see Figure 1.22). This waveguide structure is filled with a material that has permittivity $\varepsilon_{r1}$ and conductivity $\sigma_1$. Perturbing the parameter $L$, in the forward direction in a symmetric way, changes the dielectric properties of the cells outside the dielectric, marked with squares. Their field values are thus stored for both the original and adjoint simulations. We demonstrate the calculation of the system matrices $\Delta_{p_n}^+R$ in this case for a point $(j, k)$ in Table 1.1.

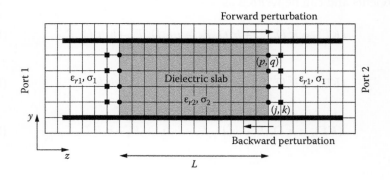

**FIGURE 1.22**
Illustration of the stored points used to calculate the sensitivities with respect to the width of dielectric slab inside a waveguide with metallic boundaries using CAVM approach [40].

**TABLE 1.1**

The Variations in the System Matrices Due to Forward Perturbation

| | $(j,k)$ | $(j+1,k)$ | $(j-1,k)$ | $(j,k+1)$ | $(j,k-1)$ |
|---|---|---|---|---|---|
| $K(p_n = L)$ | $-2(h_y^2 + h_z^2)$ | $h_y^2$ | $h_y^2$ | $h_z^2$ | $h_z^2$ |
| $K(p_n = L+\Delta L)$ | $-2(h_y^2 + h_z^2)$ | $h_y^2$ | $h_y^2$ | $h_z^2$ | $h_z^2$ |
| $\Delta K$ | $0$ | $0$ | $0$ | $0$ | $0$ |
| $E$ | $E \neq 0$ | $E \neq 0$ | $E \neq 0$ | $E \neq 0$ | $E \neq 0$ |
| $M\,(p_n = L)$ | $-\mu_r \varepsilon_{r1}(\Delta h/c\Delta t)^2$ | $0$ | $0$ | $0$ | $0$ |
| $M\,(p_n = L+\Delta L)$ | $-\mu_r \varepsilon_{r2}(\Delta h/c\Delta t)^2$ | $0$ | $0$ | $0$ | $0$ |
| $\Delta M$ | $-\mu_r \Delta\varepsilon_r(\Delta h/c\Delta t)^2$ | $0$ | $0$ | $0$ | $0$ |
| $D_{tt}E$ | $\neq 0$ | $\neq 0$ | $\neq 0$ | $\neq 0$ | $\neq 0$ |
| $N\,(p_n = L)$ | $-\mu_r \mu_o \sigma_1 \Delta h^2/\Delta t$ | $0$ | $0$ | $0$ | $0$ |
| $N\,(p_n = L+\Delta L)$ | $-\mu_r \mu_o \sigma_2 \Delta h^2/\Delta t$ | $0$ | $0$ | $0$ | $0$ |
| $\Delta N$ | $-\mu_r \mu_o \Delta\sigma(\Delta h^2/\Delta t)$ | $0$ | $0$ | $0$ | $0$ |
| $D_t E$ | $\neq 0$ | $\neq 0$ | $\neq 0$ | $\neq 0$ | $\neq 0$ |

$$\Delta^+_{p_n} R_{j,k} = \Delta_{p_n} R_{j,k} = -\mu_r \Delta\varepsilon_r (\Delta h/c\Delta t)^2 D_{tt}\, E(j,k) - \mu_r \mu_o \Delta\sigma (\Delta h^2/\Delta t) D_t\, E(j,k),$$
$$\Delta\varepsilon_r = \varepsilon_{r2} - \varepsilon_{r1} \text{ and } \Delta\sigma = \sigma_2 - \sigma_1$$

*Source:*   M. A. Swillam, M. H. Bakr, N. K. Nikolova, and X. Li, "Adjoint sensitivity analysis of dielectric discontinuities using FDTD," *Electromagnetics* 27, 123–140 (Feb. 2007).

Our experience shows that the sensitivities obtained using this approach are very similar to those obtained using forward finite difference approximation applied at the response level. A better accuracy can be obtained, especially for a highly nonlinear objective function, if a central approach is adopted. A possible central AVM (CAVM) approach is presented in the following chapter.

The excitation of the adjoint simulation is dependent on the objective function as shown before and is given by

$$\frac{\partial J^\lambda}{\partial t} = \frac{1}{\mu}\frac{\partial f}{\partial E}^T \tag{1.33}$$

For some special objective functions such as the scattering parameters, the adjoint field values can be deduced from the original. It follows that the adjoint simulation can be eliminated. Here, we apply this self-adjoint approach to the estimation of the reflection coefficient sensitivities.

Consider $\Gamma(\lambda_o)$, the reflection coefficient at wavelength $\lambda_o$. It can be written as

$$\Gamma(\lambda_o) = \frac{\displaystyle\int_\Omega \tilde{E}_r(\lambda_o)\cdot E_{inc}\, d\Omega}{\displaystyle\int_\Omega E_{inc}\cdot E_{inc}\, d\Omega} \tag{1.34}$$

where $F_{inc}$ is the modal field distribution of the fundamental mode. The vector $\tilde{E}_r(\lambda_o)$ is the reflected electric field at the wavelength $\lambda_o$. The reflection coefficient can also be written as

$$\Gamma(\omega_o) = \frac{\displaystyle\int_0^{T_{max}} \int_\Omega E_r(t) \cdot E_{inc} \, d\Omega \, e^{-j\omega_o t} \, dt}{\displaystyle\int_\Omega E_{inc} \cdot E_{inc} \, d\Omega} \tag{1.35}$$

where $\omega_o = 2\pi c/\lambda_o$ is the angular frequency, and $c$ is the velocity of light in free space. The expression given in (1.35) can be written in a form similar to (1.33):

$$\Gamma(\omega_o) = \frac{1}{S} F(\omega_o) = \frac{1}{S} \int_0^{T_{max}} \int_\Omega f(\Omega, t) \, d\Omega \, dt \tag{1.36}$$

where

$$f(\Omega, t) = E_r(t) \cdot E_{inc} \, e^{-j\omega_o t} \tag{1.37}$$

and

$$S = \int_\Omega E_{inc} \cdot E_{inc} \cdot d\Omega \tag{1.38}$$

The reflection coefficient is thus a complex function and its sensitivities with respect to the $n$th parameter, $n = 1, 2, \ldots, N$, are given by

$$\frac{\partial \Gamma}{\partial p_n} = \frac{1}{S} \frac{\partial F(\omega_o)}{\partial p_n} = \frac{1}{S} \left( \mathrm{Re}\left\{ \frac{\partial F(\omega_o)}{\partial p_n} \right\} + j \, \mathrm{Im}\left\{ \frac{\partial F(\omega_o)}{\partial p_n} \right\} \right) \tag{1.39}$$

Accordingly, the adjoint excitation in (1.33) can be written as

$$J^\lambda = \frac{1}{\gamma} \int_0^{T_{max}} \frac{\partial f}{\partial E}^T \, dt \tag{1.40}$$

where

$$\gamma = \mu_o \mu_r \Delta h^2 \tag{1.41}$$

Following a similar procedure to that in [42], the adjoint field is deduced through the expression [43]

$$\left( \lambda_{p_n}^{\pm T} \right)_{i,j,k} = \frac{\left| \left( E_{p_n}^{\pm T}(\omega_o) \right)_{i,j,k} \right|}{\gamma J_o \omega_o |H(\omega_o)|} \cdot \left( \begin{array}{l} \cos(\omega_o t + \varphi_o - \left( \phi_{E^\pm} \right)_{i,j,k} - \beta d - \pi/2) \\[2mm] + j\cos(\omega_o t + \varphi_o - \left( \phi_{E^\pm} \right)_{i,j,k} - \beta d) \end{array} \right) \tag{1.42}$$

where $|E_{p_n}^{\pm^T}(\omega_o)_{i,j,k}|$ and $(\phi_{E^\pm})_{i,j,k}$ are the magnitude and the phase of the original field of the point $(i, j, k)$ at frequency $\omega_o$, respectively. $|H(\omega_o)|$ and $\varphi_o$ in (1.42) are the magnitude and the phase of the spectral component of the wideband excitation function $h(t)$ at $\omega_o$, respectively. $J_o$ is a scaling factor that accounts for the normalization of the incident field. In (1.42), $\beta$ is the propagation constant of the fundamental mode, and $d$ is the distance between the excitation plane and the reference plane.

In order to evaluate the sensitivity using CAVM approach, the following discretization is used:

$$\frac{\partial F}{\partial p_n} = -\frac{\Delta s \Delta t}{2} \sum_{m=1}^{m_{max}} \sum_{\Omega} \frac{(\lambda_{p_n}^{+^T} \cdot \Delta_{p_n}^+ R + \lambda_{p_n}^{-^T} \cdot \Delta_{p_n}^- R)}{\Delta p_n}, \quad n = 1, \dots, N \qquad (1.43)$$

where $\Delta_s$ is the unit cell area. The adjoint field values are given in (1.42) with $t = m\Delta t$, and $T_{max} = m_{max}\Delta t$ where

$$\Delta_{p_n}^+ R = \Delta_{p_n}^+ M \cdot \ddot{E} + \Delta_{p_n}^+ N \cdot \dot{E} + \Delta_{p_n}^+ K \cdot E - \Delta_{p_n}^+ Q \qquad (1.44)$$

$$\Delta_{p_n}^- R = \Delta_{p_n}^- M \cdot \ddot{E} + \Delta_{p_n}^- N \cdot \dot{E} + \Delta_{p_n}^- K \cdot E - \Delta_{p_n}^- Q \qquad (1.45)$$

Thus, (1.43) can be obtained efficiently by obtaining the forward and backward perturbation from (1.44) and (1.45).

### 1.5.2 Examples

In this section, we illustrate our CAVM approach through two examples. Different response functions are used to illustrate the universality of our CAVM algorithm. We also apply the self-adjoint approach discussed hereafter to estimate the sensitivities of the reflection coefficient without performing any FDTD adjoint simulations. Hence, the computational overhead is practically removed. In these examples, the perfect matched layer (PML) [44] is used as an absorbing boundary condition for the computational domain.

#### 1.5.2.1 Multimode Interference 3dB Power Splitter

In this example, a compact multimode interference (MMI) device (1×2) is used as a 3 dB power splitter as shown in Figure 1.23. Strong guided structure is used with core index of 3.0. The width of the multimode section is taken to be 1.65 μm, which supports six guided modes at $\lambda_o = 1.55$ μm. The lengths of the access waveguides at the input and the output are 3.75 μm.

In this example, the sensitivities of an output energy function with respect to the dimensions of the structure are studied using our CAVM technique.

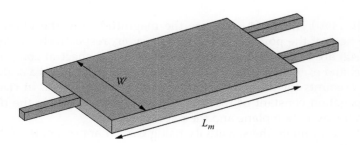

**FIGURE 1.23**
Schematic diagram of an MMI 3 dB power splitter [43].

Here, the TE case is assumed where the electric field has only one component, $E_y$.

The energy function is given in the form

$$F = \int_\Omega \int_0^{T_{\max}} \frac{E_y^2}{2} \, dt \, d\Omega \tag{1.46}$$

For evaluating the objective function in (1.46), the following discretization is used:

$$F = \sum_{m=1}^{m_{\max}} \sum_{j=1}^{j_{\max}} \frac{E_y^2}{2} \Delta x \Delta t \tag{1.47}$$

where $\Delta x$ and $\Delta t$ are the special step size and time step size, respectively. According to (1.40), the adjoint current of excitation is given by

$$J^\lambda = \frac{\Delta t}{\gamma} \sum_{m=1}^{m_{\max}} \frac{\partial f}{\partial E_y} \tag{1.48}$$

where

$$f = \frac{E_y^2}{2}$$

The discretized form given in (1.43) is used to calculate the sensitivities of the energy function with respect to the parameters $p = [W \, L_m]^T$ using the CAVM approach. The cell size is $\Delta x = \Delta z = 25.0$ nm. The sensitivities of the energy function with respect to the length and the width of the multimode section are calculated using the CAVM approach and central finite differences (CFD). The results are shown in Figures 1.24 and 1.25.

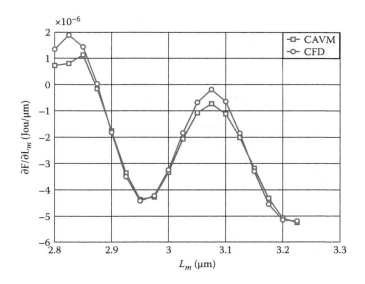

**FIGURE 1.24**
The sensitivity of the energy function with respect to the length of the MMI 3dB power splitter for $W = 1.65\ \mu m$ at wavelength $1.55\ \mu m$ [43].

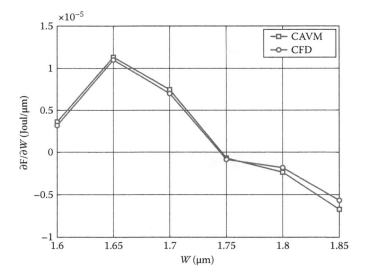

**FIGURE 1.25**
The sensitivity of the energy function with respect to the width of the MMI 3dB power splitter for $L_m = 3.2\ \mu m$ at wavelength $1.55\ \mu m$ [43].

Good agreement is obtained between our CAVM approach and the expensive central finite-difference approximation (CFD). The CAVM approach requires one extra simulation only while the CFD requires four additional FDTD simulations.

### 1.5.2.2 Deeply Etched Waveguide Terminator

A deeply etched waveguide was proposed to minimize the reflected power from waveguide terminators [45,46]. In this design, a single deep trench is etched at the end of the waveguide as shown in Figure 1.26. The design parameters are the width of the trench area and the width of the residue ($p = [L_l \; L_h]^T$). Here, we apply the CAVM approach to get the optimal design parameters that minimize the power reflectivity. The SA-CAVM approach discussed hereafter is applied to get the gradient of the objective function with no adjoint simulations. The power reflectivity can be written as

$$R(\lambda_o) = |\Gamma(\lambda_o)|^2 \tag{1.49}$$

Using the reflectivity expression in (1.49) as our objective function requires an adjoint simulation. However, the same optimal design parameters can be obtained if the magnitude of the reflection coefficient is taken as our objective function, i.e.,

$$\min_{L_l,L_h} \; |\Gamma(\lambda_o)|^2 \equiv \min_{L_l,L_h} \; |\Gamma(\lambda_o)| \tag{1.50}$$

The expression in (1.50) is valid as long as $\Gamma(\lambda_o)$ is a scalar.

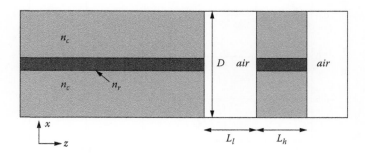

**FIGURE 1.26**
Schematic diagram of the deeply etched waveguide terminator [45].

The problem given in the right hand side of (1.50) has a self-adjoint form. Thus, the sensitivity of the objective function ($|\Gamma(\lambda_o)|$) is given by

$$\frac{\partial|\Gamma(\lambda_o)|}{\partial p_n} = \frac{1}{|S|}\frac{\partial|F(\lambda_o)|}{\partial p_n} = \frac{1}{|S|}\cdot\frac{1}{|F|}\left(F_r\cdot\mathrm{Re}\left\{\frac{\partial F(\lambda_o)}{\partial p_n}\right\} + F_m\cdot\mathrm{Im}\left\{\frac{\partial F(\lambda_o)}{\partial p_n}\right\}\right) \quad (1.51)$$

where $F_r$ and $F_m$ are the real and the imaginary parts of the function $F$ given in (1.36).

For FDTD simulations, the cell sizes are 1.2 nm and 5.5 nm in the $z$- and $x$-directions, respectively. The width of the high index region (core) of the waveguide is 0.11 μm and the depth of the trench is $D = 3.0$ μm [45]. The material model of GaAsP-InP has refractive indices of $n_r = 3.524$ and $n_c = 3.17$. The parameters of the perfectly matched layer (PML) are the same as those used in [45]. The gradient of the reflection coefficient with respect to the design parameters is evaluated using (1.51). The calculated sensitivities for the TE case are shown in Figures 1.27 and 1.28. It is clear from the displayed results that the optimal parameter values, at which minimum reflection occurs, are approximately $p = [119.55 \quad 280.0]^T$ nm. These results are very close to those

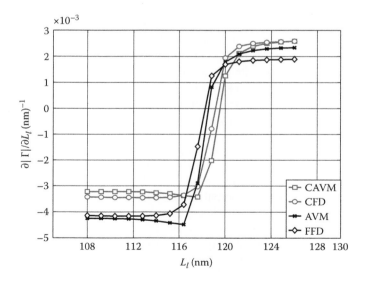

**FIGURE 1.27**
The sensitivity of the reflection coefficient with respect to the width of the trench $L_l$ of the deeply etched waveguide terminator for $L_h = 280$ nm at $\lambda = 1.55$ μm [43].

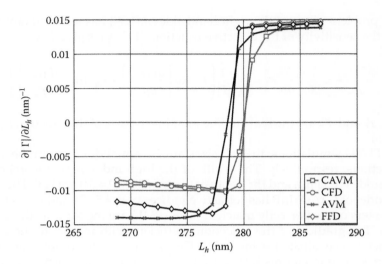

**FIGURE 1.28**
The sensitivity of the reflection coefficient with respect to the width of the residue $L_h$ of the deeply etched waveguide terminator for $L_l = 120$ nm at $\lambda = 1.55$ μm [43].

obtained in [45,46]. The sensitivities obtained using CFD are also shown for comparison. Good agreement is achieved between our SA-CAVM and the CFD. We also compare our SA-CAVM with the self-adjoint approach (AVM) suggested in [41]. The results are also shown in Figures 1.27 and 1.28. It is clear that SA-CAVM produces more accurate sensitivities that are comparable to the CFD values. In this example, the sensitivities of the response with respect to all the design parameters are obtained without any additional simulation.

In all previous examples, we addressed narrow-band sensitivity estimation. For many devices, such as semiconductor optical amplifier (SOA) or superluminescent light-emitting diode (SLED), it is desirable to obtain the wideband response and its sensitivities. The FDTD-CAVM approach can supply such wideband sensitivities. They are obtained by applying a wideband adjoint excitation and performing discrete Fourier transform (DFT). Here, the additional cost is mainly that of the DFT, which is negligible compared to the FDTD simulation cost. We demonstrate this approach by estimating the sensitivity of the reflection coefficient with respect to the width of the trench $L_l$ over a bandwidth of 100 nm centered on 1.55 μm as shown in Figure 1.29. Good match is obtained between our CAVM wideband sensitivity estimates and those obtained using CFD.

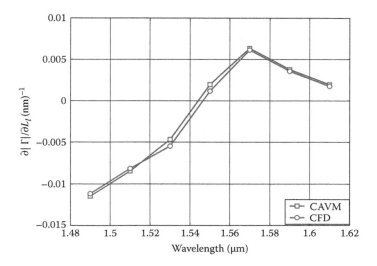

**FIGURE 1.29**
The sensitivity of the reflection coefficient with respect to the width of the trench $L_l$ over a wavelength band of the deeply etched waveguide terminator at $[L_l \; L_h] = [120.0 \; 280.0]$ nm [43].

## 1.6 Conclusion

In this chapter, the FDTD method for photonics is presented. The applications of FDTD in nanophotonics have been presented. Various examples that cover a wide range of applications in the nanophotonic field demonstrate the powerfulness of this method in the area of nano-optics. An efficient approach is proposed for obtaining second-order accurate sensitivities using the adjoint variable method applied on FDTD. This technique is utilized to obtain different responses. It has also been applied to calculate the sensitivity of the power reflectivity using no additional simulations that use a self-adjoint version of this approach. The sensitivity approach is accurate and efficient and proves an efficient method for design optimization of various photonics and nanophotonics structures.

## References

1. K. S. Yee, "Numerical solution of initial boundary value problems involving Maxwell's equations in isotropic media," *IEEE Trans. Antenna Propagat.* 14, 302–307 (May 1966).

2. A. Taflov, *Computational Electrodynamics: The Finite Difference Time Domain Method*. Norwood, MA: Artech House, 1995.
3. J. P. Berenger, "A perfectly matched layer for the absorption of electromagnetic-waves," *J. of Comp. Phys.* 114(2), 185–200 (Oct. 1994).
4. W. C. Chew and W. H. Weedon, "A 3-D perfectly matched medium from modi-fied Maxwell's equations with stretched coordinates," *Microw. Opt. Technol. Lett.* 7, 599–603 (Sept. 1994).
5. Z. S. Sacks, D. M. Kingsland, R. Lee, and J. F. Lee, "A perfectly matched anisotro-pic absorber for use as an absorbing boundary condition," *IEEE Trans. Antennas Propagat.* 43, 1460–1463 (Dec. 1995).
6. S. A. Maier, *Plasmonics: Fundamentals and Applications*. Springer, 2007.
7. D. M. Sullivan, *Electromagnetic Simulation Using the FDTD Method*. John Wiley, 2000.
8. M. A. Swillam, M. H. Bakr, and X. Li, "Efficient adjoint sensitivity analysis exploiting the FD-BPM," *J. Lightwave Technol.* 25, 1861–1869 (2007).
9. http://www.optiwave.com/products/fdtd_overview.html.
10. http://www.lumerical.com/tcad-products/fdtd/.
11. E. Ozbay, "Plasmonics: Merging photonics and electronics at nanoscale dimen-sions," *Science* 311(5758), 189–193, 2006.
12. P. Berini, "Bulk and surface sensitivity of surface plasmon waveguide," *New J. of Physics* 10, 105010 (2008).
13. Berini, P., "Long-range surface plasmon-polaritons," *(OSA) Advances in Optics and Photonics* 1, 484–588 (2009).
14. G. Veronis and S. Fan, "Bends and splitters in metal-dielectric-metal subwave-length plasmonic waveguides," *Appl. Phys. Lett.* 87, 131102 (2005).
15. M. A. Swillam and A. S. Helmy, "Feedback effects in plasmonic slot waveguides examined using a closed-form model," *Photon. Technol. Lett.* 24, 497–499 (2012).
16. Charles Lin, Mohamed A. Swillam, and Amr S. Helmy, "Analytical model for metal–insulator–metal mesh waveguide architectures," *J. Opt. Soc. Am. B* 29, 3157–3169 (2012).
17. B. Lau, M. A. Swillam, and A. S. Helmy, "Hybrid orthogonal junctions: wide-band plasmonic slot–silicon wire couplers," *Optics Express* 18(26), 27048–27059 (Dec. 2010).
18. C. Lin, H. K. Wang, B. Lau, M. A. Swillam, and A. S. Helmy, "Efficient broad-band energy transfer via momentum matching at hybrid guided-wave junc-tions," *Applied Physics Letters* 101, 123115 (2012).
19. J. D. Joannopoulos, R. D. Meade, and J. N. Winn, *Photonic Crystals*. Princeton, NJ: Princeton University Press, 1995.
20. S. Noda and T. Baba, *Roadmap on Photonic Crystals*. Dordrecht: Kluwer Academic Publishers, 2003.
21. K. Inoue and K. Ohtaka, *Photonic Crystals: Physics, Fabrication and Applications*. New York: Springer-Verlag, 2004.
22. M. Florescu, H. Lee, I. Puscasu, M. Pralle, L. Florescu, D. Z. Ting, and J. P. Dowling, "Improving solar cell efficiency using photonic band-gap materials," *Solar Energy Materials & Solar Cells* 91, 1599–1610 (2007).
23. T. W. Ebbesen, H. J. Lezec, H. F. Ghaemi, T. Thio, and P. A. Wolff, "Extraordinary optical transmission through sub-wavelength hole arrays," *Nature* 391, 667–669 (1998).
24. H. A. Bethe, "Theory of diffraction by small holes," *Phys. Rev.* 66, 163–182 (1944).

25. A. Lesuffleur, H. Im, N. C. Lindquist, and S. Oh, "Periodic nanohole arrays with shape-enhanced plasmon resonance as real-time biosensors," *Appl. Phys. Lett.* 90, 243110 (2007).

26. T. W. Ebbesen, H. J. Lezec, H. F. Ghaemi, T. Thio, and P. A. Wolff, "Extraordinary optical transmission through sub-wavelength hole arrays," *Nature* (London) 391, 667–669 (1998).

27. B. Guo, Q. Gan, G. Song, J. Gao, and L. Chen, "Numerical study of a high-resolution far-field scanning optical microscope via a surface plasmon-modulated light source," *J. Lightwave Technol.* 25, 830–833 (2007).

28. N. Yu, Q J. Wang, M. A. Kats, J. A. Fan, S. P. Khanna, L. Li, A. Giles Davies, E. H. Linfield, and F. Capasso, "Designer spoof surface plasmon structures collimate terahertz laser beams," *Nature Materials* 9, 730–735 (2010).

29. T. Ishi, J. Fujikata, K. Makita, T. Baba, and K. Ohashi, "Si nano-photodiode with a surface plasmon antenna," *Jpn. J. Appl. Phys., Part 1* 44, L364–L366 (2005).

30. Mohamed A. Swillam, Nir Rotenberg, and Henry M. van Driel, "All-optical ultrafast control of beaming through a single sub-wavelength aperture in a metal film," *Opt. Express* 19, 7856–7864 (2011).

31. H. A. Atwater, A. Polman, "Plasmonics for improved photovoltaics devices," *Nature Material* 9, 205–213 (2010).

32. Vivian E. Ferry, Marc A. Verschuuren, Hongbo B. T. Li, Ewold Verhagen, Robert J. Walters, Ruud E. I. Schropp, Harry A. Atwater, and Albert Polman, "Light trapping in ultrathin plasmonic solar cells," *Opt. Express* 18, A237–A245 (2010).

33. N. K. Nikolova, R. Safian, E. A. Soliman, M. H. Bakr, and J. W. Bandler, "Accelerated gradient based optimization using adjoint sensitivities," *IEEE Trans. Antennas Propagat.* 52, 2147–2157 (Aug. 2004).

34. M. H. Bakr and N. K. Georgieva, "An adjoint variable method for frequency domain TLM problems with conducting boundaries," *IEEE Microwave Wireless Comp. Lett.* 13, 408–410 (Nov. 2003).

35. M. H. Bakr and N. K. Nikolova, "An adjoint variable method for time domain TLM with wideband Johns matrix boundaries," *IEEE Trans. Microwave Theory Tech.* 52, 678–685 (Feb. 2004).

36. N. K. Nikolova, H. W. Tam, and M. H. Bakr, "Sensitivity analysis with the FDTD method on structured grids," *IEEE Trans. Microwave Theory Tech.* 52, 1207–1216 (Apr. 2004).

37. G. Veronis, R. W. Dutton, and S. Fan, "Method for sensitivity analysis of photonic crystal devices," *Opt. Lett.* 29, 2288–2290 (Oct. 2004).

38. E. J. Haug, K. K. Choi, and V. Komkov, *Design Sensitivity Analysis of Structural Systems*. Orlando, FL: Academic, 1986.

39. N. K. Nikolova, J. W. Bandler, and M. H. Bakr, "Adjoint techniques for sensitivity analysis in high-frequency structure CAD," *IEEE Trans. Microwave Theory Tech. (Special Issue)* 52, 403–419 (Jan. 2004).

40. M. A. Swillam, M. H. Bakr, N. K. Nikolova, and X. Li, "Adjoint sensitivity analysis of dielectric discontinuities using FDTD," *Electromagnetics* 27, 123–140 (Feb. 2007).

41. N. K. Nikolova, Ying Li, Yan Li, and M. H. Bakr, "Sensitivity analysis of scattering parameters with electromagnetic time-domain simulators," *IEEE Trans. Microwave Theory Tech.* 54, 1598–1610 (Apr. 2006).

42. M. H. Bakr, N. K. Nikolova, and P. A. W. Basl, "Self-adjoint *S*-parameter sensitivities for lossless homogeneous TLM problems," *Int. J. of Numerical Modelling: Electronic Networks, Devices and Fields* 18, 441–455 (Nov. 2005).

43. M. A. Swillam, M. H. Bakr, and X. Li, "Accurate sensitivity analysis of photonic devices exploiting the finite-difference time-domain central adjoint variable method," *Appl. Opt.* 46, 1492–1499 (Mar. 2007).
44. J. P. Berenger, "A perfectly matched layer for the absorption of electromagnetic waves," *J. Comput. Phys.* 114, 185–200 (Oct. 1994).
45. G.-R. Zhou, X. Li, and N.-N. Feng, "Design of deeply-etched antireflective waveguide terminators," *IEEE J. Quantum Electron.* 39, 384–391 (Feb. 2003).
46. G.-R. Zhou, X. Li and N. -N.Feng, "Optimization of deeply-etched antireflective waveguide terminators by space mapping technique," *Integrated Photonics Research (IPR)*, IThH2, San Francisco (2004).

# 2

# The FDTD Method: Essences, Evolutions, and Applications to Nano-Optics and Quantum Physics

**Xiaoyan Y.Z. Xiong and Wei E.I. Sha**

*Department of Electrical and Electronic Engineering,*
*University of Hong Kong, Hong Kong, China*

## CONTENTS

## 2.1 Introduction

This chapter aims to introduce the recent developments, basic theories, core techniques, real-world applications, and future directions for the finite-difference time-domain (FDTD) method. The unified theoretical frameworks of the FDTD method and its advances (Runge Kutta-FDTD, symplectic FDTD, alternative direction implicit-FDTD, high-order FDTD, multiresolution-TD, pseudospectral-TD, etc.) for solving Maxwell's equations are systematically reviewed in Section 2.1. Next, we will briefly describe core techniques of the FDTD method (involving the basic update equations, material averaging technique, perfectly matched layer, source excitation, near-to-far-field transformation, periodic boundary condition, and treatment of dispersive media) and particularly focus on those for nano-optics applications in Section 2.2. In Section 2.3, we will demonstrate powerful capabilities of the FDTD method to model versatile physical problems in the nano-optics field. The case studies on plasmonic solar cells, nanoantennas, spontaneous emissions, and meta-materials will be discussed with detailed physical understandings. Then, the numerical analyses and implementations of the FDTD method to simu-late the Schrödinger equation are presented in Section 2.4. In Section 2.5, we will show several simple examples on the numerical solution of quantum physics problems with the FDTD method. Finally, the conclusion and future direction are summarized in Section 2.6.

The concept of Yee grid was proposed in 1966 [1]. Until now, the traditional finite-difference time-domain (FDTD) method [2–4] has been widely applied to broad-band, transient, and full-wave analyses for solving Maxwell's equa-tions owing to its simplicity, generality, and facility for parallel computing. However, the FDTD method has two main drawbacks. One is the undesir-able numerical results caused by the significant accumulated errors from numerical instability, dispersion, and anisotropy, especially for long-term simulations and electrically large objects. Another is the inability to accu-rately model curved conducting surfaces and material discontinuities by using the staircase model with structured grids.

To overcome the two shortcomings, a variety of improved methods were proposed. Reviews of recent advances in the FDTD method will not only facilitate developing fast and efficient solvers in the computational electromagnetics field but also gain physical and mathematical insights to solve real-world engineering problems.

Any field components in Maxwell's equations can be defined as

$$F(x,y,z,t) = F^{n+l/m}(i\Delta_x, j\Delta_y, k\Delta_z, (n+\tau_l)\Delta_t) \qquad (2.1)$$

where $\Delta_x$, $\Delta_y$, and $\Delta_z$ are the space steps respectively along the $x$-, $y$-, and $z$-directions, $\Delta_t$ is the time step, $i, j, k, n, l$, and $m$ are integers, $n+l/m$ denotes the $l$th stage iteration after $n$ time steps, $m$ is the number of stages in each time step, and $\tau_l\Delta_t$ is the time increment corresponding to the $l$th stage.

The Maxwell's equations can be written as the following form

$$\frac{\partial}{\partial t}\begin{pmatrix} \mathbf{H} \\ \hat{\mathbf{E}} \end{pmatrix} = L\begin{pmatrix} \mathbf{H} \\ \hat{\mathbf{E}} \end{pmatrix} \qquad (2.2)$$

$$L = \begin{pmatrix} \{0\}_{3\times3} & -\dfrac{1}{\sqrt{\mu_0\varepsilon_0}}\mathfrak{R}_{3\times3} \\ \dfrac{1}{\sqrt{\mu_0\varepsilon_0}}\mathfrak{R}_{3\times3} & \{0\}_{3\times3} \end{pmatrix} \qquad \hat{\mathbf{E}} = \sqrt{\dfrac{\varepsilon_0}{\mu_0}}\mathbf{E} \qquad (2.3)$$

$$\mathfrak{R} = \begin{pmatrix} 0 & -\dfrac{\partial}{\partial z} & \dfrac{\partial}{\partial y} \\ \dfrac{\partial}{\partial z} & 0 & -\dfrac{\partial}{\partial x} \\ -\dfrac{\partial}{\partial y} & \dfrac{\partial}{\partial x} & 0 \end{pmatrix} = \nabla\times \qquad (2.4)$$

where $\mathbf{E} = (E_x, E_y, E_z)^T$ is the electric field, $\mathbf{H} = (H_x, H_y, H_z)^T$ is the magnetic field, $T$ denotes the transpose, $\{0\}_{3\times3}$ is the $3 \times 3$ null matrix, $\mathfrak{R}$ is the curl operator, and $\varepsilon_0$ and $\mu_0$ are the permittivity and permeability of free space.

The analytical solution of (2.2) from $t = 0$ to $t = \Delta_t$ is

$$\begin{pmatrix} \mathbf{H} \\ \hat{\mathbf{E}} \end{pmatrix}(\Delta_t) = \exp(\Delta_t L)\begin{pmatrix} \mathbf{H} \\ \hat{\mathbf{E}} \end{pmatrix}(0) \qquad (2.5)$$

### 2.1.1 Discretization Strategies in the Time Domain

Regarding the time domain, to approximate the time evolution matrix $\exp(\Delta_t L)$, various discretization methods are proposed.

### 2.1.1.1 Runge-Kutta Method [3, 5, 6]

Using Taylor series to expand the time evolution matrix, we have

$$\exp(\Delta_t L) = I + \Delta_t L + (\Delta_t L)^2 / 2! + (\Delta_t L)^3 / 3! + (\Delta_t L)^4 / 4! + O(\Delta_t^5) \qquad (2.6)$$

A multistage strategy can be adopted to approach (2.6). The numerical implementation of the Runge-Kutta (R-K) method from the $n$th time step to the $(n + 1)$th time step is given by

$$\mathbf{F}^{n+1} = \mathbf{F}^n + (\Delta_t / 6) \times (\mathbf{F}_1 + 2\mathbf{F}_2 + 2\mathbf{F}_3 + \mathbf{F}_4) \qquad (2.7)$$

$$\mathbf{F}_1 = L(t, \mathbf{F}^n), \quad \mathbf{F}_2 = L\left(t + \Delta_t / 2, \mathbf{F}^n + (\Delta_t/2) \cdot \mathbf{F}_1\right) \qquad (2.8)$$

$$\mathbf{F}_3 = L\left(t + \Delta_t/2, \mathbf{F}^n + (\Delta_t/2) \cdot \mathbf{F}_2\right), \quad \mathbf{F}_4 = L\left(t + \Delta_t, \mathbf{F}^n + \Delta_t \cdot \mathbf{F}_3\right) \qquad (2.9)$$

where $\mathbf{F} = (H_x, H_y, H_z, E_x, E_y, E_z)^T$.

The R-K method is high-order accurate and has a good stability in the time direction. However, it introduces amplitude error and consumes a large quantity of memory. To save the memory, E. Turkel improved the traditional R-K method [3] with a reduction of half memory. For nonlinear partial differential equations (PDEs), the improved R-K method is only second-order accurate. But it remains fourth-order accuracy for the linear Maxwell's equations.

### 2.1.1.2 Second-Order Leap-Frog (Staggered) Time-Stepping Approach [1]

The matrix $L$ can be split into the summation of the matrices $U$ and $V$, i.e.,

$$L = U + V \qquad (2.10)$$

$$U = \begin{pmatrix} \{0\}_{3\times3} & -\dfrac{1}{\sqrt{\mu_0\varepsilon_0}}\Re_{3\times3} \\ \{0\}_{3\times3} & \{0\}_{3\times3} \end{pmatrix}, \quad V = \begin{pmatrix} \{0\}_{3\times3} & \{0\}_{3\times3} \\ \dfrac{1}{\sqrt{\mu_0\varepsilon_0}}\Re_{3\times3} & \{0\}_{3\times3} \end{pmatrix} \qquad (2.11)$$

where $U$ and $V$ satisfy $U^\gamma = 0, \gamma \geq 2$, $V^\gamma = 0, \gamma \geq 2$, and $UV \neq VU$.

According to the matrix decomposition technique, we can rewrite the time evolution matrix $\exp(\Delta_t(U + V))$ as a product of matrices

$$\exp(\Delta_t(U+V)) \approx \exp\left(\frac{U\Delta_t}{2}\right) \cdot \exp(V\Delta_t) \cdot \exp\left(\frac{U\Delta_t}{2}\right) \qquad (2.12)$$

We employ the Taylor series to expand each term of (2.12) and then arrive at

$$\exp\left(\frac{U\Delta_t}{2}\right) = I + \frac{U\Delta_t}{2}, \quad \exp(V\Delta_t) = I + V\Delta_t \tag{2.13}$$

where $I$ is the unit matrix.

So (2.12) can be rewritten as

$$\exp(\Delta_t(U+V)) = \left(I + \frac{U\Delta_t}{2}\right) \cdot (I + V\Delta_t) \cdot \left(I + \frac{U\Delta_t}{2}\right) + O(\Delta t^2) \tag{2.14}$$

By recombining the first with the third terms of (2.14) and adopting the staggered time-stepping strategy, we obtain the well-known leap-frog time-stepping approach used in the traditional FDTD method. The approach is simple and requires only one iteration or stage in each time step. It has second-order accuracy, and thus the error will accumulate under a long-term simulation.

### 2.1.1.3 Unconditionally Stable Algorithms [7–11]

Similar to (2.12), we recast the time evolution matrix into

$$\exp(\Delta_t(U+V)) \approx \exp\left(\frac{V\Delta_t}{2}\right) \cdot \exp\left(\frac{U\Delta_t}{2}\right) \cdot \exp\left(\frac{U\Delta_t}{2}\right) \cdot \exp\left(\frac{V\Delta_t}{2}\right) \tag{2.15}$$

The Padé approximation is of form

$$\exp\left(\frac{V\Delta_t}{2}\right) = \frac{I}{I - (V\Delta_t)/2} \quad \exp\left(\frac{U\Delta_t}{2}\right) = \frac{I}{I - (U\Delta_t)/2} \tag{2.16}$$

Alternatively employing Taylor and Padé expansions, (2.15) can be changed to

$$\exp(\Delta_t(U+V)) = \left(\frac{I}{I - (V\Delta_t)/2}\right) \cdot \left(I + \frac{U\Delta_t}{2}\right) \cdot \left(\frac{I}{I - (U\Delta_t)/2}\right) \cdot \left(I + \frac{V\Delta_t}{2}\right) + O(\Delta t^2) \tag{2.17}$$

The alternating direction implicit (ADI) algorithm [7, 8] can efficiently simulate electrically small objects with a large time step and a small space step. Unfortunately, the computational complexity of the algorithm increases due to the implicit matrix inversion in each time step compared to other explicit methods. Furthermore, a larger time step will make the numerical dispersion of the ADI algorithm worse, which can be improved by the unconditionally stable Crank-Nicolson scheme [10] but with higher computational complexity. Recently, an unconditionally stable one-step algorithm [9] has been proposed based on the accurate solution of the time evolution matrix in the space or spectral domain. Although it saves a lot of CPU time, the one-step algorithm is very difficult to handle inhomogeneous boundary conditions.

### 2.1.1.4 High-Order Symplectic Integration Scheme [12–15]

With the aid of the high-order decomposition technique of the exponential matrix, the time evolution matrix can be approximately reconstructed by the $m$-stage $p$th-order symplectic integrator. Here, the matrix corresponding to each stage is called the elementary symplectic mapping

$$\exp(\Delta_t(U+V)) = \prod_{l=1}^{m} \exp(d_l \Delta_t V) \exp(c_l \Delta_t U) + O\left(\Delta_t^{p+1}\right) \qquad (2.18)$$

where $c_l$ and $d_l$ are the symplectic integrators or propagators.

The high-order symplectic integration scheme has desired numerical precision and high numerical stability but needs multiple stages in every time step. Compared with the R-K method, it has the energy-preserving property and saves memory.

### 2.1.2 Space Discretization Methods

Besides discretization strategies in the time domain, versatile space discretization methods are also proposed to approximate the spatial first-order derivatives.

### 2.1.2.1 Second-Order Leap-Frog (Staggered) Difference [1]

$$\left.\frac{\partial F}{\partial \delta}\right|_{\delta=h\Delta_\delta} = \frac{F(h+1/2) - F(h-1/2)}{\Delta_\delta} + \left(\Delta_\delta^2\right) \quad \delta = x, y, z \quad h = i, j, k \quad (2.19)$$

where $F(h+1/2)$ denotes the value of $F$ at $\delta = (h+1/2)\Delta_\delta$, and $\partial F / \partial \delta|_{\delta=h\Delta_\delta}$ denotes the first derivatives of $F$ at $\delta = h\Delta_\delta$.

The second-order leap-frog (staggered) difference has advantages in terms of low complexity and natural parallelism. Meanwhile, it can model electromagnetic responses of inhomogeneous materials with curved boundaries by using local subgridding [16] and conformal techniques [17–21]. In view of a long-term or large-scale simulation, the method produces a significant numerical dispersion. Consequently, to accurately capture the wave physics of inhomogeneous scatterers, fine spatial grids should be adopted, which consumes a great number of computer resources.

### 2.1.2.2 Fourth-Order Staggered Difference [22–25]

$$\left.\frac{\partial F}{\partial \delta}\right|_{\delta=h\Delta_\delta} = \frac{9}{8} \times \frac{F(h+1/2) - F(h-1/2)}{\Delta_\delta} - \frac{1}{24} \times \frac{F(h+3/2) - F(h-3/2)}{\Delta_\delta} + \left(\Delta_\delta^4\right)$$

$$(2.20)$$

The fourth-order staggered difference achieves much lower numerical dispersion compared to the second-order staggered difference and shows potential advantages in large-scale electromagnetic simulations. The main pitfalls of the method involve low stability and difficult treatments of inhomogeneous boundaries. The former can be improved by introducing the R-K method or the symplectic integration scheme. The latter can be improved by recently developed high-order conformal and subgridding techniques [14,26–29].

### 2.1.2.3 Fourth-Order Implicit Compact Difference Algorithm [5, 6]

$$\frac{\left.\frac{\partial F}{\partial \delta}\right|_{\delta=(h+1)\Delta_\delta} + \left.\frac{\partial F}{\partial \delta}\right|_{\delta=(h-1)\Delta_\delta}}{24} + \frac{11}{12}\left.\frac{\partial F}{\partial \delta}\right|_{\delta=h\Delta_\delta} = \frac{F(h+1/2)-F(h-1/2)}{\Delta_\delta} + \left(\Delta_\delta^4\right)$$

(2.21)

The fourth-order implicit compact difference algorithm, which is more accurate than the fourth-order explicit staggered difference method, could employ the same conformal and subgridding techniques as the second-order explicit staggered difference. However, a tridiagonal matrix inversion is required for each time step, resulting in low computational efficiency.

### 2.1.2.4 Multiresolution Expansion Method [30]

$$\left(\frac{\partial F}{\partial \delta}\right)_h \approx \frac{1}{\Delta_\delta}\int \psi_h(\delta)\frac{\partial \psi_{h'+1/2}(\delta)}{\partial \delta} F_{h'+1/2}\, d\delta$$

(2.22)

where $(\partial F/\partial \delta)_h$ denotes the derivatives of $F$ along the $\delta$ direction, $\psi$ is the scaling function in the wavelet theory, and $F_{h'+1/2}$ denotes the sampling of $F$ at the staggered grids.

The multiresolution expansion method saves a large quantity of memory and CPU time with drastically reduced sampling points per wavelength. Similar to the fourth-order staggered difference, it has low numerical stability and is not good at modeling inhomogeneous boundaries. The multiple image technique (MIT) [31] partially overcomes the difficulties in boundary treatments. However, the high-order precision of the multiresolution expansion method cannot be maintained.

### 2.1.2.5 Pseudo-Spectral Scheme [32]

$$\left(\frac{\partial F}{\partial \delta}\right)_h \approx \frac{2\pi}{N_\delta \Delta_\delta}\xi_\delta^{-1}\left[j_0 h \cdot \xi_\delta\left(F_h\right)\right]$$

(2.23)

where $\xi$ and $\xi^{-1}$ respectively denote the forward and inverse operators of the centered discrete Fourier transform, $F_h$ denotes the sampling of $F$ at the collocated grids, $N_\delta$ is the number of sampling points, and $j_0$ is the imaginary unit.

The pseudo-spectral scheme achieves the exponential convergence rate and remarkably lowers the complexity with coarse grids. The intrinsic weakness of the scheme lies at the Gibbs phenomena occurring at the inhomogeneous boundaries, which limits the application of the scheme.

From our personal views, a stable, accurate, fast, and efficient time-domain solver for Maxwell's equations can be proposed based on the following principles. First, one can design new methods to approximate the time evolution matrix or spatial first-order derivatives. Second, the proposed time-domain and space-domain algorithms can be recombined with each other. Third, these developed algorithms can be hybridized [33–35] with the finite-element time-domain [36], finite-volume time-domain [37] or discontinuous Galerkin time-domain methods [38].

## 2.2 Core Techniques

Nanoscience and technology become more and more important in cutting-edge industry. The advances in nanoscience and nanotechnology are attributed to newly acquired abilities to measure, fabricate, and manipulate individual structures on the nanometer scale. Controlling the light-matter interaction at the nanoscale is of paramount importance for emerging nanodevices and quantum devices. Photonics [39,40], plasmonics [41–47], and metamaterials [48–55] unprecedentedly change the traditional views and tools to control the propagation, radiation, and scattering of electromagnetic fields and to some extent break the diffraction limit in optics. Characterizing unique features, exploring new functionalities, and optimizing performances of nanostructures and nanodevices strongly depend on the rigorous solution of Maxwell's equations. As an accurate, fast, and efficient full-wave solver, the FDTD method can help to predict electromagnetic responses, understand working principles, reduce experimental costs, and shorten development periods of nano-optical designs. On one hand, the random, multilayered, and periodic nanostructures of interest have the inhomogeneous, dispersive, and anisotropic characteristics, presenting many challenges in developing the FDTD method. On the other hand, the large-scale simulation with broadband and wide-angle excitations requires huge computer resources and has to be tackled with the parallel FDTD technique. In this section, we will briefly describe core techniques of the FDTD method and particularly focus on those for nano-optics applications.

### 2.2.1 Basic Update Equations and Material Averaging Technique

Considering homogenous and isotropic media, the basic update equation of the FDTD method for the $x$ component of the scaled electric field (E-field) is given by

$$\hat{E}_x^{n+1}\left(i+\frac{1}{2},j,k\right)=\hat{E}_x^{n}\left(i+\frac{1}{2},j,k\right)+\frac{1}{\varepsilon_r\left(i+\frac{1}{2},j,k\right)}$$

$$\times\left\{\begin{array}{l}\alpha_y\times\left[H_z^{n+1/2}\left(i+\frac{1}{2},j+\frac{1}{2},k\right)-H_z^{n+1/2}\left(i+\frac{1}{2},j-\frac{1}{2},k\right)\right]\\[2mm]-\alpha_z\times\left[H_y^{n+1/2}\left(i+\frac{1}{2},j,k+\frac{1}{2}\right)-H_y^{n+1/2}\left(i+\frac{1}{2},j,k-\frac{1}{2}\right)\right]\end{array}\right\}$$

$$(2.24)$$

$$\alpha_y=\frac{1}{\sqrt{\mu_0\varepsilon_0}}\frac{\Delta_t}{\Delta_y},\quad \alpha_z=\frac{1}{\sqrt{\mu_0\varepsilon_0}}\frac{\Delta_t}{\Delta_z}\qquad(2.25)$$

where $\varepsilon_r$ is the relative permittivity. Similarly, the update equation for the $y$ component of the magnetic field (H-field) is of form

$$H_y^{n+1/2}\left(i+\frac{1}{2},j,k+\frac{1}{2}\right)=H_y^{n-1/2}\left(i+\frac{1}{2},j,k+\frac{1}{2}\right)+\frac{1}{\mu_r\left(i+\frac{1}{2},j,k+\frac{1}{2}\right)}$$

$$\times\left\{\begin{array}{l}\alpha_x\times\left[\hat{E}_z^{n}\left(i+1,j,k+\frac{1}{2}\right)-\hat{E}_z^{n}\left(i,j,k+\frac{1}{2}\right)\right]\\[2mm]-\alpha_z\times\left[\hat{E}_x^{n}\left(i+\frac{1}{2},j,k+1\right)-\hat{E}_x^{n}\left(i+\frac{1}{2},j,k\right)\right]\end{array}\right\}$$

$$(2.26)$$

$$\alpha_x=\frac{1}{\sqrt{\mu_0\varepsilon_0}}\frac{\Delta_t}{\Delta_x}\qquad(2.27)$$

where $\mu_r$ is the relative permeability. For the lossy media (take $\hat{E}_x$ component as an example), the iteration equation can be rewritten as

$$\hat{E}_x^{n+1}\left(i+\frac{1}{2},j,k\right) = \exp(-\xi) \times \hat{E}_x^n\left(i+\frac{1}{2},j,k\right)$$

$$+ \frac{1-\exp(-\xi)}{\xi} \times \frac{1}{\varepsilon_r\left(i+\frac{1}{2},j,k\right)}$$

$$\times \left\{ \begin{array}{l} \alpha_y \times \left[ H_z^{n+1/2}\left(i+\frac{1}{2},j+\frac{1}{2},k\right) - H_z^{n+1/2}\left(i+\frac{1}{2},j-\frac{1}{2},k\right) \right] \\ -\alpha_z \times \left[ H_y^{n+1/2}\left(i+\frac{1}{2},j,k+\frac{1}{2}\right) - H_y^{n+1/2}\left(i+\frac{1}{2},j,k-\frac{1}{2}\right) \right] \end{array} \right\}$$

(2.28)

where

$$\xi = \frac{\Delta_t \sigma\left(i+\frac{1}{2},j,k\right)}{\varepsilon\left(i+\frac{1}{2},j,k\right)},$$

and $\sigma$ is the electric conductivity. For the exponential functions, we can employ the Padé expansion to access their values, i.e.,

$$\exp(-\xi) = \frac{1-\dfrac{\xi}{2}+\dfrac{\xi^2}{12}}{1+\dfrac{\xi}{2}+\dfrac{\xi^2}{12}} + O(\xi^5), \quad \frac{1-\exp(-\xi)}{\xi} = \frac{1-\dfrac{\xi}{10}+\dfrac{\xi^2}{60}}{1+\dfrac{2\xi}{5}+\dfrac{\xi^2}{20}} + O(\xi^5)$$

(2.29)

For inhomogeneous nonmagnetic media, which are always modeled in nano-optics problems, the above constitutive parameters can be averaged as

$$\bar{\varepsilon}_r = \frac{1}{A_S} \iint_S \varepsilon_r dS, \quad \bar{\sigma} = \frac{1}{A_S} \iint_S \sigma dS$$

(2.30)

where $S$ is the surface corresponding to the E-field component enclosed by the four H-field components, $A_S$ is the area of $S$, and $\bar{\varepsilon}_r$ and $\bar{\sigma}$ are the averaged relative permittivity and electric conductivity, respectively. The integrals (2.30) can be converted into the summation forms according to the subgridding techniques [4,14].

### 2.2.2 Absorbing Boundary Condition

Because of limited computer resources, an absorbing boundary condition is compulsory to truncate the infinite free space (or air region) for simulating the optical radiation and scattering from nanoscatterers. The perfectly matched layer (PML) [56,57] is able to absorb the outgoing waves without spurious reflections and to perfectly simulate unbounded wave propagations.

Using the split-field technique, we decompose the $x$ polarized E-field into two subcomponents propagating along the $y$- and $z$-directions, respectively.

$$\hat{E}_x = \hat{E}_{xy} + \hat{E}_{xz} \tag{2.31}$$

The update equation of the subcomponent propagating along the $y$-direction is written as

$$\hat{E}_{xy}^{n+1}\left(i+\frac{1}{2},j,k\right) = \exp(-\xi) \times \hat{E}_{xy}^{n}\left(i+\frac{1}{2},j,k\right) + \frac{1-\exp(-\xi)}{\xi}$$

$$\times \alpha_y \times \left[ H_z^{n+1/2}\left(i+\frac{1}{2},j+\frac{1}{2},k\right) - H_z^{n+1/2}\left(i+\frac{1}{2},j-\frac{1}{2},k\right) \right] \tag{2.32}$$

where

$$\xi = \frac{\Delta_t \sigma_y\left(i+\frac{1}{2},j,k\right)}{\varepsilon_0}$$

Likewise, the update equation of the subcomponent propagating along the $z$-direction is given by

$$\hat{E}_{xz}^{n+1}\left(i+\frac{1}{2},j,k\right) = \exp(-\xi) \times \hat{E}_{xz}^{n}\left(i+\frac{1}{2},j,k\right) - \frac{1-\exp(-\xi)}{\xi}$$

$$\times \alpha_z \times \left[ H_y^{n+1/2}\left(i+\frac{1}{2},j,k+\frac{1}{2}\right) - H_y^{n+1/2}\left(i+\frac{1}{2},j,k-\frac{1}{2}\right) \right] \tag{2.33}$$

where

$$\xi = \frac{\Delta_t \sigma_z\left(i+\frac{1}{2},j,k\right)}{\varepsilon_0}$$

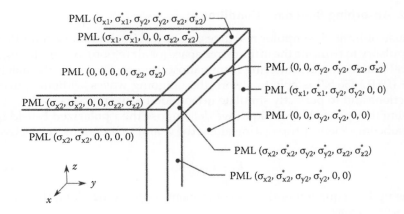

**FIGURE 2.1**
The settings of electric and magnetic conductivities in the PMLs.

To reduce the spurious numerical reflection, the electric conductivity can be set as the following polynomial form, i.e.,

$$\sigma(\Lambda) = \sigma_{max} \left( \frac{\Lambda}{\Gamma} \right)^{\kappa} \tag{2.34}$$

where $\Lambda$ is the distance from the PML-air interface, $\Gamma$ is the thickness of the PMLs, $\sigma_{max}$ is the maximum electric conductivity, and $\kappa$ is the order of the polynomial. Besides the $\hat{E}_x$ component, all the other components should use the split-field technique also. To absorb the outgoing waves from different directions, the settings of the electric and magnetic conductivities in the PMLs are shown in Figure 2.1. Additionally, an easier and more elegant technique to implement the PML, called convolution PML (CPML), can be found in [58].

### 2.2.3 Source Excitations

The typical source excitations in the FDTD solution include the plane wave and point (dipole) sources.

A rigorous way of introducing the plane wave excitation is to use the total-field and scattered-field (TF-SF) technique [2]. We assume that the E-field is $x$-polarized and propagates along the $z$-direction. The total-field region occupies $[i_1, i_2] \times [j_1, j_2] \times [k_1, k_2]$. The additional update equation of $\hat{E}_x$ field at the $k = k_1$ plane is of form

$$\hat{E}_x^{n+1} \left( i + \frac{1}{2}, j, k_1 \right) = \hat{E}_x^n \left( i + \frac{1}{2}, j, k_1 \right) + \alpha_z \times H_{y,inc}^{n+1/2} \left( k_1 - \frac{1}{2} \right) \tag{2.35}$$

At the $k = k_2$ plane, the additional update equation of $\hat{E}_x$ field is given by

$$\hat{E}_x^{n+1}\left(i+\frac{1}{2},j,k_2\right) = \hat{E}_x^n\left(i+\frac{1}{2},j,k_2\right) - \alpha_z \times H_{y,inc}^{n+1/2}\left(k_2+\frac{1}{2}\right) \qquad (2.36)$$

where $i_1 \leq i \leq i_2 - 1$, $j_1 \leq j \leq j_2$, and $H_{y,inc}$ is the incident H-field. We can modify the iterations of $\hat{E}_z$, $H_y$, and $H_z$ components at the TF-SF interfaces accordingly through introducing the equivalent incident sources. In view of the incident source, the one-dimensional (1D) FDTD method can be employed. For example, $\hat{E}_{x,inc}$ can be updated by

$$\hat{E}_{x,inc}^{n+1}(k) = \hat{E}_{x,inc}^n(k) - \alpha_z \times \left[ H_{y,inc}^{n+1/2}\left(k+\frac{1}{2}\right) - H_{y,inc}^{n+1/2}\left(k-\frac{1}{2}\right) \right] \qquad (2.37)$$

At the $k = k_s$ point, the source $\zeta(t)$ is added as

$$\hat{E}_{x,inc}^{n+1}(k_s) = \zeta^{n+1}((n+1)\Delta_t) \qquad (2.38)$$

where

$$\zeta(t) = -\cos(\omega t)\exp\left(-\frac{4\pi(t-T_0)^2}{W^2}\right) \qquad (2.39)$$

with an effective frequency range $f \in \left[ \omega/2\pi - 2/W, \omega/2\pi + 2/W \right]$ and $T_0 = 9\pi/2\omega$. Alternatively, the plane wave incidence can be approximately realized by exciting a transparent or soft source at a planar surface. One can place the planar source along one of the inner PML boundaries (see Figure 2.2). Make sure that the size of the source is the same as that of the entire simulation cell (including the PML thickness) along its planar dimension. Under this situation,

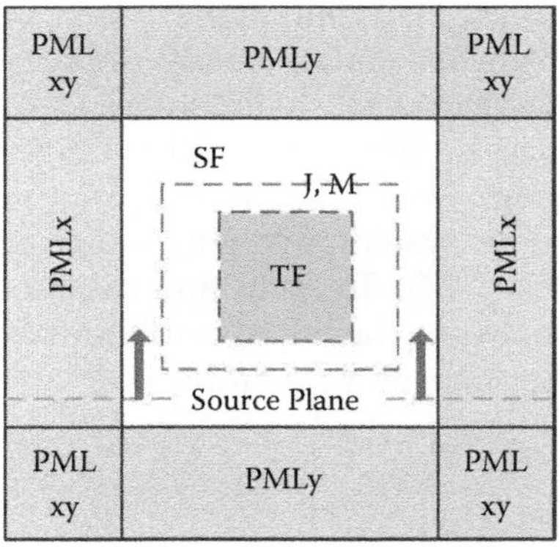

**FIGURE 2.2**
The 2D FDTD configurations involving the perfectly matched layers (PMLs), total-field (TF) region, scattered-field (SF) region, planar source plane, and equivalent currents $J$ and $M$ for the near-to-far-field (NFF) transformation.

the entire computational domain becomes the total-field region, and thus there is no scattered-field region. The scattered fields can be obtained by extracting the incident fields from the total fields by running the FDTD program twice.

The dipole source is essential to predict the emission of atoms, molecules, and quantum dots in inhomogeneous environments. It can be used for analyzing the spontaneous emission [59–62], near-field heat transfer [63–65], Casimir force [66–68], solar cells [69], and so on. The electromagnetic radiation of a point source relates to the local density of states (LDOS) concept [42], which plays a fundamental role in modern physics. The dipole source is implemented as follows:

$$\hat{E}_x^{n+1}(k_s) = \hat{E}_x^n(k_s) - \frac{\Delta_t}{\varepsilon_0 \delta^3} \left[ \frac{dp}{dt} \right]^{n+1/2} \tag{2.40}$$

where $p = A \exp[-4\pi((t - T_0)/W)^2]$ corresponds to the dipole momentum and $\delta^3 = \Delta_x \Delta_y \Delta_z$ is the volume of a Yee cell. In fact, the transparent source adopted is the differential Gaussian pulse without the direct current (DC) component. The LDOS, which counts the number of electromagnetic modes where photons can be emitted at the specific location of the emitter, can be calculated by the superposition of the projected LDOS. The projected (perpolarization) LDOS is exactly proportional to the power radiated by an $l$-oriented point-dipole current $J(\omega)$ at the given position in space [70]:

$$\text{LDOS}(\mathbf{r}, \omega) = \frac{-2}{\pi} \varepsilon(\mathbf{r}) \frac{\text{Re}\left[ \tilde{E}_l(\mathbf{r}, \omega) \tilde{J}(\omega) \right]}{\left| \tilde{J}(\omega) \right|^2} \tag{2.41}$$

where the normalization of $|\tilde{J}(\omega)|^2$ is necessary for obtaining the power exerted by a unit-amplitude Hertzian dipole.

The eigenvalue problem is fundamentally important in the nano-optics field. On one hand, the excitation solution by the plane wave or dipole sources can be expanded in terms of dominant eigenmodes. On the other hand, the dispersion diagram or relation generated provides key characteristics of an electromagnetic system involving the group velocity, quality factor, and density of states (DOS). To excite all possible resonant modes in the electromagnetic system, we set the initial condition for all the E-field components, taking the $z$ component as an example [71]:

$$\hat{E}_z^0\left( i, j, k + \frac{1}{2} \right) = \exp\left[ -\frac{(i - i_c)^2 + (j - j_c)^2}{2\tau_g^2} \right] \tag{2.42}$$

where $i_c$, $j_c$, $k_c$ are the center points of the system, $\tau_g$ is the width of the spatial pulse, and $i,j,k$ can be set as $i > i_c$, $j > j_c$, and $k > k_c$. The asymmetric setup enables all the eigenmodes to be excited.

### 2.2.4 Near-to-Far-Field Transformation

To obtain far-field scattering or radiation information, the near-to-far-field (NFF) transformation [2] should be implemented at a closed surface enclosing the scatterers. The E-field and H-field values at the closed surface in the frequency domain should be evaluated via the discrete Fourier transform or fast Fourier transform before the NFF transformation.

The discrete Fourier transform for the scaled $\hat{E}_x$ field is described as

$$\tilde{\hat{E}}_x(f) = \sum_{n=0}^{n_{max}} \hat{E}_x^n(n\Delta_t)\exp(-j_0 2\pi f n\Delta_t) \tag{2.43}$$

where $\tilde{\hat{E}}$ is the scaled E-field in the frequency domain, and $n_{max}$ is the required time steps before the steady state is reached. The equivalent electric current $\tilde{\mathbf{J}}$ and magnetic current $\hat{\tilde{\mathbf{j}}}_m$ are respectively defined by

$$\tilde{\mathbf{J}} = \mathbf{n} \times \tilde{\mathbf{H}}, \quad \hat{\tilde{\mathbf{j}}}_m = -\mathbf{n} \times \tilde{\hat{\mathbf{E}}} \tag{2.44}$$

where $\mathbf{n}$ is the outer normal vector of the closed surface. With the help of the equivalent principle, the E-field in the far-field zone can be calculated by

$$\tilde{\hat{E}}_\theta = -j_0 k_0 \frac{\exp(-j_0 k_0 r)}{4\pi r}\left[\left(\tilde{f}_x \cos\theta\cos\varphi + \tilde{f}_y \cos\theta\sin\varphi - \tilde{f}_z \sin\theta\right)\right.$$
$$\left. + \left(-\tilde{f}_{mx}\sin\varphi + \tilde{f}_{my}\cos\varphi\right)\right] \tag{2.45}$$

$$\tilde{\hat{E}}_\varphi = j_0 k_0 \frac{\exp(-j_0 k_0 r)}{4\pi r}\left[\left(\tilde{f}_{mx}\cos\theta\cos\varphi + \tilde{f}_{my}\cos\theta\sin\varphi - \tilde{f}_{mz}\sin\theta\right)\right.$$
$$\left. + (\tilde{f}_x \sin\varphi - \tilde{f}_y \cos\varphi)\right] \tag{2.46}$$

$$\tilde{f}_\delta = \iint_A \tilde{J}_\delta(r')\exp[j_0 k_0(x'\sin\theta\cos\varphi + y'\sin\theta\sin\varphi + z'\cos\theta)]ds' \tag{2.47}$$

$$\tilde{\hat{f}}_{m\delta} = \iint_A \hat{\tilde{J}}_{m\delta}(r')\exp[j_0 k_0(x'\sin\theta\cos\varphi + y'\sin\theta\sin\varphi + z'\cos\theta)]ds' \tag{2.48}$$

where $k_0$ is the wave number of free space, $r$ is the distance from the source to the field points, $\iint_A ds'$ denotes the area integral in the whole closed surface, and $\theta$ and $\varphi$ are the spherical angles.

### 2.2.5 Periodic Structures

A periodic structure has profound theoretical meanings and practical uses in nanoscience and technology. It can be made from dielectrics, metals, and their hybrids with structured lattices. The periodic structure is crucial for optical components and devices, such as nanoantennas [72], nanocircuits (waveguides, polarizers, filters) [40, 73–75], and optoelectronics (solar cells, light-emitting diodes, and lasers) [45, 76–79]. First, due to the constructive and destructive interferences, the period structures could open up a band gap where the photon emission is forbidden or inhibited and form a band edge where the light intensity is extremely enhanced [39, 40]. Second, the group delay of modulated optical signals is also highly tunable in the slow-wave structure [80,81]. Third, periodic structures could control the emission direction of atoms, molecules, and quantum dots by the diffraction effect of Floquet modes [82,83]. Finally, the interferences between the quasi-guided mode of periodic structures and incident light induce many interesting physical phenomena involving the Fano-resonance or Wood's anomaly [84–86], electromagnetically induced transparency [87], and so on. In sum, modeling periodic structures and unveiling relevant physical mechanisms have a high impact in the nano-optics field. The challenges in the FDTD method to model periodic structures can be understood from the Bloch theorem:

$$\tilde{E}(r+R) = \tilde{E}(r)\exp(-j_0 k_B \cdot R) \tag{2.49}$$

$$\tilde{E}(r) = \tilde{E}(r+R)\exp(j_0 k_B \cdot R) \tag{2.50}$$

where $k_B$ is the Bloch wave vector.

In the time domain, (2.49) can be numerically implemented because $\tilde{E}(r+R)$ has a time delay (or retardation) with respect to $\tilde{E}(r)$. However, the anti-causal property can be found in (2.50). The value of $\tilde{E}(r)$ at the current time step depends on that of $\tilde{E}(r+R)$ at the following time step, which is unknown physically. Here we implement periodic boundary conditions in the time domain by the constant horizontal wave number approach [88] with complex field values. Although the memory cost of the FDTD method will double, the phase delay in periodic boundary conditions (PBCs) can be incorporated conveniently.

For example, to update $\hat{E}_x$ field on the periodic boundary, the value of $H_z$ component outside the unit cell is needed. Fortunately, assuming the periodicity in the $y$-direction, one can use the $H_z$ component of interest inside the unit cell to update the E-field, such that

$$\hat{E}_x^{n+1}\left(i+\frac{1}{2},0,k\right) = \hat{E}_x^n\left(i+\frac{1}{2},0,k\right) + \frac{1}{\varepsilon_r\left(i+\frac{1}{2},0,k\right)}$$

$$\times \left\{ \begin{array}{l} \alpha_y \times \left[ H_z^{n+1/2}\left(i+\frac{1}{2},\frac{1}{2},k\right) - H_z^{n+1/2}\left(i+\frac{1}{2},n_y-\frac{1}{2},k\right) \times \exp(jk_y P_y) \right] \\ -\alpha_z \times \left[ H_y^{n+1/2}\left(i+\frac{1}{2},0,k+\frac{1}{2}\right) - H_y^{n+1/2}\left(i+\frac{1}{2},0,k-\frac{1}{2}\right) \right] \end{array} \right\} \tag{2.51}$$

where $P_y$ and $k_y$ are, respectively, the periodicity and Bloch wave number along the $y$-direction. The TF-SF technique is not applicable to implement the plane wave source for periodic structures, and the pure total-field technique should be employed with a transparent source at the excitation plane, which has been described in the "source excitations" subsection. The incident angle of the plane wave can be controlled by changing the value of $k_y$. There exists a problem of horizontal resonance [88], where fields do not decay to zero over time. To avoid this problem, the proper frequency range for the excitation waveform must be chosen as follows:

$$f_c = \frac{k_y c}{2\pi} + \frac{BW}{2} \qquad (2.52)$$

where $f_c$ is the center frequency of the Gaussian pulse and $BW$ is the corresponding bandwidth. The constant horizontal wave number approach can be naturally extended to the skewed grid periodic structures with a simple linear interpolation [88].

### 2.2.6 Dispersive Media

In the nano-optics field, most materials are dispersive with a frequency-dependent complex permittivity. Debye, Lorentz, and Drude media are three main classes of dispersive materials and have different frequency-dependent behaviors. Various techniques have been developed to model these dispersive materials, such as the recursive convolution (RC) [89], the auxiliary differential equation (ADE) [90], piecewise linear recursive convolution [91], and the Z-transform methods [92]. Mathematically, we can use several Lorentz terms with different harmonic resonances to represent the permittivity of an arbitrary dispersive material, i.e.,

$$\varepsilon(\omega) = \varepsilon_\infty + \sum_n \frac{\beta_n \omega_n^2}{\omega_n^2 - \omega^2 + j\omega\gamma_n} \qquad (2.53)$$

where the first term is the instantaneous dielectric function corresponding to the infinite-frequency response and the Lorentz terms are related to the frequency-dependent polarization density in the material

$$\mathbf{D} = \varepsilon\mathbf{E} = \varepsilon_\infty\mathbf{E} + \mathbf{P} \qquad (2.54)$$

$$\mathbf{P} = \sum_n \mathbf{P}_n \qquad (2.55)$$

$$\frac{d^2\mathbf{P}_n}{dt^2} + \gamma_n \frac{d\mathbf{P}_n}{dt} + \omega_n^2\mathbf{P}_n = \beta_n \omega_n^2 \mathbf{E} \qquad (2.56)$$

For a practical curve fitting process, we must force $\omega_n \Delta t/2 \leq 1$ to ensure the stability of the algorithm and $\beta_n$ also cannot be too large.

The Ampere's law in the time domain can be expressed as

$$\nabla \times H = \varepsilon_\infty \frac{d}{dt} \mathbf{E}(t) + \sigma \mathbf{E}(t) + \sum_p \mathbf{J}_p(t) \tag{2.57}$$

where $\mathbf{J}_p$ are the polarization currents satisfying

$$\frac{d^2 \mathbf{J}_p}{dt^2} + \gamma_n \frac{d\mathbf{J}_p}{dt} + \omega_n^2 \mathbf{J}_p = \beta_n \omega_n^2 \frac{d\mathbf{E}}{dt} \tag{2.58}$$

Using the ADE technique, we discretize the differential equation (2.58) with centered differences

$$\frac{\mathbf{J}_p^{n+1} - 2\mathbf{J}_p^n + \mathbf{J}_p^{n-1}}{\Delta t^2} + \gamma_n \frac{\mathbf{J}_p^{n+1} - \mathbf{J}_p^{n-1}}{2\Delta t} + \omega_n^2 \mathbf{J}_p^n = \beta_n \omega_n^2 \frac{\mathbf{E}^{n+1} - \mathbf{E}^{n-1}}{2\Delta t} \tag{2.59}$$

From the ADE, we can obtain the polarization current $\mathbf{J}_p$ at the $(n+1)$th time step by those at the $n$th and $(n-1)$th. Considering that we require the polarization current at the $(n+1/2)$th time step from (2.57), a simple linear interpolation can be adopted:

$$\mathbf{J}_p^{n+1/2} = \frac{1}{2}\left[\mathbf{J}_p^n + \mathbf{J}_p^{n+1}\right] \tag{2.60}$$

Substituting (2.59) and (2.60) into (2.57), we can get the update equation of E-field components. Starting with the known (stored) component values of $\mathbf{E}^{n-1}$, $\mathbf{E}^n$, $\mathbf{J}_p^{n-1}$, $\mathbf{J}_p^n$ and $\mathbf{H}^{n+1/2}$, we calculate the new $\mathbf{E}^{n+1}$ components. Then, we get the new $\mathbf{J}_p^{n+1}$ components by using the just-computed $\mathbf{E}^{n+1}$ components.

---

## 2.3 Numerical Examples for Nano-Optics Applications

Having unique features of tunable resonance and unprecedented near-field concentration, plasmon is an enabling technique for light manipulation and management [41–47]. By altering the metallic nanostructure, the properties of plasmons, in particular their interactions with light, can be tailored, which offers the potential for developing emerging optical components and devices. Meanwhile, the use of metallic materials with a negative permittivity is one of the most feasible ways of circumventing the fundamental (half-wavelength) limit and achieving localization of electromagnetic energy (at optical frequencies) into nanoscale. In this section, we will investigate and explore plasmonic effects in various nano-optics applications.

### 2.3.1 Thin-Film Solar Cells

Solar cells [93,94], which can provide renewable and clean energy by converting the sunlight to electrical power, have attracted much attention in the past few years. Despite the growing importance, we need to reduce the cost of solar cells and increase the power conversion efficiency (PCE) before they can successfully replace fossil fuels for electrical power generation. A light-trapping scheme can realize the above two goals simultaneously and thus is useful for emerging solar cell technology meeting clean energy demands. As an efficient light-trapping technique [45], surface plasmons are collective oscillations of free electrons in metals that are confined to the surface between the metal and dielectric and interact strongly with light. The extremely near-field enhancement inherent from surface plasmons allows us to significantly improve the optical absorption of thin-film solar cells.

Here, we use the FDTD method to simulate a typical thin-film solar cell nanostructure in the literature [78]. Figure 2.3 shows the unit cell of the plasmonic thin-film solar cell. The incident wave is a *p*-polarized plane wave propagating vertically into the plasmonic solar cell. The ADE technique and Lorenz models are adopted for simulating the dispersive materials involving the absorbing material (amorphous silicon, A-Si) and metal (Au). The frequency-domain values of E-fields are obtained by the discrete (fast) Fourier transform. The real and imaginary parts of permittivities for the A-Si and Au are depicted in Figure 2.4. The complex permittivity of Au can be expressed by the Brendel-Bormann model [95] and that of A-Si is taken from [96]. The relative permittivities of ITO and SiO$_2$ are 4.0 and 2.1, respectively.

The electron-hole pair generation of solar cells depends on the photon energy absorbed by the absorbing material per unit time per unit area, i.e.,

$$\eta = \frac{\int_S \sigma(\omega)\left|\tilde{E}(\omega,r)\right|^2 dS}{\Delta_S} = \frac{-\omega\varepsilon_0 \int_S \text{Im}\left[\varepsilon_r(\omega)\right]\left|\tilde{E}(\omega,r)\right|^2 dS}{\Delta_S} \tag{2.61}$$

where $\eta$ is the power density, $S$ denotes the region of the absorbing material, $\Delta_S$ is the area of $S$, and $\sigma(\omega) = -\omega\varepsilon_0 \text{Im}[\varepsilon_r(\omega)]$ is the conductivity of the absorbing material. Figure 2.6 shows the absorbed power density of the A-Si layer. Using the planar Au layer, the non-strip (planar) structure is also modeled. For the non-strip structure, $d_2 = 140$ nm is adopted for achieving the same A-Si area while other parameters are unchanged. The momentum of surface plasmons [41] is given by

$$k_{sp} = k_0 \sqrt{\frac{\varepsilon_m \varepsilon_a}{\varepsilon_m + \varepsilon_a}} \tag{2.62}$$

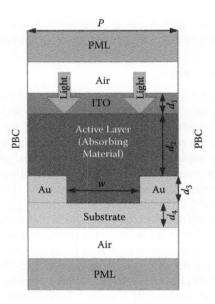

**FIGURE 2.3**
The unit cell of the plasmonic thin-film solar cell. The four-layered structure includes the indium tin oxide (ITO), active layer (absorbing material of amorphous silicon), metal (Au) electrodes, and substrate (SiO$_2$) with thicknesses of $d_1$, $d_2$, $d_3$, and $d_4$, respectively. The distance between two adjacent strips is $w$ and the periodicity is $P$. The incident light propagates into the structure through the ITO. The PMLs are employed at the top and the bottom of the solar cell structure. The PBCs at the left and right sides of the unit cell are imposed. The geometric parameters of the device are set as $d_1 = 25$ nm, $d_2 = 120$ nm, $d_3 = 40$ nm, $d_4 = 30$ nm, $w = 100$ nm, and $P = 200$ nm.

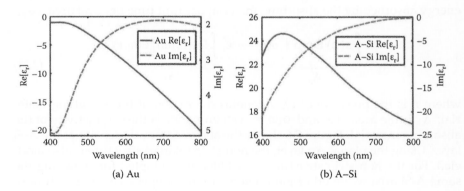

**FIGURE 2.4**
The real and imaginary parts of relative permittivities for (a) Au and (b) A-Si.

where $\varepsilon_m$ and $\varepsilon_a$ are the permittivities of metal and absorbing material, respectively. It is well known that surface plasmons will exist if the condition $Re(-\varepsilon_m) > Re(\varepsilon_a)$ is satisfied. Thus the momentum of surface plasmons $k_{sp}$ is larger than free space momentum $k_0$ of the plane wave (sunlight). Hence, additional momentum should be provided for exciting the surface plasmon. From Figure 2.6, the periodic strip incorporated solar cell shows much stronger absorption due to the excited surface plasmons. However, for the non-strip structure, the surface plasmons cannot be excited due to the momentum mismatch. The reflection coefficient results as presented in Figure 2.5 not only validate the FDTD results but also confirm the strong optical absorption of A-Si from 660 to 800 nm. The surface plasmon is successfully excited by the nanostrip as shown in Figure 2.7(b) in comparison with Figure 2.7(d). According to the mode conversion theory, the subwavelength strip can excite evanescent wave components, which may provide additional momentum. Furthermore, the Floquet modes supported by the periodic strip structure can also overcome the momentum mismatch problem. The waveguide mode also enhances the optical absorption of solar cells as illustrated in Figures 2.7(a) and (c). Particularly, the surface plasmon mode can be coupled to the waveguide mode or Floquet mode, which has a great help for boosting the absorption (Figure 2.7[b]).

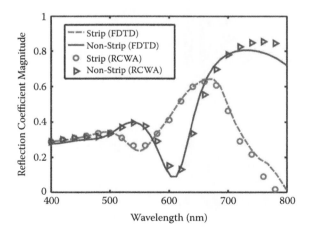

**FIGURE 2.5**
The reflection coefficients of the strip and non-strip solar cell architectures calculated by the FDTD method and rigorous coupled-wave analysis (RCWA) [97]. The pattern of the strip structure is shown in Figure 2.3. By using the planar Au layer, the non-strip structure has the same geometric size as the strip one except that $d_2 = 140$ nm is adopted for achieving the same A-Si area.

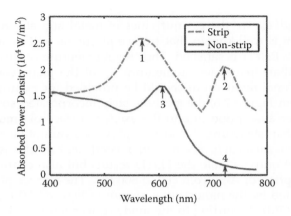

**FIGURE 2.6**
The absorbed power density of the active (A-Si) layer.

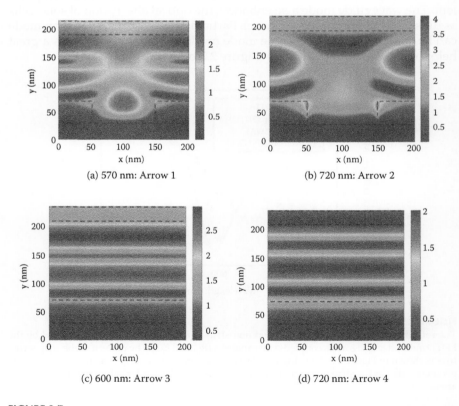

(a) 570 nm: Arrow 1

(b) 720 nm: Arrow 2

(c) 600 nm: Arrow 3

(d) 720 nm: Arrow 4

**FIGURE 2.7**
The near-field distributions for the absorption peaks of A-Si material denoted in Figure 2.6.
(See color figure.)

## 2.3.2 Nanoantennas

Nanoantennas [72, 75, 98–103] play a fundamental role in nanotechnology due to their capabilities to confine and enhance the light through converting the localized to propagating electromagnetic fields, and vice versa. The nanoantenna is a direct analogue and extended technology of the radio wave and microwave antenna. But the nanoantenna possesses lots of individual and novel features mainly owing to the existence of the electron gas oscillations in metals. The behavior of strongly coupled plasmas and the capability of manipulating light on the nanometer scale make nanoantenna particularly useful in microscopy and spectroscopy [104], fluorescence enhancement [101], surface-enhanced Raman spectroscopy [105], and photovoltaics [106,107]. The above applications mainly rely on the characteristics of nanoantennas, such as the resonance frequency, bandwidth, directivity, far-field radiation pattern, near-field distribution, and local density of states. Here, we use the FDTD method to simulate a typical dipole nanoantenna with the schematic pattern shown in Figure 2.8. The relative permittivity of $Al_2O_3$ is 3.065 and the excitation source is a plane wave polarized along the arm direction of the antenna. Figure 2.9 demonstrates that the resonance frequency of the antenna can be tunable in a wide range by changing the arm length of the antenna. The absorption cross section can be defined by

$$\sigma_a = \frac{-\int_S \frac{1}{2} \mathrm{Re}\left\{\tilde{\mathbf{E}}(\omega,\mathbf{r}) \times \mathrm{conj}\left[\tilde{\mathbf{H}}(\omega,\mathbf{r})\right]\right\} \cdot d\mathbf{S}}{\left|\tilde{\mathbf{S}}_i\right|} = \frac{-\int_S k_0 \,\mathrm{Im}\left[\varepsilon_r(\omega)\right]\left|\tilde{\mathbf{E}}(\omega,\mathbf{r})\right|^2 dS}{\left|\tilde{\mathbf{E}}_i\right|^2}$$

(2.63)

where *conj* denotes the complex conjugation, $S$ is an arbitrary surface enclosing the nanoantenna, and $\mathbf{S}_i = \frac{1}{2}\mathrm{Re}[\tilde{\mathbf{E}}_i \times \mathrm{conj}(\tilde{\mathbf{H}}_i)]$ is the incident energy flux. The radar cross sections—i.e., radiation patterns from the reciprocal theorem—at the resonance and off the resonance of the antenna are shown in Figure 2.10. Both the absorption cross section and radar cross section are

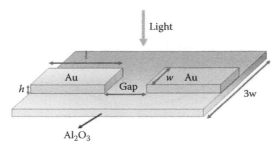

**FIGURE 2.8**
The schematic pattern of the metallic (Au) dipole nanoantenna with the $Al_2O_3$ as a substrate. The geometric size of the antenna is $l = 60$ nm, $w = 40$ nm, $h = 40$ nm, and Gap = 20 nm.

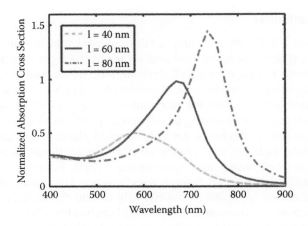

**FIGURE 2.9**
The normalized absorption cross sections of the dipole nanoantenna with varying arm lengths. The plasmonic resonance will be red-shifted and becomes stronger with the increasing arm length.

normalized with the geometric cross section area of the metallic dipole antenna ($2 \times l \times w$). At the resonance frequency, the radiation pattern of the dipole antenna is shaped like the well-known cosine square law, which is different from an asymmetric radiation pattern off the resonance. Moreover, the concentrated E-fields at the gap of the nanoantenna can be clearly observed

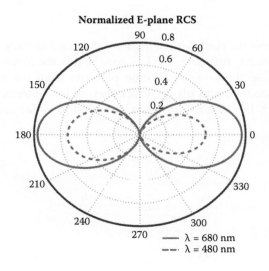

**FIGURE 2.10**
The normalized E-plane radar cross sections (RCSs) of the antenna respectively at the resonance (680 nm) and off the resonance (480 nm).

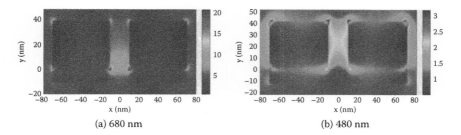

**FIGURE 2.11**
The near-field distributions of the antenna respectively at the resonance (680 nm) and off the resonance (480 nm). (See color figure.)

in Figure 2.11. The amplitude of the E-field at the resonance is significantly larger than that off the resonance (see colorbar).

### 2.3.3 Spontaneous Emissions

Control of spontaneously emitted light lies at the heart of quantum optics. It is also essential for diverse applications ranging from lasers, light-emitting diodes, and quantum information [59–62, 108]. It is well known that the radiation dynamics of an atom strongly depends on its environment, which was first discovered by Purcell [109], and the spontaneous emission (SE) can be enhanced if the emitting atom is coupled to a cavity resonator. According to the quantum electrodynamics theory, the SE of an atom can be a weak-coupling radiation process due to the vacuum fluctuations of electromagnetic fields. A suitable modification of inhomogeneous environment is required so that the vacuum fluctuations controlling the SE can be manipulated. Inhibiting unwanted SEs and boosting desired ones will promote the novel optoelectronic designs tailored to industrial standards. The local density of states (LDOS) [42] counts the number of electromagnetic modes where photons can be emitted at the specific location of the emitter, and can be interpreted as the density of vacuum fluctuations. The inhibition or enhancement of SE boils down to how the LDOS of photons is controlled. The SE rate is also proportional to the LDOS. Regarding laser and light-emitting diode applications, enhancing SE enables the improved photoluminescence, low threshold current, and fast turn-on time. Meanwhile, SE can be redirected with a high directionality in the Yagi-Uda nanoantenna system, which is quite useful in the molecular detection and sensing [103]. As an efficient tool to control the SE, plasmonic effects are successfully explored to design various optical elements and devices for enhancing and redirecting the emission [82, 83]. Here, we employ the FDTD method to investigate the LDOS or SE in a hybrid plasmonic system [108] as plotted in Figure 2.12. The LDOS can be calculated by using (2.41). The localized plasmon from the nanosphere will strongly interact with the surface plasmon from the plate substrate resulting in a strong confinement and large spontaneous decay rate. Particularly,

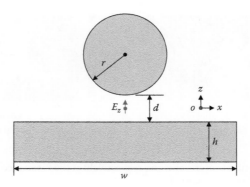

**FIGURE 2.12**
A schematic pattern for the SE in a hybrid plasmonic system. A gold sphere is located above the square gold substrate excited by the $z$-polarized dipole, where $r = 20$ nm, $d = 20$ nm, $h = 30$ nm, and $w = 100$ nm.

constructive or coherent interferences by the evanescent wave coupling in the case of $z$-polarized dipole make the normalized SE rate of the hybrid system stronger than the summation of those of the single nanosphere and the single substrate (Figure 2.13). Figures 2.14(a) and (b) show the scattered near-field distributions for the hybrid plasmonic system and single metallic nanosphere, respectively. The two hot spots at the opposite edges of the sphere and plate can be seen clearly in Figure 2.14(a).

### 2.3.4 Metamaterials

Artificially engineered metamaterials (MMs) have attracted much attention due to their interesting properties not attainable in naturally occurring

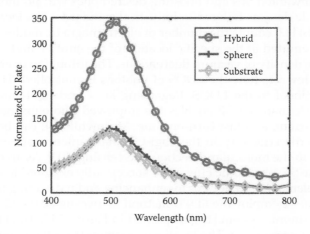

**FIGURE 2.13**
The normalized SE rate of a $z$-polarized emitter versus wavelength in a hybrid, a single sphere, and a single substrate plasmonic system.

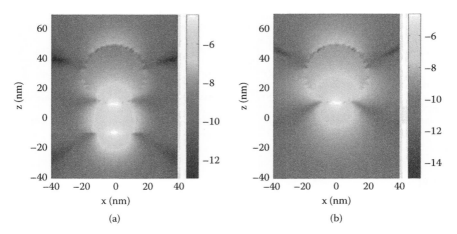

**FIGURE 2.14**
The scattered near-field distributions $E_z$ of (a) a hybrid plasmonic system and (b) a single Au sphere excited with a z-polarized dipole (in the logarithmic scale). The dipole is located at $z = 0$ nm. (See color figure.)

materials, such as negative refraction, cloaking, and electromagnetically induced transparency [48–55]. Since MMs have a subwavelength scale, effective material parameters (EMPs) (involving permittivity/permeability or refractive index/impedance) [110] can be retrieved by a homogenization procedure to significantly simplify the description of MM properties. Here, the EMPs are obtained with the recently developed non-asymptotic homogenization theory [111,112], where the electromagnetic field is approximated with a suitably chosen set of basis functions (modes).

The EMPs, which can be generally represented by a 6×6 matrix, are used to describe the linear constitutive relationship of the four coarse-grained (macroscopic) electromagnetic (EM) fields $\tilde{\mathbf{E}}, \tilde{\mathbf{H}}, \tilde{\mathbf{D}}, \tilde{\mathbf{B}}$ in the frequency domain, i.e.,

$$\begin{pmatrix} \tilde{\mathbf{D}} \\ \tilde{\mathbf{B}} \end{pmatrix} = \begin{pmatrix} \{\varepsilon\}_{3\times3} & \{\chi\}_{3\times3} \\ \{\xi\}_{3\times3} & \{\mu\}_{3\times3} \end{pmatrix} \begin{pmatrix} \tilde{\mathbf{E}} \\ \tilde{\mathbf{H}} \end{pmatrix} \tag{2.64}$$

The EMPs can be retrieved by inverting (2.64) with a sufficient basis set of macroscopic EM fields. In view of periodic structures, the macroscopic EM fields $\tilde{\mathbf{E}}, \tilde{\mathbf{H}}, \tilde{\mathbf{D}}, \tilde{\mathbf{B}}$ can be obtained by averaging the microscopic ones $\tilde{\mathbf{e}}, \tilde{\mathbf{h}}, \tilde{\mathbf{d}}, \tilde{\mathbf{b}}$ with the line and area integrals along the edge and surface of the unit cell of the periodic structure, respectively. Here, the FDTD method is adopted to obtain the microscopic EM fields with different excitations as basis functions [113].

An array of spherical gold particles with the radius of 20 nm in a cubic unit cell with the size of 80 nm is analyzed (see Figure 2.16[a]). The number

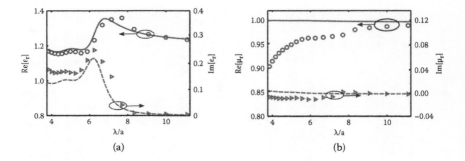

**FIGURE 2.15**
The diagonal elements of the EMPs for a periodic lattice of gold spheres with a radius of 20 nm and unit cell size of 80 nm. The number of lattice layers is five. (a) Relative permittivity; (b) relative permeability. Lines: the Lewin theory. Markers: Homogenized EMPs based on the FDTD simulation.

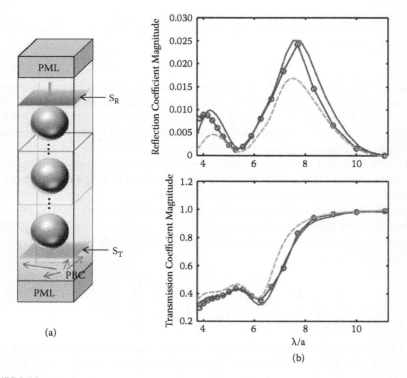

**FIGURE 2.16**
(a) A schematic pattern for the EMP retrieval. $S_R$ and $S_T$ are the planar surfaces for extracting the reflection and transmission coefficients. PML and PBC denote perfectly matched layer and periodic boundary condition, respectively. (b) The reflection and transmission coefficients. Solid red lines: real model simulation; blue circles: an equivalent slab with the EMPs obtained by the homogenization from FDTD solutions; dashed green lines: an equivalent slab with the EMPs obtained from the Lewin's theory. (See color figure.)

of lattice layers is five. The basis functions are constructed by illuminating the structure with incident waves at varying angles. PMLs are used as absorbing boundaries at the top and bottom (z-direction) of the MM slab, while PBCs are applied in the x- and y-directions. For a broadband simulation with a Gaussian pulse excitation, the constant horizontal wave number approach described in Section 2.2 is adopted especially for handling the oblique incidence case. Figure 2.15 displays the calculated EMPs compared to the Lewin's theory results [114]. Figure 2.16(b) illustrates the reflection (R) and transmission (T) coefficients used to check the validity of the EMPs. We compare the T/R coefficients calculated from the real model simulation and those calculated by an equivalent slab with the EMPs. From Figure 2.16(b), the EMPs by the homogenization from FDFD solutions are more accurate than those obtained from the Lewin's theory.

## 2.4 Extended to a Time-Dependent Schrödinger Equation

For conventional devices, the applied or built-in potentials vary slowly in comparison to the crystal potential, so that wave phenomena such as reflections and tunneling are absent, and therefore the electron motion can be described by classical physics. The classical approach will break down in ultra-small devices. Because most ultra-small devices contain quantum wells, the carriers within such wells clearly display their wave nature. Quantum confinements alter the wave function of electrons confined in potential wells, but the transport outside the confined region can often be described semiclassically. As a result, wave phenomena in ultra-small quantum devices, governed by the Schrödinger equation, strongly affect the device performances.

The time-dependent Schrödinger equation is given by [115]

$$i\hbar\frac{\partial\psi(\mathbf{r},t)}{\partial t} = -\frac{\hbar^2}{2m^*}\nabla^2\psi(\mathbf{r},t) + V(\mathbf{r})\psi(\mathbf{r},t) \qquad (2.65)$$

where $\psi$ is the wave function that is a probability amplitude in quantum mechanics describing the quantum state of a particle at the position $r$ and time $t$, $m^*$ is the (effective) mass of the particle, $-(\hbar^2/2m^*)\nabla^2$ is the kinetic energy operator, $V(\mathbf{r})$ is the time-independent potential energy, and $-(\hbar^2/2m^*)\nabla^2 + V$ is the Hamiltonian operator.

To avoid using complex numbers, one can separate the variable $\psi(r,t)$ into its real and imaginary parts as

$$\psi(r,t) = \psi_R(r,t) + i\psi_I(r,t) \tag{2.66}$$

Inserting (2.66) into (2.65), we get the following coupled set of equations [116,117]:

$$\hbar\frac{\partial\psi_R(r,t)}{\partial t} = -\frac{\hbar^2}{2m^*}\left[\frac{\partial^2\psi_I(r,t)}{\partial x^2} + \frac{\partial^2\psi_I(r,t)}{\partial y^2} + \frac{\partial^2\psi_I(r,t)}{\partial z^2}\right] + V(r)\psi_I(r,t) \tag{2.67}$$

$$\hbar\frac{\partial\psi_I(r,t)}{\partial t} = \frac{\hbar^2}{2m^*}\left[\frac{\partial^2\psi_R(r,t)}{\partial x^2} + \frac{\partial^2\psi_R(r,t)}{\partial y^2} + \frac{\partial^2\psi_R(r,t)}{\partial z^2}\right] - V(r)\psi_R(r,t) \tag{2.68}$$

A mesh is defined as a discrete set of grid points that sample the wave function in space and time. The discretized real and imaginary parts of the wave function can be represented as

$$\psi_R(\mathbf{r},t) \approx \psi_R^n(i,j,k) = \psi_R(i\Delta_x, j\Delta_y, k\Delta_z, n\Delta_t) \tag{2.69}$$

$$\psi_I(\mathbf{r},t) \approx \psi_I^n(i,j,k) = \psi_I(i\Delta_x, j\Delta_y, k\Delta_z, n\Delta_t) \tag{2.70}$$

where $\Delta_x$, $\Delta_y$, and $\Delta_z$ are, respectively, the space steps in the $x$-, $y$-, and $z$-directions, $\Delta_t$ is the time step, and $i, j, k$, and $n$ are integers. The first-order time derivatives can be discretized by a second-order centered difference scheme. The second-order Laplace operator in (2.67) and (2.68) is discretized by using the second-order collocated difference, which distinguishes from the Yee (staggered) cell in the FDTD method for Maxwell's equations [1]. Accordingly, the update equations of the real and imaginary parts of the wave function are of the forms

$$
\begin{aligned}
\psi_R^{n+1}(i,j,k) = {}& \psi_R^n(i,j,k) \\
& -\frac{\hbar\Delta t}{2m^*\Delta x^2}\left[\psi_I^{n+1/2}(i+1,j,k) - 2\psi_I^{n+1/2}(i,j,k) + \psi_I^{n+1/2}(i-1,j,k)\right] \\
& -\frac{\hbar\Delta t}{2m^*\Delta y^2}\left[\psi_I^{n+1/2}(i,j+1,k) - 2\psi_I^{n+1/2}(i,j,k) + \psi_I^{n+1/2}(i,j-1,k)\right] \\
& -\frac{\hbar\Delta t}{2m^*\Delta z^2}\left[\psi_I^{n+1/2}(i,j,k+1) - 2\psi_I^{n+1/2}(i,j,k) + \psi_I^{n+1/2}(i,j,k-1)\right] \\
& +\frac{V(i,j,k)\Delta t}{\hbar} \times \psi_I^{n+1/2}(i,j,k)
\end{aligned}
\tag{2.71}
$$

$$\psi_I^{n+1/2}(i,j,k) = \psi_I^{n-1/2}(i,j,k) + \frac{\hbar \Delta t}{2m^* \Delta x^2} \Big[ \psi_R^n(i+1,j,k) - 2\psi_R^n(i,j,k) + \psi_R^n(i-1,j,k) \Big]$$

$$+ \frac{\hbar \Delta t}{2m^* \Delta y^2} \Big[ \psi_R^n(i,j+1,k) - 2\psi_R^n(i,j,k) + \psi_R^n(i,j-1,k) \Big]$$

$$+ \frac{\hbar \Delta t}{2m^* \Delta z^2} \Big[ \psi_R^n(i,j,k+1) - 2\psi_R^n(i,j,k) + \psi_R^n(i,j,k-1) \Big]$$

$$- \frac{V(i,j,k)\Delta t}{\hbar} \times \psi_R^n(i,j,k) \qquad (2.72)$$

To absorb the outgoing waves, the stretched coordinate PML [118,119] is adopted. By using the case of a 1D Schrödinger equation, we have

$$i\hbar \frac{\partial \psi(x,t)}{\partial t} = -\frac{\hbar^2}{2m^*} \left[ \frac{1}{1+R\sigma(x)} \frac{\partial \psi(x,t)}{\partial x} \left( \frac{1}{1+R\sigma(x)} \frac{\partial \psi(x,t)}{\partial x} \right) \right] + V(\mathbf{r})\psi(\mathbf{r},t) \quad (2.73)$$

where $\sigma(x) = 0.005(x - x_r)^2$ is the parameter that is stretched as it approaches the edge of the problem space and $R = e^{i\pi/4}$.

According to the von Neumann stability analysis, the solution of the wave function can be represented as a superposition of plane waves

$$\psi(x,y,z) = \psi(i\Delta_x, j\Delta_y, k\Delta_z) = A_0 \exp(-j_0(i\Delta_x k_x + j\Delta_y k_y + k\Delta_z k_z))$$

$$k_x = k_0 \sin\theta \cos\varphi, \quad k_y = k_0 \sin\theta \sin\varphi, \quad k_z = k_0 \cos\theta \qquad (2.74)$$

where $k_x = p_m/\hbar$ is the wave number, $p_m$ is the momentum, and $\theta$ and $\varphi$ are the spherical angles. The second-order collocated differences are used to discretize the second-order spatial derivatives, i.e.,

$$\frac{\partial^2 \psi}{\partial z^2} \approx \sum_{r=-1}^{1} W_r \frac{\psi(i,j,k+r)}{\Delta_z^2} = \sum_{r=-1}^{1} W_r \frac{\exp(-j_0 r k_z \Delta_z)}{\Delta_z^2} \psi(i,j,k) = \eta_z \psi \quad (2.75)$$

where

$$\eta_z = \sum_{r=-1}^{1} W_r \frac{\exp(-j_0 r k_z \Delta_z)}{\Delta_z^2} \quad \text{and} \quad W = \{1, -2, 1\}.$$

For simplicity, we consider a 1D Schrödinger equation with zero potential energy

$$\frac{\partial}{\partial t}\begin{pmatrix} \psi_R \\ \psi_I \end{pmatrix} = \begin{pmatrix} 0 & -\dfrac{\hbar}{2m^*}\dfrac{\partial^2}{\partial z^2} \\ \dfrac{\hbar}{2m^*}\dfrac{\partial^2}{\partial z^2} & 0 \end{pmatrix}\begin{pmatrix} \psi_R \\ \psi_I \end{pmatrix} \tag{2.76}$$

and corresponding spatial discretization form is given by

$$\frac{\partial}{\partial t}\begin{pmatrix} \psi_R \\ \psi_I \end{pmatrix} = \begin{pmatrix} 0 & -\dfrac{\hbar}{2m^*}\eta_z \\ \dfrac{\hbar}{2m^*}\eta_z & 0 \end{pmatrix}\begin{pmatrix} \psi_R \\ \psi_I \end{pmatrix} \tag{2.77}$$

It is trivial to access the discretized evolution matrix $L^d$ with the second-order staggered time-stepping method

$$L^d = \begin{bmatrix} l_{11} & l_{12} \\ l_{21} & l_{22} \end{bmatrix} = \prod_{l=1}^{m}\begin{pmatrix} 1 & 0 \\ \dfrac{\hbar}{2m^*}\eta_z d_l \Delta_t & 1 \end{pmatrix}\begin{pmatrix} 1 & -\dfrac{\hbar}{2m^*}\eta_z c_l \Delta_t \\ 0 & 1 \end{pmatrix} \tag{2.78}$$

where $m = 2$, $c_1 = c_2 = 0.5$, $d_1 = 1$, and $d_2 = 0$.

The eigenvalues $\lambda$ of the discretized evolution matrix satisfy the following eigen-equation

$$\lambda^2 - tr(L^d)\lambda + \det(L^d) = 0 \tag{2.79}$$

where $tr(L^d)$ and $\det(L^d)$ are the trace and determinant of the evolution matrix, respectively. Considering that the determinant of the discretized evolution matrix is 1, the eigen-equation can be simplified as

$$\lambda^2 - tr(L^d)\lambda + 1 = 0 \tag{2.80}$$

and its solutions are

$$\lambda_{1,2} = \frac{tr(L^d) \pm j_0\sqrt{4 - [tr(L^d)]^2}}{2}$$

A stable algorithm requires $|\lambda_{1,2}| = 1$, and thus $|tr(L^d)| \le 2$. Through terms of matrix multiplications, we have

$$tr(L^d) = 2 + \sum_{l=1}^{m}(-1)^l g_l\left(\left(\frac{\hbar}{2m^*}\right)^2 \Delta_t^2 \eta_z^2\right)^l \tag{2.81}$$

$$g_l = \sum_{1 \leq i_1 \leq j_1 < i_2 \leq j_2 < \cdots < i_l \leq j_l \leq m} c_{i_1} d_{j_1} c_{i_2} d_{j_2} \cdots c_{i_l} d_{j_l} + \sum_{1 \leq i_1 < j_1 \leq i_2 < j_2 \leq \cdots \leq i_l < j_l \leq m} d_{i_1} c_{j_1} d_{i_2} c_{j_2} \cdots d_{i_l} c_{j_l}$$

(2.82)

The above results can be generalized to a three-dimensional (3D) Schrödinger equation with zero potential energy, i.e.,

$$tr(L^d) = 2 + \sum_{l=1}^{m} (-1)^l g_l \left( \left( \frac{\hbar}{2m^*} \right)^2 \Delta_t^2 (\eta_x + \eta_y + \eta_z)^2 \right)^l$$

(2.83)

Finally, we can get

$$\sqrt{\frac{\hbar}{m^*} \frac{\Delta_t}{\Delta_\delta^2}} \leq CFL$$

(2.84)

where $CFL = 1/\sqrt{d}$ is the Courant-Friedrichs-Levy (CFL) number and $d = 1,2,3$ is the dimension number.

The dispersion relation of free space photon described by Maxwell's equations is

$$\omega = c|\mathbf{k}_0|$$

(2.85)

where $c$ is the speed of light and $\mathbf{k}_0 = (k_x, k_y, k_z)$ is the wave vector with the amplitude of $k_0$. Critically different from free space photon with the cone-shaped dispersion relation, the dispersion relation of free electron is a paraboloid, i.e.,

$$\omega = \left( \frac{\hbar}{2m^*} \right) |\mathbf{k}_0|^2$$

(2.86)

We can define a dummy velocity of the Schrödinger equation as $v_0 = (\hbar/2m^*)$, analogous to Maxwell's equations, so (2.86) can be rewritten as

$$\omega = v_0 |\mathbf{k}_0|^2$$

(2.87)

Due to the energy-conserving property of the second-order staggered time-stepping method, the dispersion relation of free electron can be written as

$$\omega \Delta_t = a \cos \left[ \frac{tr(L^d)}{2} \right]$$

(2.88)

The relative error of phase velocity is given by

$$\eta = \left| \frac{v_p - v_0}{v_0} \right|$$

(2.89)

where $v_p = \omega/k_0^2$, and $\omega$ can be obtained by (2.88). The above analyses for the numerical stability and dispersion can be extended to the high-order quantum FDTD approaches [120].

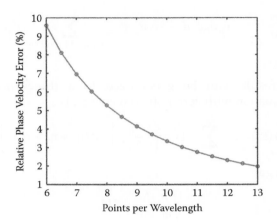

**FIGURE 2.17**
The relative dispersion error as a function of the spatial resolution (points per wavelength). Here, the spherical angles are set to $\theta = 0°$, $\varphi = 0°$, and the stability constant is $S_\delta = \frac{\Delta_t}{\Delta_\delta^2} \frac{\hbar}{2m^*} = 0.125$.

We set the stability criterion to be

$$S_\delta = \frac{\Delta_t}{\Delta_\delta^2} \frac{\hbar}{2m^*} = 0.125$$

Figure 2.17 shows the relative phase velocity error as a function of points per wavelength (PPW) for a plane wave traveling at $\theta = 0°$ and $\varphi = 0°$. Next, the spatial resolution is set to be 7 points per wavelength. We redraw the relative error at $\theta = 30°$ versus the propagating angle $\varphi$ as shown in Figure 2.18.

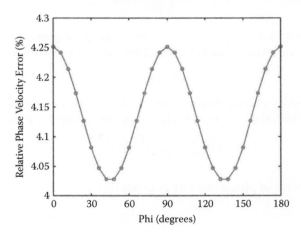

**FIGURE 2.18**
The relative dispersion error as a function of the spherical angle $\varphi$. The spherical angle is $\theta = 30°$, the spatial resolution is 7 points per wavelength, and the stability constant is $S_\delta = \frac{\Delta_t}{\Delta_\delta^2} \frac{\hbar}{2m^*} = 0.125$.

## 2.5 Numerical Examples for Quantum Physics

### 2.5.1 1D Propagation Problem

We consider a simulation of a particle interacting with a barrier of 0.1 eV. The spatial step is $\Delta_\delta = 0.1\,\text{nm}$ and

$$S_\delta = \frac{\Delta_t}{\Delta_\delta^2}\frac{\hbar}{2m^*} = 0.1$$

The simulation domain occupies 400 grids. Figure 2.19 shows the propagation of the particle near the barrier with the initial wave function of a modulated

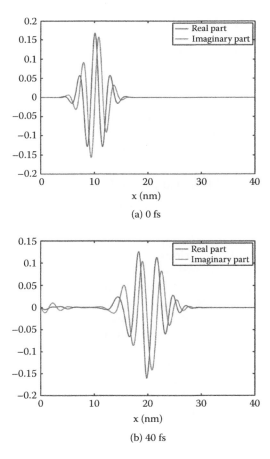

(a) 0 fs

(b) 40 fs

**FIGURE 2.19**
A particle is initiated in free space and strikes a barrier with a potential of 0.1 eV. The time evolutions of the wave function are recorded. (*Continued*)

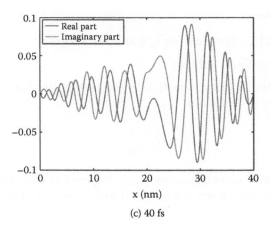

(c) 40 fs

**FIGURE 2.19** (*Continued*)
A particle is initiated in free space and strikes a barrier with a potential of 0.1 eV. The time evolutions of the wave function are recorded.

Gaussian pulse. After 40 fs, the particle reaches the barrier. After 90 fs, the wave function is reflected back and meanwhile penetrated into the barrier.

## 2.5.2 Two-Dimensional Eigenvalue Problem

The eigenvalue problem of the Schrödinger equation is fundamentally important for the quantum transport and nanodevice modeling. The ballistic electron transport strongly depends on the transverse eigenstates of the conducting channel [121]. Various intriguing quantum phenomena for the microscopic electron transport, such as the resonant tunneling effect [115], Fano-resonance [122], and so on, contribute to the excitation of eigenstates or interplay of different eigenstates. Thus, an accurate and efficient method to calculate the eigenstates and eigenfrequencies is crucial to understand the fundamental and device physics. Moreover, eigenstates and eigenfrequencies extraction tailored to industrial requirements is also indispensable in the quantum computer-aided design (CAD). One of the commonly adopted algorithms to solve the eigenvalue problem of the time-dependent Schrödinger equation is the FDTD method [116,117]. Here we investigate the eigenvalue problem of a particle in a two-dimensional (2D) quantum well with the area of $40 \times 30$ nm$^2$. The spatial steps are set to $\Delta_x = \Delta_y = 1$ nm, and

$$S_\delta = \frac{\Delta_t}{\Delta_\delta^2} \frac{\hbar}{2m^*} = 0.1$$

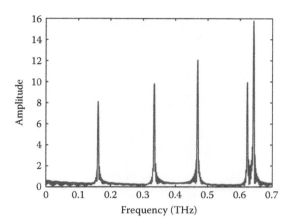

**FIGURE 2.20**
The eigenfrequencies of a 2D quantum well calculated by the quantum FDTD method and discrete (fast) Fourier transform.

The simulation time step is set to 200,000 to force all the eigenstates completely excited. The initial value of the wave function is taken as (2.42). Figure 2.20 shows the amplitude-frequency spectrum at an arbitrary point of the quantum well from 0 to 0.7 THz calculated by the quantum FDTD method and discrete (fast) Fourier transform. Table 2.1 lists the eigenfrequencies related to the amplitude peaks in Figure 2.20. The calculated eigenfrequencies by the quantum FDTD method agree with the analytical solutions well. The eigenstate corresponding to one of the eigenfrequencies is illustrated in Figure 2.21.

**TABLE 2.1**

Comparisons of Eigenfrequencies of a 2D Quantum Well between the FDTD Method and the Analytical Solution

| Eigenfrequencies | Numerical (THz) | Analytical (THz) |
|---|---|---|
| $f_{11}$ | 0.161 | 0.158 |
| $f_{21}$ | 0.335 | 0.328 |
| $f_{12}$ | 0.469 | 0.461 |
| $f_{31}$ | 0.623 | 0.613 |
| $f_{22}$ | 0.642 | 0.631 |

*Note:* The analytical solution is

$$f_{pq} = \frac{\hbar\pi}{4m^*}\left(\frac{p^2}{a^2} + \frac{q^2}{b^2}\right),$$

where $p,q$ are integers, and $a$ and $b$ are the length and width of the well.

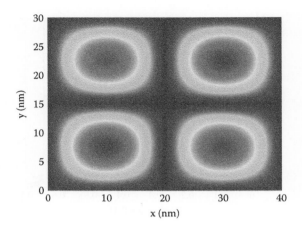

**FIGURE 2.21**
The eigenstate corresponding to the eigenfrequency $f_{22}$ for a 2D quantum well obtained by the quantum FDTD method and discrete Fourier transform. (See color figure.)

## 2.6 Conclusion

The FDTD method and its advances tremendously promote the development of the computational electromagnetics field and play more and more fundamental roles in nanotechnology applications. Due to the features of extreme flexibility and easy implementation, the FDTD method is an indispensable tool in modeling inhomogeneous, anisotropic, and dispersive media with random, multilayered, and periodic fundamental (or device) nanostructures. Regarding future possible research directions, the FDTD method is expected to be widely adopted to simulate multiphysics problems, through which the optical, electrical, thermal, mechanical, and quantum properties of multifunctional components and devices can be investigated. Based on authors' knowledge, some interesting and hot topics for the multiphysics simulation are summarized as follows:

1. Nonlinear optics problem, such as the second harmonic generation, Kerr effect, and four-wave mixing [123]. The Maxwell's equations and hydrodynamic equation can be solved self-consistently [124].

2. Optoelectronic device simulation for light-emitting diodes, photodetectors, solar cells, and lasers. The Maxwell's equations, semiconductor equations (Poisson, drift-diffusion, continuity, heat conduction, and energy balance equations), and thermal stress field equations will be coupled with each other [125–127]. The Maxwell's equations also could be interacted with the rate equations [128,129].

3. Quantum optics and quantum transport problems. The self-consistent solutions to Maxwell's equations and effective-mass Schrödinger
equation, von Neumann equation, or density functional theory
method are required [130,131].

Exploring start-of-the-art techniques and emerging applications for the
FDTD method will open up a fantastic, fruitful, and challenging research
area in the near future. Are you ready?

## Acknowledgment

The authors acknowledge the support of the grants from the National Natural Science Foundation of China (No. 61201122 and No. 60931002).

## References

1. K. S. Yee, "Numerical solution of initial boundary value problems involving Maxwell's equations in isotropic media," *IEEE Trans. Antennas Propag.* 14, 302–307 (1966).
2. A. Taflove, *Computational Electrodynamics: The Finite-Difference Time-Domain Method.* Norwood, MA: Artech House, 1995.
3. A. Taflove, et al., *Advances in Computational Electrodynamics: The Finite-Difference Time-Domain Method.* Norwood, MA: Artech House, 1998.
4. D. M. Sullivan, *Electromagnetic Simulation Using the FDTD Method.* New York: IEEE Press, 2000.
5. J. L. Young, D. Gaitonde, and J. S. Shang, "Toward the construction of a fourth-order difference scheme for transient EM wave simulation: Staggered grid approach," *IEEE Trans. Antennas Propag.* 45, 1573–1580 (1997).
6. J. S. Shang, "High-order compact-difference schemes for time-dependent Maxwell equations," *J. Comput. Phys.* 153, 312–333 (1999).
7. T. Namiki, "New FDTD algorithm based on alternating-direction implicit method," *IEEE Trans. Microw. Theory Tech.* 47, 2003–2007 (1999).
8. F. H. Zhen, Z. Z. Chen, and J. Z. Zhang, "Toward the development of a three-dimensional unconditionally stable finite-difference time-domain method," *IEEE Trans. Microw. Theory Tech.* 48, 1550–1558 (2000).
9. J. S. Kole, M. T. Figge, and H. De Raedt, "Higher-order unconditionally stable algorithms to solve the time-dependent Maxwell equations," *Phys. Rev. E* 65, 066705 (2002).
10. G. Sun and C. W. Trueman, "Unconditionally stable Crank-Nicolson scheme for solving two-dimensional Maxwell's equations," *Electron. Lett.* 39, 595–597 (2003).

11. J. Shibayama, M. Muraki, J. Yamauchi, and H. Nakano, "Efficient implicit FDTD algorithm based on locally one-dimensional scheme," *Electron. Lett.* 41, 1046–1047 (2005).

12. H. Yoshida, "Construction of higher order symplectic integrators," *Physica D* 46, 262–268 (1990).

13. T. Hirono, W. Lui, S. Seki, and Y. Yoshikuni, "A three-dimensional fourth-order finite-difference time-domain scheme using a symplectic integrator propagator," *IEEE Trans. Microw. Theory Tech.* 49, 1640–1648 (2001).

14. W. E. I. Sha, Z. X. Huang, X. L. Wu, and M. S. Chen, "Application of the symplectic finite-difference time-domain scheme to electromagnetic simulation," *J. Comput. Phys.* 225, 33–50 (2007).

15. W. E. I. Sha, Z. X. Huang, M. S. Chen, and X. L. Wu, "Survey on symplectic finite-difference time-domain schemes for Maxwell's equations," *IEEE Trans. Antennas Propag.* 56, 493–500 (2008).

16. M. Okoniewski, E. Okoniewska, and M. A. Stuchly, "Three-dimensional subgridding algorithm for FDTD," *IEEE Trans. Antennas Propag.* 45, 422–429 (1997).

17. S. Dey and R. Mittra, "A locally conformal finite-difference time-domain algorithm for modeling three-dimensional perfectly conducting objects," *IEEE Microwave and Guided Wave Letters* 7, 273–275 (1997).

18. T. G. Jurgens, A. Taflove, K. Umashankar, and T. G. Moore, "Finite-difference time-domain modeling of curved surfaces," *IEEE Trans. Antennas Propag.* 40, 357–366 (1992).

19. W. H. Yu and R. Mittra, "A conformal FDTD algorithm for modeling perfectly conducting objects with curve-shaped surfaces and edges," *Microw. Opt. Technol. Lett.* 27, 136–138 (2000).

20. I. A. Zagorodnov, R. Schuhmann, and T. Weiland, "A uniformly stable conformal FDTD-method in Cartesian grids," *Int. J. Numer. Model.-Electron. Netw. Device Fields* 16, 127–141 (2003).

21. T. Xiao and Q. H. Liu, "Enlarged cells for the conformal FDTD method to avoid the time step reduction," *IEEE Microw. Wirel. Compon. Lett.* 14, 551–553 (2004).

22. J. Fang, "Time domain finite difference computation for Maxwell's equations," Ph.D. dissertation, Univ. California, Berkeley, CA, 1989.

23. A. Yefet and P. G. Petropoulos, "A staggered fourth-order accurate explicit finite difference scheme for the time-domain Maxwell's equations," *J. Comput. Phys.* 168, 286–315 (2001).

24. S. V. Georgakopoulos, C. R. Birtcher, C. A. Balanis, and R. A. Renaut, "Higher-order finite-difference schemes for electromagnetic radiation, scattering, and penetration, Part I: Theory," *IEEE Antennas Propag. Mag.* 44, 134–142 (2002).

25. S. V. Georgakopoulos, C. R. Birtcher, C. A. Balanis, and R. A. Renaut, "Higher-order finite-difference schemes for electromagnetic radiation, scattering, and penetration, Part 2: Applications," *IEEE Antennas Propag. Mag.* 44, 92–101 (2002).

26. W. E. I. Sha, X. L. Wu, M. S. Chen, and Z. X. Huang, "Application of the high-order symplectic FDTD scheme to the curved three-dimensional perfectly conducting objects," *Microw. Opt. Technol. Lett.* 49, 931–934 (2007).

27. W. E. I. Sha, X. L. Wu, Z. X. Huang, and M. S. Chen, "A new conformal FDTD(2,4) scheme for modeling three-dimensional curved perfectly conducting objects," *IEEE Microw. Wirel. Compon. Lett.* 18, 149–151 (2008).

28. J. A. Wang, W. Y. Yin, P. G. Liu, and Q. H. Liu, "High-order interface treatment techniques for modeling curved dielectric objects," *IEEE Trans. Antennas Propag.* 58, 2946–2953 (2010).

29. B. A. Al-Zohouri and M. F. Hadi, "Conformal modelling of perfect conductors in the high-order M24 finite-difference time-domain algorithm," *IET Microw. Antennas Propag.* 5, 583–587 (2011).

30. M. Krumpholz and L. P. B. Katehi, "MRTD: New time-domain schemes based on multiresolution analysis," *IEEE Trans. Microw. Theory Tech.* 44, 555–571 (1996).

31. Q. S. Cao, Y. C. Chen, and R. Mittra, "Multiple image technique (MIT) and anisotropic perfectly matched layer (APML) in implementation of MRTD scheme for boundary truncations of microwave structures," *IEEE Trans. Microw. Theory Tech.* 50, 1578–1589 (2002).

32. Q. H. Liu, "PSTD algorithm: A time-domain method requiring only two cells per wavelength," *Microw. Opt. Technol. Lett.* 15, 158–165 (1997).

33. F. Edelvik and G. Ledfelt, "A comparison of time-domain hybrid solvers for complex scattering problems," *Int. J. Numer. Model.-Electron. Netw. Device Fields* 15, 475–487 (2002).

34. A. Monorchio, A. R. Bretones, R. Mittra, G. Manara, and R. G. Martin, "A hybrid time-domain technique that combines the finite element, finite difference and method of moment techniques to solve complex electromagnetic problems," *IEEE Trans. Antennas Propag.* 52, 2666–2674 (2004).

35. T. Rylander and A. Bondeson, "Stable FEM-FDTD hybrid method for Maxwell's equations," *Comput. Phys. Commun.* 125, 75–82 (2000).

36. J. F. Lee, R. Lee, and A. Cangellaris, "Time-domain finite-element methods," *IEEE Trans. Antennas Propag.* 45, 430–442 (1997).

37. V. Shankar, A. H. Mohammadian, and W. F. Hall, "A time-domain, finite-volume treatment for the Maxwell equations," *Electromagnetics* 10, 127–145 (1990).

38. J. S. Hesthaven and T. Warburton, "Nodal high-order methods on unstructured grids—I. Time-domain solution of Maxwell's equations," *J. Comput. Phys.* 181, 186–221 (2002).

39. K. Sakoda, *Optical Properties of Photonic Crystals*. Berlin: Springer, 2005.

40. J. D. Joannopoulos, S. G. Johnson, J. N. Winn, and R. D. Meade, *Photonic Crystals Molding the Flow of Light*. Woodstock: Princeton University Press, 2008.

41. W. L. Barnes, A. Dereux, and T. W. Ebbesen, "Surface plasmon subwavelength optics," *Nature* 424, 824–830 (2003).

42. L. Novotny, and B. Hecht, *Principles of Nano-Optics*. Cambridge University Press, 2006.

43. S. Kawata, and V. M. Shalaev, *Nanophotonics with Surface Plasmons*. Elsevier, 2007.

44. S. A. Maier, *Plasmonics: Fundamentals and Applications*. Springer, 2007.

45. H. A. Atwater, and A. Polman, "Plasmonics for improved photovoltaic devices," *Nat. Mater.* 9, 205–213 (2010).

46. D. K. Gramotnev, and S. I. Bozhevolnyi, "Plasmonics beyond the diffraction limit," *Nat. Photonics* 4, 83–91 (2010).

47. J. A. Schuller, E. S. Barnard, W. S. Cai, Y. C. Jun, J. S. White, and M. L. Brongersma, "Plasmonics for extreme light concentration and manipulation," *Nat. Mater.* 9, 193–204 (2010).

48. J. B. Pendry, A. J. Holden, D. J. Robbins, and W. J. Stewart, "Magnetism from conductors and enhanced nonlinear phenomena," *IEEE Trans. Microw. Theory Tech.* 47, 2075–2084 (1999).

49. J. B. Pendry, "Negative refraction makes a perfect lens," *Phys. Rev. Lett.* 85, 3966–3969 (2000).

50. D. R. Smith, W. J. Padilla, D. C. Vier, S. C. Nemat-Nasser, and S. Schultz, "Composite medium with simultaneously negative permeability and permittivity," *Phys. Rev. Lett.* 84, 4184–4187 (2000).

51. D. R. Smith, J. B. Pendry, and M. C. K. Wiltshire, "Metamaterials and negative refractive index," *Science* 305, 788–792 (2004).

52. V. M. Shalaev, W. S. Cai, U. K. Chettiar, H. K. Yuan, A. K. Sarychev, V. P. Drachev, and A. V. Kildishev, "Negative index of refraction in optical metamaterials," *Opt. Lett.* 30, 3356–3358 (2005).

53. N. Engheta and R. W. Ziolkowski, *Metamaterials: Physics and Engineering Explorations.* Wiley-IEEE Press, 2006.

54. V. M. Shalaev, "Optical negative-index metamaterials," *Nat. Photonics* 1, 41–48 (2007).

55. C. M. Soukoulis and M. Wegener, "Past achievements and future challenges in the development of three-dimensional photonic metamaterials," *Nat. Photonics* 5, 523–530 (2011).

56. J. P. Berenger, "Three-dimensional perfectly matched layer for the absorption of electromagnetic waves," *J. Comput. Phys.* 127, 363–379 (1996).

57. S. D. Gedney, "An anisotropic PML absorbing media for the FDTD simulation of fields in lossy and dispersive media," *Electromagnetics* 16, 399–415 (1996).

58. J. A. Roden and S. D. Gedney, "Convolution PML (CPML): An efficient FDTD implementation of the CFS-PML for arbitrary media," *Microw. Opt. Technol. Lett.* 27, 334–339 (2000).

59. M. Francardi, L. Balet, A. Gerardino, N. Chauvin, D. Bitauld, L. H. Li, B. Alloing, and A. Fiore, "Enhanced spontaneous emission in a photonic-crystal light-emitting diode," *Appl. Phys. Lett.* 93, 143102 (2008).

60. M. A. Noginov, H. Li, Y. A. Barnakov, D. Dryden, G. Nataraj, G. Zhu, C. E. Bonner, M. Mayy, Z. Jacob, and E. E. Narimanov, "Controlling spontaneous emission with metamaterials," *Opt. Lett.* 35, 1863–1865 (2010).

61. C. Walther, G. Scalari, M. I. Amanti, M. Beck, and J. Faist, "Microcavity laser oscillating in a circuit-based resonator," *Science* 327, 1495–1497 (2010).

62. P. F. Qiao, W. E. I. Sha, W. C. H. Choy, and W. C. Chew, "Systematic study of spontaneous emission in a two-dimensional arbitrary inhomogeneous environment," *Phys. Rev. A* 83, 043824 (2011).

63. J. J. Greffet, R. Carminati, K. Joulain, J. P. Mulet, S. P. Mainguy, and Y. Chen, "Coherent emission of light by thermal sources," *Nature* 416, 61–64 (2002).

64. K. Joulain, J. P. Mulet, F. Marquier, R. Carminati, and J. J. Greffet, "Surface electromagnetic waves thermally excited: Radiative heat transfer, coherence properties and Casimir forces revisited in the near field," *Surf. Sci. Rep.* 57, 59–112 (2005).

65. Y. De Wilde, F. Formanek, R. Carminati, B. Gralak, P. A. Lemoine, K. Joulain, J. P. Mulet, Y. Chen, and J. J. Greffet, "Thermal radiation scanning tunnelling microscopy," *Nature* 444, 740–743 (2006).

66. L. D. Landau, E. M. Lifshitz, and L. P. Pitaevskii, *Statistical Physics. Part 2.* New York: Pergamon, 1980.

67. J. L. Xiong and W. C. Chew, "Efficient evaluation of Casimir force in z-invariant geometries by integral equation methods," *Appl. Phys. Lett.* 95, 154102 (2009).

68. P. R. Atkins, W. C. Chew, Q. I. Dai, and W. E. I. Sha, "The Casimir force for arbitrary three-dimensional objects with low frequency methods," *Progress In Electromagnetics Research Symposium*, 738–741 (2012).

69. D. M. Callahan, J. N. Munday, and H. A. Atwater, "Solar cell light trapping beyond the ray optic limit," *Nano Lett.* 12, 214–218 (2012).

70. A. F. Oskooi, D. Roundy, M. Ibanescu, P. Bermel, J. D. Joannopoulos, and S. G. Johnson, "MEEP: A flexible free-software package for electromagnetic simulations by the FDTD method," *Comput. Phys. Commun.* 181, 687–702 (2010).

71. S. M. Rao, *Time Domain Electromagnetics*. New York: Academic Press, 1999.

72. L. Novotny and N. van Hulst, "Antennas for light," *Nat. Photonics* 5, 83–90 (2011).

73. A. Yariv, Y. Xu, R. K. Lee, and A. Scherer, "Coupled-resonator optical waveguide: A proposal and analysis," *Opt. Lett.* 24, 711–713 (1999).

74. T. Kawanishi and M. Izutsu, "Coaxial periodic optical waveguide," *Opt. Express* 7, 10–22 (2000).

75. A. Alu and N. Engheta, "Tuning the scattering response of optical nanoantennas with nanocircuit loads," *Nat. Photonics* 2, 307–310 (2008).

76. H. Ichikawa and T. Baba, "Efficiency enhancement in a light-emitting diode with a two-dimensional surface grating photonic crystal," *Appl. Phys. Lett.* 84, 457–459 (2004).

77. M. C. Y. Huang, Y. Zhou, and C. J. Chang-Hasnain, "A surface-emitting laser incorporating a high-index-contrast subwavelength grating," *Nat. Photonics* 1, 119–122 (2007).

78. W. E. I. Sha, W. C. H. Choy, and W. C. Chew, "A comprehensive study for the plasmonic thin-film solar cell with periodic structure," *Opt. Express* 18, 5993–6007 (2010).

79. W. E. I. Sha, W. C. H. Choy, and W. C. Chew, "Angular response of thin-film organic solar cells with periodic metal back nanostrips," *Opt. Lett.* 36, 478–480 (2011).

80. Y. A. Vlasov, M. O'Boyle, H. F. Hamann, and S. J. McNab, "Active control of slow light on a chip with photonic crystal waveguides," *Nature* 438, 65–69 (2005).

81. T. Baba, "Slow light in photonic crystals," *Nat. Photonics* 2, 465–473 (2008).

82. H. Aouani, O. Mahboub, E. Devaux, H. Rigneault, T. W. Ebbesen, and J. Wenger, "Plasmonic antennas for directional sorting of fluorescence emission," *Nano Lett.* 11, 2400–2406 (2011).

83. Y. C. Jun, K. C. Y. Huang, and M. L. Brongersma, "Plasmonic beaming and active control over fluorescent emission," *Nat. Commun.* 2, 283 (2011).

84. A. Hessel and A. A. Oliner, "A new theory of Wood's anomalies on optical gratings," *Appl. Optics* 4, 1275–1297 (1965).

85. B. Luk'yanchuk, N. I. Zheludev, S. A. Maier, N. J. Halas, P. Nordlander, H. Giessen, and C. T. Chong, "The Fano resonance in plasmonic nanostructures and metamaterials," *Nat. Mater.* 9, 707–715 (2010).

86. A. E. Miroshnichenko, S. Flach, and Y. S. Kivshar, "Fano resonances in nanoscale structures," *Rev. Mod. Phys.* 82, 2257–2298 (2010).

87. M. D. Lukin and A. Imamoglu, "Controlling photons using electromagnetically induced transparency," *Nature* 413, 273–276 (2001).

88. K. ElMahgoub, F. Yang, and A. Elsherbeni, *Scattering Analysis of Periodic Structures Using Finite-Difference Time-Domain Method*. Morgan & Claypool Publishers, 2012.
89. R. Luebbers, F. P. Hunsberger, K. S. Kunz, R. B. Standler, and M. Schneider, "A frequency-dependent finite-difference time-domain formulation for dispersive materials," *IEEE Trans. Electromagn. Compat.* 32, 222–227 (1990).
90. T. Kashiwa and I. Fukai, "A treatment by the FD-TD method of the dispersive characteristics associated with electronic polarization," *Microw. Opt. Technol. Lett.* 3, 203–205 (1990).
91. D. F. Kelley and R. J. Luebbers, "Piecewise linear recursive convolution for dispersive media using FDTD," *IEEE Trans. Antennas Propag.* 44, 792–797 (1996).
92. D. M. Sullivan, "Frequency-dependent FDTD methods using z-transforms," *IEEE Trans. Antennas Propag.* 40, 1223–1230 (1992).
93. J. Nelson, *The Physics of Solar Cells*. London: Imperial College Press, 2003.
94. K. L. Chopra, P. D. Paulson, and V. Dutta, "Thin-film solar cells: An overview," *Prog. Photovoltaics* 12, 69–92 (2004).
95. A. D. Rakic, A. B. Djurisic, J. M. Elazar, and M. L. Majewski, "Optical properties of metallic films for vertical-cavity optoelectronic devices," *Appl. Optics* 37, 5271–5283 (1998).
96. E. D. Palik, *Handbook of Optical Constants of Solids*. London: Academic Press, 1998.
97. M. G. Moharam, E. B. Grann, D. A. Pommet, and T. K. Gaylord, "Formulation for stable and efficient implementation of the rigorous coupled-wave analysis of binary gratings," *J. Opt. Soc. Am. A* 12, 1068–1076 (1995).
98. P. Muhlschlegel, H. J. Eisler, O. J. F. Martin, B. Hecht, and D. W. Pohl, "Resonant optical antennas," *Science* 308, 1607–1609 (2005).
99. J. J. Li, A. Salandrino, and N. Engheta, "Shaping light beams in the nanometer scale: A Yagi-Uda nanoantenna in the optical domain," *Phys. Rev. B* 76, 245403 (2007).
100. T. H. Taminiau, F. D. Stefani, and N. F. van Hulst, "Enhanced directional excitation and emission of single emitters by a nano-optical Yagi-Uda antenna," *Opt. Express* 16, 10858–10866 (2008).
101. A. Kinkhabwala, Z. F. Yu, S. H. Fan, Y. Avlasevich, K. Mullen, and W. E. Moerner, "Large single-molecule fluorescence enhancements produced by a bowtie nano-antenna," *Nat. Photonics* 3, 654–657 (2009).
102. Y. G. Liu, Y. Li, and W. E. I. Sha, "Directional far-field response of a spherical nanoantenna," *Opt. Lett.* 36, 2146–2148 (2011).
103. Y. G. Liu, W. C. H. Choy, W. E. I. Sha, and W. C. Chew, "Unidirectional and wavelength-selective photonic sphere-array nanoantennas," *Opt. Lett.* 37, 2112–2114 (2012).
104. T. Kalkbrenner, U. Hakanson, A. Schadle, S. Burger, C. Henkel, and V. Sandoghdar, "Optical microscopy via spectral modifications of a nanoantenna," *Phys. Rev. Lett.* 95, 200801 (2005).
105. J. Grand, M. L. de la Chapelle, J. L. Bijeon, P. M. Adam, A. Vial, and P. Royer, "Role of localized surface plasmons in surface-enhanced Raman scattering of shape-controlled metallic particles in regular arrays," *Phys. Rev. B* 72, 033407 (2005).
106. W. E. I. Sha, W. C. H. Choy, Y. P. Chen, and W. C. Chew, "Optical design of organic solar cell with hybrid plasmonic system," *Opt. Express* 19, 15908–15918 (2011).

107. W. E. I. Sha, W. C. H. Choy, Y. G. Liu, and W. C. Chew, "Near-field multiple scattering effects of plasmonic nanospheres embedded into thin-film organic solar cells," *Appl. Phys. Lett.* 99, 113304 (2011).

108. Y. P. Chen, W. E. I. Sha, W. C. H. Choy, L. J. Jiang, and W. C. Chew, "Study on spontaneous emission in complex multilayered plasmonic system via surface integral equation approach with layered medium Green's function," *Opt. Express* 20, 20210–20221 (2012).

109. E. M. Purcell, "Spontaneous emission probabilities at radio frequencies," *Phys. Rev. Lett.* 69, 681 (1946).

110. D. R. Smith, S. Schultz, P. Markos, and C. M. Soukoulis, "Determination of effective permittivity and permeability of metamaterials from reflection and transmission coefficients," *Phys. Rev. B* 65, 195104 (2002).

111. A. Pors, I. Tsukerman, and S. I. Bozhevolnyi, "Effective constitutive parameters of plasmonic metamaterials: Homogenization by dual field interpolation," *Phys. Rev. E* 84, 016609 (2011).

112. I. Tsukerman, "Effective parameters of metamaterials: a rigorous homogenization theory via Whitney interpolation," *J. Opt. Soc. Am. B* 28, 577–586 (2011).

113. X. Y. Z. Xiong, L. J. Jiang, V. A. Markel, and I. Tsukerman, "Surface waves in three-dimensional electromagnetic composites and their effect on homogenization," *Opt. Express* 21, 10412–10421 (2013).

114. L. Lewin, "The electrical constants of a material loaded with spherical particles," *Journal of the Institution of Electrical Engineers* 94, 65–68 (1947).

115. D. J. Griffiths, *Introduction to Quantum Mechanics*, 2nd ed. Boston: Addison-Wesley, 2004.

116. A. Soriano, E. A. Navarro, J. A. Porti, and V. Such, "Analysis of the finite difference time domain technique to solve the Schrodinger equation for quantum devices," *J. Appl. Phys.* 95, 8011–8018 (2004).

117. D. M. Sullivan, and D. S. Citrin, "Determining quantum eigenfunctions in three-dimensional nanoscale structures," *J. Appl. Phys.* 97, 104305 (2005).

118. W. C. Chew, J. M. Jin, and E. Michielssen, "Complex coordinate stretching as a generalized absorbing boundary condition," *Microw. Opt. Technol. Lett.* 15, 363–369 (1997).

119. D. M. Sullivan, and P. M. Wilson, "Time-domain determination of transmission in quantum nanostructures," *J. Appl. Phys.* 112, 064325 (2012).

120. J. Shen, W. E. I. Sha, Z. X. Huang, M. S. Chen, and X. L. Wu, "High-order symplectic FDTD scheme for solving a time-dependent Schrödinger equation," *Comput. Phys. Commun.* 184, 480–492 (2013).

121. S. Datta, *Quantum Transport: Atom to Transistor*. New York: Cambridge University Press, 2005.

122. Y. S. Joe, A. M. Satanin, and C. S. Kim, "Classical analogy of Fano resonances," *Phys. Scr.* 74, 259–266 (2006).

123. M. Kauranen, and A. V. Zayats, "Nonlinear plasmonics," *Nat. Photonics* 6, 737–748 (2012).

124. J. J. Liu, M. Brio, Y. Zeng, A. R. Zakharian, W. Hoyer, S. W. Koch, and J. V. Moloney, "Generalization of the FDTD algorithm for simulations of hydrodynamic nonlinear Drude model," *J. Comput. Phys.* 229, 5921–5932 (2010).

125. W. E. I. Sha, W. C. H. Choy, and W. C. Chew, "The roles of metallic rectangular-grating and planar anodes in the photocarrier generation and transport of organic solar cells," *Appl. Phys. Lett.* 101, 223302 (2012).

126. W. E. I. Sha, W. C. H. Choy, Y. Wu, and W. C. Chew, "Optical and electrical study of organic solar cells with a 2D grating anode," *Opt. Express* 20, 2572–2580 (2012).

127. F. Z. Kong, W. Y. Yin, J. F. Mao, and Q. H. Liu, "Electro-thermo-mechanical characterizations of various wire bonding interconnects illuminated by an electromagnetic pulse," *IEEE Trans. Adv. Packag.* 33, 729–737 (2010).

128. A. Fang, T. Koschny, and C. M. Soukoulis, "Self-consistent calculations of loss-compensated fishnet metamaterials," *Phys. Rev. B* 82, 121102 (2010).

129. Z. X. Huang, T. Koschny, and C. M. Soukoulis, "Theory of Pump-Probe experiments of metallic metamaterials coupled to a gain medium," *Phys. Rev. Lett.* 108, 187402 (2012).

130. L. Pierantoni, D. Mencarelli, and T. Rozzi, "A new 3-D transmission line matrix scheme for the combined Schrodinger-Maxwell problem in the electronic/electromagnetic characterization of nanodevices," *IEEE Trans. Microw. Theory Tech.* 56, 654–662 (2008).

131. K. Lopata, and D. Neuhauser, "Multiscale Maxwell-Schrodinger modeling: A split field finite-difference time-domain approach to molecular nanopolaritonics," *J. Chem. Phys.* 130, 104707 (2009).

# 3

# Modeling of Optical Metamaterials Using the FDTD Method

**Yan Zhao**

*Faculty of Engineering, Chulalongkorn University, Thailand*

## CONTENTS

## 3.1 Introduction

Recently, the investigation and design of metamaterials have attracted considerable attention. Metamaterials are artificially engineered periodic structures that possess desirable electromagnetic properties that cannot be found in naturally occurring materials. For instance, metamaterials allow one to create negative permittivity and/or permeability, and consequently a negative refractive index to construct the so-called left-handed material (LHM, or negative index material) [1]. LHMs can be used to build a perfect lens [2], through which subwavelength imaging with unlimited resolutions can be achieved. Potential applications of metamaterials in optical frequencies include optical cloaking structures for reducing scattering cross sections [3], optical hyperlens for simultaneous subwavelength image transmission and magnification [4], plasmonic waveguides for highly efficient light transmission [5], and plasmon solar cells for enhancing light scattering and absorption.

In the study of the aforementioned novel electromagnetic structures, numerical techniques have always played an important role. This is due to the complexity of these structures such that analytical techniques are often incapable of providing valid solutions without certain approximations. The finite-difference time-domain (FDTD) method is one of the most popular numerical techniques in modeling electromagnetic structures, thanks to its ability of dealing with inhomogeneous, anisotropic, and frequency dispersive materials [6]. Moreover, since FDTD calculations are performed in the time domain, solutions at all frequencies of interest can be simultaneously obtained in a single simulation run. Nonetheless, for the modeling of optical metamaterials, due to the high contrast between the material properties of the structure and the free space, numerical simulations of these structures become very challenging, and the accuracy of the conventional FDTD method is usually insufficient [7]. Although using extremely fine mesh in FDTD simulations can improve the simulation accuracy, it may result in the requirement for excessive computational resources.

In this chapter, the modeling of three types of optical metamaterials using the FDTD method is introduced. To characterize these metamaterials accurately, either the conventional FDTD method is improved, or new FDTD schemes are developed. Details of the developed FDTD schemes are provided in the following sections.

## 3.2 FDTD Modeling of Cloaking Structures

Recently, a great deal of attention has been paid to the analysis and design of electromagnetic cloaking structures, first proposed by Pendry et al. [8]. The specially designed cloak is able to guide waves to propagate around its central region, rendering the objects placed inside invisible to external electromagnetic radiations. The ideal cloak in [8] requires inhomogeneous and anisotropic media, with both permittivity and permeability independently controlled and radially dependent, making its practical realization very difficult. Therefore it has been proposed to use simplified material parameters for both transverse electric (TE) [9] and transverse magnetic (TM) [10] cases. To reduce the scattering due to the impedance mismatch introduced by the simplified cloaks, an improved linear cloak [11], a high-order transformation based cloak [12], and a "square root" transformation-based cloak [13] have also been proposed.

The coordinate transformation technique used in [8,14] has also been applied to the design of magnifying perfect and super lenses [15], electromagnetic field rotators [16], the reflectionless complex media for shifting and splitting optical beams [17], and conformal antennas [18]. The experimental demonstration of a simplified cloak consisting of split-ring resonators (SRRs) has been reported at microwave frequencies [19]. For the optical frequency range, the cloak can be constructed by embedding silver wires in a dielectric medium [10], or using a gold-dielectric concentric layered structure [20,21].

The modeling of Pendry's invisible cloak has been performed by using both analytical and numerical methods. Besides the widely used coordinate transformation technique [8,14–18,22], a cylindrical wave expansion technique [23] and a method based on the full-wave Mie scattering model [24,25] have also been applied. In addition, the full-wave finite element method (FEM)–based commercial simulation software COMSOL Multiphysics™ has been extensively used to model different cloaks and validate theoretical predictions [9,10,12,16,26], due to its ability to deal with anisotropic and radial dependent material parameters. However, similar to other frequency domain techniques, the FEM may become inefficient when wideband solutions are needed. So far, the time-domain techniques that have been developed to model the cloaking structures include the time-dependent scattering theory [27], the transmission line method (TLM) [28], and the FDTD method [29]. Due to its simplicity in implementation and the ability of treating anisotropic, inhomogeneous, and nonlinear materials, the FDTD method has been extremely popular for the analysis of electromagnetic structures. In the following, we develop a parallel dispersive FDTD method to model three-dimensional (3D) cloaking structures and reveal their extraordinary behavior.

### 3.2.1 Parallel Dispersive FDTD Modeling of 3D Cloaks

A complete set of material parameters of the ideal cloak in spherical coordinate is given by [8]

$$\varepsilon_r = \mu_r = \frac{R_2}{R_2 - R_1}\left(\frac{r - R_1}{r}\right)^2,$$

$$\varepsilon_\theta = \mu_\theta = \frac{R_2}{R_2 - R_1}, \tag{3.1}$$

$$\varepsilon_\phi = \mu_\phi = \frac{R_2}{R_2 - R_1},$$

where $R_1$ and $R_2$ are the inner and outer radii of the cloak, respectively, and $r$ is the distance from a spatial point within the cloak to the center of the cloak. Since the values of $\varepsilon_r$ and $\mu_r$ are less than one (between 0 and $(R_2 - R_1)/R_2$, same as the case for LHMs, the cloak cannot be directly modeled using the conventional FDTD method. However, one can map the material parameters using dispersive material models, for example, the Drude model

$$\varepsilon_r(\omega) = 1 - \frac{\omega_p^2}{\omega^2 - j\omega\gamma}, \tag{3.2}$$

where $\omega_p$ and $\gamma$ are the plasma and collision frequencies of the material, respectively. By varying the plasma frequency, the radial-dependent material parameters in (3.1) can be achieved. For example, for the ideal lossless case considered here, i.e., the collision frequency in (3.2) is equal to zero ($\gamma = 0$), the radial-dependent plasma frequency can be calculated using $\omega_p = \omega - \sqrt{1 - \varepsilon_r}$ with a given value of $\varepsilon_r$ calculated from (3.1). Note that in practice, the plasma frequency of the material depends on the periodicity of the split ring resonators (SRRs) [19] or wires [10], which varies along the radial direction. It should be noted that other dispersive material models (e.g., Debye, Lorentz) can be also considered for the modeling of electromagnetic cloaks. However, the Drude model has the simplest form when implemented using the dispersive FDTD method and has been widely used in the modeling of metamaterials. The Lorentz model can be also used in FDTD simulations with some additional modifications to the iterative equations. The Debye model may be less accurate to characterize the dispersion behavior of metamaterials and is rarely used in the community.

Because the conventional FDTD method [6,30] deals with frequency-independent materials, the frequency-dependent FDTD method is hence referred to as the dispersive FDTD method [31–33]. There are also different dispersive FDTD methods using different approaches to deal with the frequency-dependent material parameters: the recursive convolution (RC) method [31], the auxiliary differential equation (ADE) method [32], and the Z-transform method [33]. Due to its simplicity, we have chosen the ADE method to model the 3D cloak.

The ADE dispersive FDTD method is based on the Faraday's and Ampere's laws

$$\nabla \times \mathbf{E} = -\frac{\partial \mathbf{B}}{\partial t}, \tag{3.3}$$

$$\nabla \times \mathbf{H} = -\frac{\partial \mathbf{D}}{\partial t}, \tag{3.4}$$

as well as the constitutive relations $\mathbf{D} = \varepsilon\mathbf{E}$ and $\mathbf{B} = \mu\mathbf{H}$ where $\varepsilon$ and $\mu$ are expressed by (3.1). Equations (3.3) and (3.4) can be discretized following a standard procedure [6,30], which leads to the conventional FDTD updating equations

$$\mathbf{B}^{n+1} = \mathbf{B}^n - \Delta t \cdot \tilde{\nabla} \times \mathbf{E}^{n+\frac{1}{2}} \tag{3.5}$$

$$\mathbf{D}^{n+1} = \mathbf{D}^n - \Delta t \cdot \tilde{\nabla} \times \mathbf{H}^{n+\frac{1}{2}}, \tag{3.6}$$

where $\tilde{\nabla}$ is the discrete curl operator, $\Delta t$ is the discrete FDTD time step, and $n$ is the number of the time steps.

In addition, auxiliary differential equations need to be taken into account and they can be discretized through the following steps. For the conventional Cartesian FDTD mesh, since the material parameters given in (3.1) are in spherical coordinates, the following coordinate transformation is used [34]:

$$
\begin{bmatrix}
\varepsilon_{xx} & \varepsilon_{xy} & \varepsilon_{xz} \\
\varepsilon_{yx} & \varepsilon_{yy} & \varepsilon_{yz} \\
\varepsilon_{zx} & \varepsilon_{zy} & \varepsilon_{zz}
\end{bmatrix}
=
\begin{bmatrix}
\sin\theta\cos\phi & \cos\theta\cos\phi & -\sin\phi \\
\sin\theta\sin\phi & \cos\theta\sin\phi & \cos\phi \\
\cos\theta & -\sin\theta & 0
\end{bmatrix}
\begin{bmatrix}
\varepsilon_r & 0 & 0 \\
0 & \varepsilon_\theta & 0 \\
0 & 0 & \varepsilon_\phi
\end{bmatrix}
$$

$$
\times
\begin{bmatrix}
\sin\theta\cos\phi & \sin\theta\sin\phi & \cos\theta \\
\cos\theta\cos\phi & \cos\theta\sin\phi & -\sin\theta \\
-\sin\phi & \cos\phi & 0
\end{bmatrix}
\tag{3.7}
$$

The tensor form of the constitutive relation is given by

$$
\varepsilon_0
\begin{bmatrix}
\varepsilon_{xx} & \varepsilon_{xy} & \varepsilon_{xz} \\
\varepsilon_{yx} & \varepsilon_{yy} & \varepsilon_{yz} \\
\varepsilon_{zx} & \varepsilon_{zy} & \varepsilon_{zz}
\end{bmatrix}
\begin{bmatrix}
E_x \\
E_y \\
E_z
\end{bmatrix}
=
\begin{bmatrix}
D_x \\
D_y \\
D_z
\end{bmatrix}
\Leftrightarrow
\begin{bmatrix}
\varepsilon_{xx} & \varepsilon_{xy} & \varepsilon_{xz} \\
\varepsilon_{yx} & \varepsilon_{yy} & \varepsilon_{yz} \\
\varepsilon_{zx} & \varepsilon_{zy} & \varepsilon_{zz}
\end{bmatrix}^{-1}
\begin{bmatrix}
D_x \\
D_y \\
D_z
\end{bmatrix},
$$

$$\tag{3.8}$$

where

$$
\begin{bmatrix}
\varepsilon_{xx} & \varepsilon_{xy} & \varepsilon_{xz} \\
\varepsilon_{yx} & \varepsilon_{yy} & \varepsilon_{yz} \\
\varepsilon_{zx} & \varepsilon_{zy} & \varepsilon_{zz}
\end{bmatrix}^{-1}
=
\begin{bmatrix}
\varepsilon'_{xx} & \varepsilon'_{xy} & \varepsilon'_{xz} \\
\varepsilon'_{yx} & \varepsilon'_{yy} & \varepsilon'_{yz} \\
\varepsilon'_{zx} & \varepsilon'_{zy} & \varepsilon'_{zz}
\end{bmatrix},
\tag{3.9}
$$

and

$$\varepsilon'_{xx} = \frac{1}{\varepsilon_r}\sin^2\theta\cos^2\phi + \frac{1}{\varepsilon_\theta}\cos^2\theta\cos^2\phi + \frac{1}{\varepsilon_\phi}\sin^2\phi,$$

$$\varepsilon'_{xy} = \frac{1}{\varepsilon_r}\sin^2\theta\sin\phi\cos\phi + \frac{1}{\varepsilon_\theta}\cos^2\theta\sin\phi\cos\phi - \frac{1}{\varepsilon_\phi}\sin\phi\cos\phi,$$

$$\varepsilon'_{xz} = \frac{1}{\varepsilon_r}\sin\theta\cos\theta\cos\phi - \frac{1}{\varepsilon_\theta}\sin\theta\cos\theta\cos\phi,$$

$$\varepsilon'_{yx} = \frac{1}{\varepsilon_r}\sin^2\theta\sin\phi\cos\phi + \frac{1}{\varepsilon_\theta}\cos^2\theta\sin\phi\cos\phi - \frac{1}{\varepsilon_\phi}\sin\phi\cos\phi,$$

$$\varepsilon'_{yy} = \frac{1}{\varepsilon_r}\sin^2\theta\sin\phi + \frac{1}{\varepsilon_\theta}\cos^2\theta\sin^2\phi + \frac{1}{\varepsilon_\phi}\cos^2\phi,$$

$$\varepsilon'_{yz} = \frac{1}{\varepsilon_r}\sin\theta\cos\theta\cos\phi - \frac{1}{\varepsilon_\theta}\sin\theta\cos\theta\cos\phi,$$

$$\varepsilon'_{zx} = \frac{1}{\varepsilon_r}\sin\theta\cos\theta\cos\phi - \frac{1}{\varepsilon_\theta}\sin\theta\cos\theta\cos\phi,$$

$$\varepsilon'_{zy} = \frac{1}{\varepsilon_r}\sin\theta\cos\theta\sin\phi - \frac{1}{\varepsilon_\theta}\sin\theta\cos\theta\sin\phi,$$

$$\varepsilon'_{zz} = \frac{1}{\varepsilon_r}\cos^2\theta + \frac{1}{\varepsilon_\theta}\sin^2\theta.$$

Note that the inverse of the permittivity tensor matrix (3.9) exists only when $\varepsilon_r \neq 0$, $\varepsilon_\theta \neq 0$, and $\varepsilon_\phi \neq 0$. However, the inner boundary of the cloak does not satisfy the condition of $\varepsilon_r \neq 0$. Therefore, in our FDTD simulations, we place a perfect electric conductor (PEC) sphere with radius equal to $R_1$ inside the cloak to guarantee the validity of (3.9).

Substituting (3.9) into (3.8) gives

$$\varepsilon_0 E_x = \left(\frac{1}{\varepsilon_r}\sin^2\theta\cos^2\phi + \frac{1}{\varepsilon_\theta}\cos^2\theta\cos^2\phi + \frac{1}{\varepsilon_\phi}\sin^2\phi\right)D_x$$

$$+ \left(\frac{1}{\varepsilon_r}\sin^2\theta\sin\phi\cos\phi + \frac{1}{\varepsilon_\theta}\cos^2\theta\sin\phi\cos\phi + \frac{1}{\varepsilon_\phi}\sin\phi\cos\phi\right)D_x \qquad (3.10)$$

$$+ \left(\frac{1}{\varepsilon_r}\sin\theta\cos\theta\cos\phi - \frac{1}{\varepsilon_\theta}\sin\theta\cos\theta\cos\phi\right)D_z,$$

$$\varepsilon_0 E_y = \left( \frac{1}{\varepsilon_r} \sin^2 \theta \sin \phi \cos \phi + \frac{1}{\varepsilon_\theta} \cos^2 \theta \sin \phi \cos \phi + \frac{1}{\varepsilon_\phi} \sin \phi \cos \phi \right) D_x$$

$$+ \left( \frac{1}{\varepsilon_r} \sin^2 \theta \sin^2 \phi + \frac{1}{\varepsilon_\theta} \cos^2 \theta \sin^2 \phi + \frac{1}{\varepsilon_\phi} \cos^2 \phi \right) D_y \qquad (3.11)$$

$$+ \left( \frac{1}{\varepsilon_r} \sin \theta \cos \theta \sin \phi - \frac{1}{\varepsilon_\theta} \sin \theta \cos \theta \sin \phi \right) D_z,$$

$$\varepsilon_0 E_z = \left( \frac{1}{\varepsilon_r} \sin \theta \cos \theta \cos \phi - \frac{1}{\varepsilon_\theta} \sin \theta \cos \theta \cos \phi \right) D_x$$

$$+ \left( \frac{1}{\varepsilon_r} \sin \theta \cos \theta \sin \phi - \frac{1}{\varepsilon_\theta} \sin \theta \cos \theta \sin \phi \right) D_y \qquad (3.12)$$

$$+ \left( \frac{1}{\varepsilon_r} \cos^2 \theta + \frac{1}{\varepsilon_\theta} \sin^2 \theta \right) D_z.$$

Because the above equations have a similar form, in the following, the derivation of the updating equation is given only for the $E_x$ component. The updating equations for the $E_y$ and $E_z$ components can be derived following the same procedure.

Expressing $\varepsilon_r$ in the Drude form of (3.1), we can write Equation (3.10) as

$$\varepsilon_0 \left( \omega^2 - j\omega\gamma - \omega_p^2 \right) E_x$$

$$= \left[ (\omega^2 - j\omega\gamma) \sin^2 \theta \cos^2 \phi + (\omega^2 - j\omega\gamma - \omega_p^2) \left( \frac{\cos^2 \theta \cos^2 \phi}{\varepsilon_\theta} + \frac{\sin^2 \phi}{\varepsilon_\phi} \right) \right] D_x$$

$$+ \left[ (\omega^2 - j\omega\gamma) \sin^2 \theta \sin \phi \cos \phi \right. \qquad (3.13)$$

$$+ (\omega^2 - j\omega\gamma - \omega_p^2) \left( \frac{\cos^2 \theta \sin \phi \cos \phi}{\varepsilon_\theta} - \frac{\sin \phi \cos \phi}{\varepsilon_\phi} \right) \right] D_y$$

$$+ \left[ (\omega^2 - j\omega\gamma) \sin \theta \cos \theta \cos \phi - (\omega^2 - j\omega\gamma - \omega_p^2) \frac{\sin \theta \cos \theta \cos \phi}{\varepsilon_\theta} \right] D_z.$$

Note that $\varepsilon_\theta$ and $\varepsilon_\phi$ remain in (3.13) because their values are always greater than one and can be directly used in the conventional FDTD updating equations [6,30]. Applying the inverse Fourier transform and the following rules,

$$j\omega \to \frac{\partial}{\partial t}, \quad \omega^2 \to -\frac{\partial^2}{\partial t^2},$$

we can rewrite Equation (3.13) in the time domain as

$$\varepsilon_0 \left( \frac{\partial^2}{\partial t^2} + \gamma \frac{\partial}{\partial t} + \omega_p^2 \right) E_x$$

$$= \left[ \left( \frac{\partial^2}{\partial t^2} + \gamma \frac{\partial}{\partial t} \right) \sin^2 \theta \cos^2 \phi + \left( \frac{\partial^2}{\partial t^2} + \gamma \frac{\partial}{\partial t} + \omega_p^2 \right) \left( \frac{\cos^2 \theta \cos^2 \phi}{\varepsilon_\theta} + \frac{\sin^2 \phi}{\varepsilon_\phi} \right) \right] D_x$$

$$+ \left[ \left( \frac{\partial^2}{\partial t^2} + \gamma \frac{\partial}{\partial t} \right) \sin^2 \theta \sin \phi \cos \phi \right.$$

$$\left. + \left( \frac{\partial^2}{\partial t^2} + \gamma \frac{\partial}{\partial t} + \omega_p^2 \right) \left( \frac{\cos^2 \theta \sin \phi \cos \phi}{\varepsilon_\theta} - \frac{\sin \phi \cos \phi}{\varepsilon_\phi} \right) \right] D_y$$

$$+ \left[ \left( \frac{\partial^2}{\partial t^2} + \gamma \frac{\partial}{\partial t} \right) \sin \theta \cos \theta \cos \phi - \left( \frac{\partial^2}{\partial t^2} + \gamma \frac{\partial}{\partial t} + \omega_p^2 \right) \frac{\sin \theta \cos \theta \cos \phi}{\varepsilon_\theta} \right] D_z. \quad (3.14)$$

The FDTD simulation domain is represented by an equally spaced 3D grid with the periods $\Delta x$, $\Delta y$, and $\Delta z$ along the x-, y-, and z-directions, respectively. For the discretization of Equation (3.14), we use the central finite difference operators in time ($\delta_t$ and $\delta_t^2$) and the central average operators with respect to time ($\mu_t$ and $\mu_t^2$):

$$\frac{\partial^2}{\partial t^2} \to \frac{\delta_t^2}{(\Delta t)^2}, \quad \frac{\partial}{\partial t} \to \frac{\delta_t}{\Delta t} \mu_t, \quad 1 \to \mu_t^2,$$

where the operators $\delta_t$, $\delta_t^2$, $\mu_t$, and $\mu_t^2$ are defined as in [35]:

$$\delta_t \mathbf{F}\,|_{m_x,m_y,m_z}^n \equiv \mathbf{F}\,|_{m_x,m_y,m_z}^{n+\frac{1}{2}} - \mathbf{F}\,|_{m_x,m_y,m_z}^{n-\frac{1}{2}},$$

$$\delta_t^2 \mathbf{F}\,|_{m_x,m_y,m_z}^n \equiv \mathbf{F}\,|_{m_x,m_y,m_z}^{n+1} - 2\mathbf{F}\,|_{m_x,m_y,m_z}^{n} + \mathbf{F}\,|_{m_x,m_y,m_z}^{n-1},$$

$$\mu_t \mathbf{F}\,|_{m_x,m_y,m_z}^n \equiv \frac{\mathbf{F}\,|_{m_x,m_y,m_z}^{n+\frac{1}{2}} + \mathbf{F}\,|_{m_x,m_y,m_z}^{n-\frac{1}{2}}}{2},$$

$$\mu_t^2 \mathbf{F}\,|_{m_x,m_y,m_z}^n \equiv \frac{\mathbf{F}\,|_{m_x,m_y,m_z}^{n+1} + 2\mathbf{F}\,|_{m_x,m_y,m_z}^{n} + \mathbf{F}\,|_{m_x,m_y,m_z}^{n-1}}{4},$$

where $\mathbf{F}$ represents field components and $m_x$, $m_y$, $m_z$, are the indices corresponding to a certain discretization point in the FDTD domain. The discretized Equation (3.14) reads

$$\varepsilon_0 \left[ \frac{\delta_t^2}{(\Delta t)^2} + \gamma \frac{\delta_t}{\Delta t} \mu_t + \omega_p^2 \mu_t^2 \right] E_x$$

$$= \left\{ \left[ \frac{\delta_t^2}{(\Delta t)^2} + \gamma \frac{\delta_t}{\Delta t} \mu_t \right] \sin^2 \theta \cos^2 \phi \right.$$

$$+ \left[ \frac{\delta_t^2}{(\Delta t)^2} + \gamma \frac{\delta_t}{\Delta t} \mu_t + \omega_p^2 \mu_t^2 \right] \left( \frac{\cos^2 \theta \cos^2 \phi}{\varepsilon_\theta} + \frac{\sin^2 \phi}{\varepsilon_\phi} \right) \right\} D_x$$

$$+ \left\{ \left[ \frac{\delta_t^2}{(\Delta t)^2} + \gamma \frac{\delta_t}{\Delta t} \mu_t \right] \sin^2 \theta \sin \phi \cos \phi \right.$$

$$+ \left[ \frac{\delta_t^2}{(\Delta t)^2} + \gamma \frac{\delta_t}{\Delta t} \mu_t + \omega_p^2 \mu_t^2 \right] \left( \frac{\cos^2 \theta \sin \phi \cos \phi}{\varepsilon_\theta} - \frac{\sin \phi \cos \phi}{\varepsilon_\phi} \right) \right\} D_y$$

$$+ \left\{ \left[ \frac{\delta_t^2}{(\Delta t)^2} + \gamma \frac{\delta_t}{\Delta t} \mu_t \right] \sin \theta \cos \theta \cos \phi \right.$$

$$- \left[ \frac{\delta_t^2}{(\Delta t)^2} + \gamma \frac{\delta_t}{\Delta t} \mu_t + \omega_p^2 \mu_t^2 \right] \frac{\sin \theta \cos \theta \cos \phi}{\varepsilon_\theta} \right\} D_z.$$

$$(3.15)$$

Note that in (3.15), the discretization of the term $\omega_p^2$ of (3.14) is performed using the central average operator $\mu_t^2$ in order to guarantee the improved stability; the central average operator $\mu_t$ is used for the term containing $\gamma$

to preserve the second-order feature of the equation. Equation (3.15) can be written as

$$\varepsilon_0 \left[ \frac{E_x^{n+1} - 2E_x^n + E_x^{n-1}}{(\Delta t)^2} + \gamma \frac{E_x^{n+1} - E_x^{n-1}}{2\Delta t} + \omega_p^2 \frac{E_x^{n+1} + 2E_x^n + E_x^{n-1}}{4} \right]$$

$$= \sin^2\theta\cos^2\phi \left[ \frac{D_x^{n+1} - 2D_x^n + D_x^{n-1}}{(\Delta t)^2} + \gamma \frac{D_x^{n+1} - D_x^{n-1}}{2\Delta t} \right] + \left( \frac{\cos^2\theta\cos^2\phi}{\varepsilon_\theta} + \frac{\sin^2\phi}{\varepsilon_\phi} \right)$$

$$\times \left[ \frac{D_x^{n+1} - 2D_x^n + D_x^{n-1}}{(\Delta t)^2} + \gamma \frac{D_x^{n+1} - D_x^{n-1}}{2\Delta t} + \omega_p^2 \frac{D_x^{n+1} + 2D_x^n + D_x^{n-1}}{4} \right]$$

$$+ \sin^2\theta\sin\phi\cos\phi \left[ \frac{D_y^{n+1} - 2D_y^n + D_y^{n-1}}{(\Delta t)^2} + \gamma \frac{D_y^{n+1} - D_y^{n-1}}{2\Delta t} \right]$$

$$+ \left( \frac{\cos^2\theta\sin\phi\cos\phi}{\varepsilon_\theta} - \frac{\sin\phi\cos\phi}{\varepsilon_\phi} \right)$$

$$\times \left[ \frac{D_y^{n+1} - 2D_y^n + D_y^{n-1}}{(\Delta t)^2} + \gamma \frac{D_y^{n+1} - D_y^{n-1}}{2\Delta t} + \omega_p^2 \frac{D_y^{n+1} + 2D_y^n + D_y^{n-1}}{4} \right]$$

$$+ \sin\theta\cos\theta\cos\phi \left[ \frac{D_z^{n+1} - 2D_z^n + D_z^{n-1}}{(\Delta t)^2} + \gamma \frac{D_z^{n+1} - D_z^{n-1}}{2\Delta t} \right] - \frac{\sin\theta\cos\theta\cos\phi}{\varepsilon_\theta}$$

$$\times \left[ \frac{D_z^{n+1} - 2D_z^n + D_z^{n-1}}{(\Delta t)^2} + \gamma \frac{D_z^{n+1} - D_z^{n-1}}{2\Delta t} + \omega_p^2 \frac{D_z^{n+1} + 2D_z^n + D_z^{n-1}}{4} \right].$$

$$(3.16)$$

After simple manipulations, the updating equation for $E_x$ can be obtained as

$$E_x^{n+1} = \left[ b_{0xx}D_x^{n+1} + b_{1xx}D_x^n + b_{2xx}D_x^{n-1} + b_{0xy}\bar{D}_y^{n+1} + b_{1xy}\bar{D}_y^n + b_{2xy}\bar{D}_y^{n-1} \right.$$

$$\left. + b_{0xz}\bar{D}_z^{n+1} + b_{1xz}\bar{D}_z^n + b_{2xz}\bar{D}_z^{n-1} - \left( a_{1x}E_x^n + a_{2x}E_x^{n-1} \right) \right]/a_{0x}, \qquad (3.17)$$

where the coefficients are given by

$$a_{0x} = \varepsilon_0 \left[ \frac{1}{(\Delta t)^2} + \frac{\gamma}{2\Delta t} + \frac{\omega_p^2}{4} \right],$$

$$a_{1x} = \varepsilon_0 \left[ -\frac{2}{\Delta t^2} + \frac{\omega_p^2}{2} \right],$$

$$a_{2x} = \varepsilon_0 \left[ \frac{1}{(\Delta t)^2} - \frac{\gamma}{2\Delta t} + \frac{\omega_p^2}{4} \right],$$

$$b_{0xx} = \sin^2 \theta \cos^2 \phi \left[ \frac{1}{(\Delta t)^2} + \frac{\gamma}{2\Delta t} \right]$$
$$+ \left( \frac{\cos^2 \theta \cos^2 \phi}{\varepsilon_\theta} + \frac{\sin^2 \phi}{\varepsilon_\phi} \right) \left[ \frac{1}{(\Delta t)^2} + \frac{\gamma}{2\Delta t} + \frac{\omega_p^2}{4} \right],$$

$$b_{1xx} = -\sin^2 \theta \cos^2 \phi \frac{2}{(\Delta t)^2} + \left( \frac{\cos^2 \theta \cos^2 \phi}{\varepsilon_\theta} + \frac{\sin^2 \phi}{\varepsilon_\phi} \right) \left[ -\frac{2}{(\Delta t)^2} + \frac{\omega_p^2}{2} \right],$$

$$b_{2xx} = \sin^2 \theta \cos^2 \phi \left[ \frac{1}{(\Delta t)^2} - \frac{\gamma}{2\Delta t} \right]$$
$$+ \left( \frac{\cos^2 \theta \cos^2 \phi}{\varepsilon_\theta} + \frac{\sin^2 \phi}{\varepsilon_\phi} \right) \left[ \frac{1}{(\Delta t)^2} - \frac{\gamma}{2\Delta t} + \frac{\omega_p^2}{4} \right],$$

$$b_{0xy} = \sin^2 \theta \sin \phi \cos \phi \left[ \frac{1}{(\Delta t)^2} + \frac{\gamma}{2\Delta t} \right]$$
$$+ \left( \frac{\cos^2 \theta \sin \phi \cos \phi}{\varepsilon_\theta} - \frac{\sin \phi \cos \phi}{\varepsilon_\phi} \right) \left[ \frac{1}{(\Delta t)^2} + \frac{\gamma}{2\Delta t} + \frac{\omega_p^2}{4} \right],$$

$$b_{1xy} = -\sin^2 \theta \sin \phi \cos \phi \frac{2}{(\Delta t)^2}$$
$$+ \left( \frac{\cos^2 \theta \sin \phi \cos \phi}{\varepsilon_\theta} - \frac{\sin \phi \cos \phi}{\varepsilon_\phi} \right) \left[ -\frac{2}{(\Delta t)^2} + \frac{\omega_p^2}{2} \right],$$

$$b_{2xy} = \sin^2 \theta \sin \phi \cos \phi \left[ \frac{1}{(\Delta t)^2} - \frac{\gamma}{2\Delta t} \right]$$
$$+ \left( \frac{\cos^2 \theta \sin \phi \cos \phi}{\varepsilon_\theta} - \frac{\sin \phi \cos \phi}{\varepsilon_\phi} \right) \left[ \frac{1}{(\Delta t)^2} - \frac{\gamma}{2\Delta t} + \frac{\omega_p^2}{4} \right],$$

$$b_{0xz} = \sin \theta \cos \theta \cos \phi \left[ \frac{1}{(\Delta t)^2} + \frac{\gamma}{2\Delta t} \right] - \frac{\sin \theta \cos \theta \cos \phi}{\varepsilon_\theta} \left[ \frac{1}{(\Delta t)^2} + \frac{\gamma}{2\Delta t} + \frac{\omega_p^2}{4} \right],$$

$$b_{1xz} = -\sin \theta \cos \theta \cos \phi \frac{2}{(\Delta t)^2} - \frac{\sin \theta \cos \theta \cos \phi}{\varepsilon_\theta} \left[ -\frac{2}{(\Delta t)^2} + \frac{\omega_p^2}{2} \right],$$

$$b_{2xz} = \sin \theta \cos \theta \cos \phi \left[ \frac{1}{(\Delta t)^2} - \frac{\gamma}{2\Delta t} \right] - \frac{\sin \theta \cos \theta \cos \phi}{\varepsilon_\theta} \left[ \frac{1}{(\Delta t)^2} - \frac{\gamma}{2\Delta t} + \frac{\omega_p^2}{4} \right].$$

Following the same procedure, the updating equation for $E_y$ is

$$E_y^{n+1} = \left[ b_{0yx}\bar{D}_x^{n+1} + b_{1yx}\bar{D}_x^n + b_{2yx}\bar{D}_x^{n-1} + b_{0yy}D_y^{n+1} + b_{1yy}D_y^n + b_{2yy}D_y^{n-1} \right.$$

$$\left. + b_{0yz}\bar{D}_z^{n+1} + b_{1yz}\bar{D}_z^n + b_{2yz}\bar{D}_z^{n-1} - \left( a_{1y}E_y^n + a_{2y}E_y^{n-1} \right) \right]/a_{0y}, \tag{3.18}$$

with the coefficients given by

$$a_{0y} = \varepsilon_0 \left[ \frac{1}{(\Delta t)^2} + \frac{\gamma}{2\Delta t} + \frac{\omega_p^2}{4} \right],$$

$$a_{1y} = \varepsilon_0 \left[ -\frac{2}{(\Delta t)^2} + \frac{\omega_p^2}{2} \right],$$

$$a_{2y} = \varepsilon_0 \left[ \frac{1}{(\Delta t)^2} - \frac{\gamma}{2\Delta t} + \frac{\omega_p^2}{4} \right],$$

$$b_{0yx} = \sin^2\theta \sin\phi \cos\phi \left[ \frac{1}{(\Delta t)^2} + \frac{\gamma}{2\Delta t} \right]$$

$$+ \left( \frac{\cos^2\theta \sin\phi \cos\phi}{\varepsilon_\theta} - \frac{\sin\phi \cos\phi}{\varepsilon_\phi} \right) \left[ \frac{1}{(\Delta t)^2} + \frac{\gamma}{2\Delta t} + \frac{\omega_p^2}{4} \right],$$

$$b_{1yx} = -\sin^2\theta \sin\phi \cos\phi \frac{2}{(\Delta t)^2}$$

$$+ \left( \frac{\cos^2\theta \sin\phi \cos\phi}{\varepsilon_\theta} - \frac{\sin\phi \cos\phi}{\varepsilon_\phi} \right) \left[ -\frac{2}{(\Delta t)^2} + \frac{\omega_p^2}{2} \right],$$

$$b_{2yx} = \sin^2\theta \sin\phi \cos\phi \left[ \frac{1}{(\Delta t)^2} - \frac{\gamma}{2\Delta t} \right]$$

$$+ \left( \frac{\cos^2\theta \sin\phi \cos\phi}{\varepsilon_\theta} - \frac{\sin\phi \cos\phi}{\varepsilon_\phi} \right) \left[ \frac{1}{(\Delta t)^2} - \frac{\gamma}{2\Delta t} + \frac{\omega_p^2}{4} \right],$$

$$b_{0yy} = \sin^2\theta \sin^2\phi \left[ \frac{1}{(\Delta t)^2} + \frac{\gamma}{2\Delta t} \right]$$

$$+ \left( \frac{\cos^2\theta \sin^2\phi}{\varepsilon_\theta} + \frac{\cos^2\phi}{\varepsilon_\phi} \right) \left[ \frac{1}{(\Delta t)^2} + \frac{\gamma}{2\Delta t} + \frac{\omega_p^2}{4} \right],$$

$$b_{1yy} = -\sin^2\theta\sin^2\phi\,\frac{2}{(\Delta t)^2} + \left(\frac{\cos^2\theta\sin^2\phi}{\varepsilon_\theta} + \frac{\cos^2\phi}{\varepsilon_\phi}\right)\left[-\frac{2}{(\Delta t)^2} + \frac{\omega_p^2}{2}\right],$$

$$b_{2yy} = \sin^2\theta\sin^2\phi\left[\frac{1}{(\Delta t)^2} - \frac{\gamma}{2\Delta t}\right]$$

$$+ \left(\frac{\cos^2\theta\sin^2\phi}{\varepsilon_\theta} + \frac{\cos^2\phi}{\varepsilon_\phi}\right)\left[\frac{1}{(\Delta t)^2} - \frac{\gamma}{2\Delta t} + \frac{\omega_p^2}{4}\right],$$

$$b_{0yz} = \sin\theta\cos\theta\sin\phi\left[\frac{1}{(\Delta t)^2} + \frac{\gamma}{2\Delta t}\right] - \frac{\sin\theta\cos\theta\sin\phi}{\varepsilon_\theta}\left[\frac{1}{(\Delta t)^2} + \frac{\gamma}{2\Delta t} + \frac{\omega_p^2}{4}\right],$$

$$b_{1yz} = -\sin\theta\cos\theta\sin\phi\,\frac{2}{(\Delta t)^2} - \frac{\sin\theta\cos\theta\sin\phi}{\varepsilon_\theta}\left[-\frac{2}{(\Delta t)^2} + \frac{\omega_p^2}{2}\right],$$

$$b_{2yz} = \sin\theta\cos\theta\sin\phi\left[\frac{1}{(\Delta t)^2} - \frac{\gamma}{2\Delta t}\right] - \frac{\sin\theta\cos\theta\sin\phi}{\varepsilon_\theta}\left[\frac{1}{(\Delta t)^2} - \frac{\gamma}{2\Delta t} + \frac{\omega_p^2}{4}\right].$$

And the updating equation for $E_z$ is

$$E_z^{n+1} = \Big[b_{0zx}\bar{D}_x^{n+1} + b_{1zx}\bar{D}_x^n + b_{2zx}\bar{D}_x^{n-1} + b_{0zy}\bar{D}_y^{n+1} + b_{1zy}\bar{D}_y^n + b_{2zy}\bar{D}_y^{n-1}$$

$$+ b_{0zz}D_z^{n+1} + b_{1zz}D_z^n + b_{2zz}D_z^{n-1} - \left(a_{1z}E_z^n + a_{2z}E_z^{n-1}\right)\Big]/a_{0z}, \qquad (3.19)$$

with the coefficients given by

$$a_{0z} = \varepsilon_0\left[\frac{1}{(\Delta t)^2} + \frac{\gamma}{2\Delta t} + \frac{\omega_p^2}{4}\right],$$

$$a_{1z} = \varepsilon_0\left[-\frac{2}{(\Delta t)^2} + \frac{\omega_p^2}{2}\right],$$

$$a_{2z} = \varepsilon_0\left[\frac{1}{(\Delta t)^2} - \frac{\gamma}{2\Delta t} + \frac{\omega_p^2}{4}\right],$$

$$b_{0zx} = \sin\theta\cos\theta\cos\phi\left[\frac{1}{(\Delta t)^2} + \frac{\gamma}{2\Delta t}\right] - \frac{\sin\theta\cos\theta\cos\phi}{\varepsilon_\theta}\left[\frac{1}{(\Delta t)^2} + \frac{\gamma}{2\Delta t} + \frac{\omega_p^2}{4}\right],$$

$$b_{1zx} = -\sin\theta\cos\theta\cos\phi\,\frac{2}{(\Delta t)^2} - \frac{\sin\theta\cos\theta\cos\phi}{\varepsilon_\theta}\left[-\frac{2}{(\Delta t)^2} + \frac{\omega_p^2}{2}\right],$$

$$b_{2zx} = \sin\theta\cos\theta\cos\phi\left[\frac{1}{(\Delta t)^2} - \frac{\gamma}{2\Delta t}\right] - \frac{\sin\theta\cos\theta\cos\phi}{\varepsilon_\theta}\left[\frac{1}{(\Delta t)^2} - \frac{\gamma}{2\Delta t} + \frac{\omega_p^2}{4}\right],$$

$$b_{0zy} = \sin\theta\cos\theta\sin\phi\left[\frac{1}{(\Delta t)^2} + \frac{\gamma}{2\Delta t}\right] - \frac{\sin\theta\cos\theta\sin\phi}{\varepsilon_\theta}\left[\frac{1}{(\Delta t)^2} + \frac{\gamma}{2\Delta t} + \frac{\omega_p^2}{4}\right],$$

$$b_{1zy} = -\sin\theta\cos\theta\sin\phi\frac{2}{(\Delta t)^2} - \frac{\sin\theta\cos\theta\sin\phi}{\varepsilon_\theta}\left[-\frac{2}{(\Delta t)^2} + \frac{\omega_p^2}{2}\right],$$

$$b_{2zy} = \sin\theta\cos\theta\sin\phi\left[\frac{1}{(\Delta t)^2} - \frac{\gamma}{2\Delta t}\right] - \frac{\sin\theta\cos\theta\sin\phi}{\varepsilon_\theta}\left[\frac{1}{(\Delta t)^2} - \frac{\gamma}{2\Delta t} + \frac{\omega_p^2}{4}\right],$$

$$b_{0zz} = \cos^2\theta\left[\frac{1}{(\Delta t)^2} + \frac{\gamma}{2\Delta t}\right] + \frac{\sin^2\theta}{\varepsilon_\theta}\left[\frac{1}{(\Delta t)^2} + \frac{\gamma}{2\Delta t} + \frac{\omega_p^2}{4}\right],$$

$$b_{1zz} = -\cos^2\theta\frac{2}{(\Delta t)^2} + \frac{\sin^2\theta}{\varepsilon_\theta}\left[-\frac{2}{(\Delta t)^2} + \frac{\omega_p^2}{2}\right],$$

$$b_{2zz} = \cos^2\theta\left[\frac{1}{(\Delta t)^2} - \frac{\gamma}{2\Delta t}\right] + \frac{\sin^2\theta}{\varepsilon_\theta}\left[\frac{1}{(\Delta t)^2} - \frac{\gamma}{2\Delta t} + \frac{\omega_p^2}{4}\right].$$

Note that the field quantities $\bar{D}_x$, $\bar{D}_y$, and $\bar{D}_z$ in (3.17)–(3.19) are locally averaged values of $D_x$, $D_y$, and $D_z$, respectively, since the x-, y-, and z-components of the electric fields are in different locations in the FDTD domain [36]. However, the averaging needs to be applied along different directions depending on the updating equations. Specifically in (3.17), the averaged $D_y$ and $D_z$ can be calculated using

$$\bar{D}_y(i,j,k) = \frac{D_y(i,j,k) + D_y(i+1,j,k) + D_y(i,j-1,k) + D_y(i+1,j-1,k)}{4},$$

$$\bar{D}_z(i,j,k) = \frac{D_z(i,j,k) + D_z(i+1,j,k) + D_z(i,j,k-1) + D_z(i+1,j,k-1)}{4},$$

where $(i,j,k)$ is the coordinate of the field component. In (3.18), the averaged $D_x$, and $D_z$ can be calculated using

$$\bar{D}_x(i,j,k) = \frac{D_x(i,j,k) + D_x(i,j+1,k) + D_x(i-1,j,k) + D_x(i-1,j+1,k)}{4},$$

$$\bar{D}_z(i,j,k) = \frac{D_z(i,j,k) + D_z(i,j+1,k) + D_z(i,j,k-1) + D_z(i,j+1,k-1)}{4}.$$

And in (3.19), the averaged $D_x$ and $D_y$ can be calculated using

$$\bar{D}_x(i,j,k) = \frac{D_x(i,j,k) + D_x(i,j,k+1) + D_x(i-1,j,k) + D_x(i-1,j,k+1)}{4},$$

$$\bar{D}_y(i,j,k) = \frac{D_y(i,j,k) + D_y(i,j,k+1) + D_y(i,j-1,k) + D_y(i,j-1,k+1)}{4}.$$

The updating equations for the magnetic fields $H_x$, $D_y$, and $D_z$ are in the same form as (3.17)–(3.19) with the same coefficients and can be obtained by replacing **E** with **H** and **D** with **B**. The averaged field components can be calculated in a similar manner. Equations (3.5), (3.6), and (3.17)–(3.19), and the updating equations for **H** from **B** (not given) form the updating equation set for the modeling of 3D cloaks using the well-known leap-frog scheme [30]. If the plasma frequency in (3.2) is equal to zero, i.e., $\omega_p = 0$, and $\varepsilon_\theta = \mu_\theta = \varepsilon_\phi = \mu_\phi = 1$, the above updating equation set reduces to the updating equation set for the free space.

Because the FDTD method is inherently a numerical technique, the spatial as well as time discretization has important effects on the accuracy of simulation results. Also, because the permittivity and permeability are frequency dependent, one can expect a slight difference between the analytical and numerical material parameters due to the discrete time step in the FDTD method. From our previous analysis [7] that for the modeling of metamaterials, especially the LHMs, the numerical errors due to the time discretization will cause spurious resonances, it follows that a requirement of $\Delta x < \lambda/80$ is necessary. Following the same approach as in [29], one can find that the numerical permittivity $\tilde{\varepsilon}_r$ for the ideal 3D cloak takes the following form:

$$\tilde{\varepsilon}_r = \varepsilon_0 \left[ 1 - \frac{\omega_p^2 (\Delta t)^2 \cos^2 \frac{\omega \Delta t}{2}}{2 \sin \frac{\omega \Delta t}{2} \left( 2 \sin \frac{\omega \Delta t}{2} - j\gamma \Delta t \cos \frac{\omega \Delta t}{2} \right)} \right]. \tag{3.20}$$

Note that Equation (3.20) simplifies to the analytical Drude dispersion model (3.2) when $\Delta t \to 0$. With the expression of the numerical permittivity (3.20) available, one can correct the errors introduced by the discrepancy between the numerical and analytical material parameters. For example, if the required permittivity is $\varepsilon_r = \varepsilon_r' + j\varepsilon_r''$, after simple derivations, the corrected plasma and collision frequencies can be calculated as

$$\tilde{\omega}_p^2 = \frac{2\sin\dfrac{\omega\Delta t}{2}\left[-2(\varepsilon_r'-1)\sin\dfrac{\omega\Delta t}{2}-\varepsilon_r''\gamma\Delta t\cos\dfrac{\omega\Delta t}{2}\right]}{(\Delta t)^2\cos^2\dfrac{\omega\Delta t}{2}}, \tag{3.21}$$

$$\tilde{\gamma} = \frac{2\varepsilon_r''\sin\dfrac{\omega\Delta t}{2}}{(\varepsilon_r'-1)\Delta t\cos\dfrac{\omega\Delta t}{2}}. \tag{3.22}$$

The above averaging of field components and the correction of numerical material parameters ensure stable and accurate FDTD simulations of the 3D cloak. However, if the averaging is not applied, the field distribution becomes unsymmetrical and hence incorrect; if the numerical material parameters are not corrected, the FDTD simulations become unstable before reaching the steady state. Therefore, in our simulations of the 3D cloaks, the field averaging and corrected material parameters (3.21) and (3.22) are always used.

The FDTD method is a versatile numerical technique. However, similar to other numerical methods, it is computationally intensive. For large electromagnetic problems such as the modeling of 3D cloaks, the requirement for system resources is beyond the capability of a single personal computer (PC). One way to resolve this problem is to divide the whole computational domain into many smaller subdomains, and each subdomain can be handled by a PC. By linking the PCs together with an appropriate synchronization procedure, the original large problem can be decomposed and solved efficiently.

One of the most attractive features of the FDTD method is that it can be easily parallelized with very few modifications to the algorithm. Because it solves Maxwell's equations in the time-space domain, the parallel FDTD algorithm is based on the space decomposition technique [37,38]. The data transfer functionality between processors (PCs) is provided by the message passing interface (MPI) library. Data exchange is required only for the adjacent cells at the interface between different subdomains and is performed at each time step; hence the parallel FDTD algorithm is a self-synchronized process. Figure 3.1 shows the arrangement of the field components in different subdomains in parallel FDTD simulations. The gray arrows are the transferred field components from the neighboring subdomain during the data communication process. At the end of parallel FDTD simulations, the results calculated at each processor need to be combined to obtain the final simulation result. In comparison to the conventional parallel FDTD method, the parallelization of the dispersive FDTD method introduced in this chapter requires additional field components to be transferred between adjacent

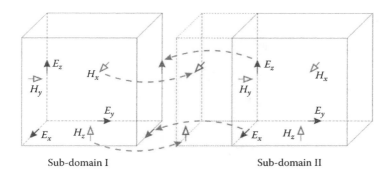

Sub-domain I                                             Sub-domain II

**FIGURE 3.1**

The arrangement of field components in different subdomains in parallel FDTD simulations. The gray arrows indicate the field components that are transferred from the neighboring domain during the data communication process and used to update the field components on the boundary of the current subdomain.

subdomains during the synchronization process, because of the applied field averaging scheme. The complexity of the algorithm further increases if the whole computational domain is divided along more than one direction, although the data communication load and the overall simulation time may be reduced.

The PC cluster used to simulate 3D cloaks consists of one head node for monitoring purposes and 15 compute nodes for performing calculation tasks. Each node has Dual Intel Xeon E5405 (Quad Core 2.0 GHz) central processing units (CPUs), and there are 128 cores and 512 GB memory in total. The nodes are connected by a 24-port gigabit switch. The GNU C compiler (GCC) and a free version of MPI, MPI Chameleon (MPICH), developed by Argonne National Laboratory [39], are used to compile the developed parallel dispersive FDTD code and handle the inter-core data communications, respectively. The above developed parallel dispersive FDTD method has been implemented to model the ideal 3D cloaks, and the simulation results and discussions are presented in the following section.

### 3.2.2 Numerical Results and Discussions

The 3D FDTD simulation domain is shown in Figure 3.2. The FDTD cell size in all simulations is $\Delta x = \Delta y = \lambda/150$, where $\lambda$ is the wavelength at the operating frequency $f = 2.0$ GHz. The time step is chosen according to the Courant stability criterion [6], i.e., $\Delta t = \Delta x/\sqrt{3}c$. The radii of the cloak are $R_1 = 0.1$ m and $R_2 = 0.2$ m. In the present work, only the ideal case (lossless) is considered, i.e., the collision frequency in (3.2) is equal to zero ($\gamma = 0$). The radial dependent plasma frequency can be calculated using (3.21) and (3.22) with a given value of $\varepsilon_r$ calculated from (3.1). The computational domain is truncated using Berenger's perfectly matched layer (PML) [40]

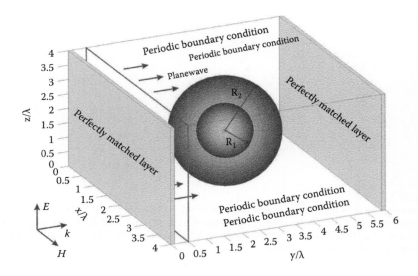

**FIGURE 3.2**
The 3D parallel dispersive FDTD simulation domain for the case of plane-wave incidence on the cloak. The red rectangle indicates the location of the source plane. (See color figure.)

in $y$-direction to absorb waves leaving the computational domain without introducing reflections, and terminated with periodic boundary conditions (PBCs) in $x$- and $z$-directions for the modeling of a plane-wave source. The plane-wave source is implemented by specifying a complete plane of FDTD cells using a certain wave function. The electric and magnetic fields of the plane wave are along the $z$- and $x$-axis, respectively, and the propagation direction is along the $y$-axis, as indicated in Figure 3.2. For simplicity, the whole simulation domain is only divided along $y$-direction into 100 sub-domains and in total 100 processors and 220 gigabyte (GB) memory were used to run the parallel dispersive FDTD simulations. Each simulation lasts around 45 hours (13,000 time steps) before reaching the steady state.

Figures 3.3 and 3.4 show the normalized steady-state field distributions for the $E_z$ and $H_x$ components in $y$-$z$ and $x$-$y$ planes, respectively. It can be seen that the plane wave is guided by the cloak to propagate around its central region and is recomposed back after leaving the cloak. There is nearly no reflection (except those tiny numerical ones due to the finite spatial resolution in FDTD simulations), since the material parameters (3.1) vary continuously in space while keeping the impedance the same as the free space one. It is also interesting to notice that the $E_z$ component in $y$-$z$ and $x$-$y$ planes in Figure 3.3 and the $H_x$ component in $x$-$y$ and $y$-$z$ planes in Figure 3.4 have the same distributions (with different amplitude), which is due to the fact that the ideal 3D cloak is a rotationally symmetric structure with respect to the electric and magnetic fields. In comparison, the two-dimensional (2D) cloaks studied previously in [29] do not have such properties because the modeled

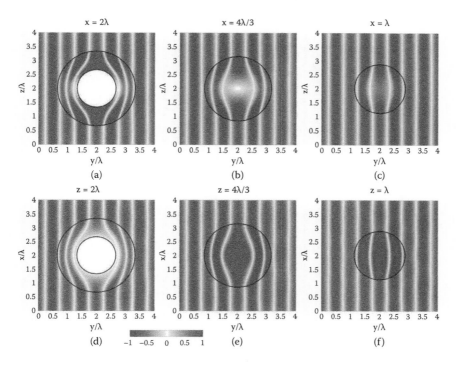

**FIGURE 3.3**
Normalized field distributions for the $E_z$ component in (a)–(c) $y$-$z$ plane and (d)–(f) $x$-$y$ plane in the steady state of the parallel dispersive FDTD simulations. The cutting planes are (see Figure 3.2) (a) $x = 2\lambda$, (b) $x = 4\lambda/3$, (c) $x = \lambda$, (d) $z = 2\lambda$, (e) $z = 4\lambda/3$, (f) $z = \lambda$. The wave propagation direction is from left to right. (See color figure.)

2D cloaks are effectively 3D infinitely long cylindrical ones, which only have cylindrical symmetry and respond differently to plane waves with different angle of incidence.

The wave behavior near the 3D cloak can be better illustrated using the power flow diagram, as plotted in Figure 3.5. It is shown that the Poynting vectors are diverted around the central area enclosed by the cloak. Therefore objects placed inside the cloak do not introduce any scattering to external radiations and hence become "invisible."

The above presented results validate the developed parallel dispersive FDTD method and demonstrate the cloaking property of the structure. However, there are some numerical issues that need to be addressed in FDTD simulations. Besides the correction of numerical material parameters introduced earlier, since the cloak is a sensitive structure, for single-frequency simulations, the switching time of the sinusoidal source also has significant impact on the convergence time. Normalized field distributions from the simulations using different switching time are plotted at the time

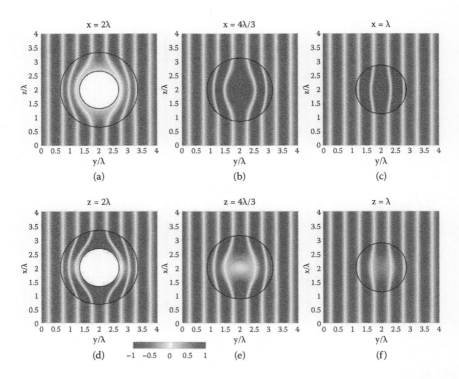

**FIGURE 3.4**

Normalized field distributions for the $H_x$ component in (a)–(c) $y$-$z$ plane and (d)–(f) $x$-$y$ plane in the steady state of the parallel dispersive FDTD simulations. The cutting planes are (see Figure 3.2) (a) $x = 2\lambda$, (b) $x = 4\lambda/3$, (c) $x = \lambda$, (d) $z = 2\lambda$, (e) $z = 4\lambda/3$, (f) $z = \lambda$. The wave propagation direction is from left to right. (See color figure.)

step $t = 1320\Delta t$ and shown in Figure 3.6. It can be seen that if the source is switched to its maximum amplitude within a short period of time, because of the multiple frequency components excited, and the cloak is essentially a narrowband structure due to its dispersive nature, the scattering from the cloak may occur, as shown in Figure 3.6(a). The scattered waves oscillate within the lossless cloak and hence it requires a very long time for the simulations to reach the steady state. It is also demonstrated that if the switching time is greater than $10T_0$ where $T_0$ is the period of the sinusoidal wave, the scattered waves can be significantly reduced and a much shorter convergence time in simulations can be achieved. Therefore in the previous simulations, the switching time of $30T_0$ was used.

Because the FDTD method is a time-domain technique, it is convenient to study the transient response of the 3D cloak. The snapshots of the field distributions for the $E_z$ component at different time steps $t = 3000\Delta t$ (5.77 ns), $t = 5000\Delta t$ (9.62 ns), and $t = 8000\Delta t$ (15.40 ns) are taken and plotted

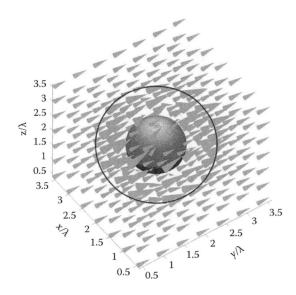

**FIGURE 3.5**
Power flow diagram of a plane wave incidence on the ideal 3D cloak calculated from parallel dispersive FDTD simulations.

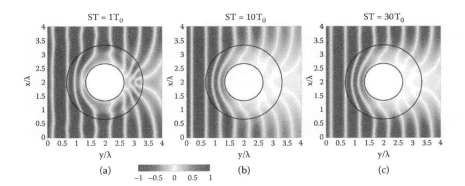

**FIGURE 3.6**
Comparison of the influence of different switching time (ST) of the sinusoidal source on the simulation results: (a) $ST = T_0$, (b) $ST = 10T_0$, (c) $ST = 30T_0$, where $T_0$ is the period of the sinusoidal wave. The wave propagation direction is from left to right and the normalized field distributions are plotted in the $x$-$y$ plane ($z = 2\lambda$; see Figure 3.2) and at the time step $t = 1320\Delta t$. (See color figure.)

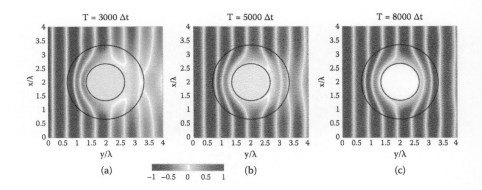

**FIGURE 3.7**
Snapshots of the field distributions for the $E_z$ component at different time steps in the parallel dispersive FDTD simulations: (a) $t = 3000\Delta t$ (5.77 ns), (b) $t = 5000\Delta t$ (9.62 ns), (c) $t = 8000\Delta t$ (15.40 ns), plotted in the $x$-$y$ plane ($z = 2\lambda$; see Figure 3.2). The wave propagation direction is from left to right. (See color figure.)

in Figure 3.7. It is shown in Figure 3.7(a) that outside the shadow region behind the cloak ($y \sim 3.5\lambda$, $x < 0.5\lambda$, and $x > 3.5\lambda$), waves propagate at the speed of light and the wave front remains the same as the one before reaching the cloak. However, due to the fact that the waves that travel through the cloak undergo a longer path compared to the free space one, and since the group velocity cannot exceed the speed of light, the wave front experiences a considerable delay in forming back to the free space one in the shadow region behind the cloak, as is illustrated by the field distributions at different time steps in FDTD simulations in Figure 3.7. In fact, the convergence of simulations is quite slow and the steady state is reached in simulations at around 13,000 time steps (25.02 ns). This delay effect has been explicitly studied in [41] for the 3D spherical cloak and numerically analyzed in [42] for the 2D cylindrical cloak. The time delay is slightly different for 2D and 3D cloaks due to their different constitutive material parameters. A detailed comparison can also be performed to analyze the difference in time delay for different cloaks.

Another advantage of the FDTD method is that a wideband frequency response can be obtained with a single run of simulations. In comparison to the previous single-frequency case, a wideband Gaussian pulse with the central frequency of 2.0 GHz and covering the frequency range of 1.5 ~ 2.5 GHz is used instead. At 5 FDTD cells away at both the front and back of the cloak along its central axis ($x = z = 2\lambda$; see Figure 3.2), the amplitude of $E_z$ is recorded during the simulation and then transformed to the frequency domain. The recorded time-domain signals and their spectra are plotted in Figure 3.8. It is found from Figure 3.8(a) that the time delay between the directly received time-domain signal in front of the cloak (gray solid line)

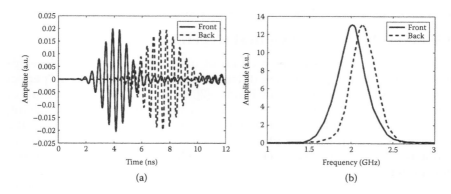

**FIGURE 3.8**
(a) The recorded time domain signals at 5 FDTD cells away at both the front and back of the cloak along its central axis ($x = z = 2\lambda$; see Figure 3.2). (b) The spectra of the recorded time domain signals.

and the received time domain signal through the cloak (gray dashed line) is approximately 3.4 ns. However, the distance between the two recording points is 0.402 m and the expected time delay for a plane wave propagating in the free space is 1.34 ns. This clearly demonstrates that the 3D cloak introduces a time delay to the waves propagating through it. It is also found from Figure 3.8(a) that the width of the pulse passing through the cloak has been broadened, which is due to the dispersive nature of the cloaking material. It is interesting to note that in Figure 3.8(b), the spectrum of the directly received time domain signal is centered at 2.0 GHz; however, the central frequency of the signal that passes through the cloak is considerably higher (~2.1 GHz). This frequency shift has been demonstrated theoretically in [43]; the explanation is that the frequency components higher than the working frequency of the cloak are enhanced and the frequency components lower than the working frequency of the cloak are weakened. The shift is found to be much more pronounced from our analysis since the observation point is taken near the back surface of the cloak in our simulations, and it was $30\lambda$ away from the cloak considered in [43]. The difference may be also due to the Lorentz dispersion model considered for the permeability of the cloak in [43], while in our simulations, both the permittivity and permeability are assumed to follow the same Drude dispersion model.

The method developed here can also be used to model cylindrical cloaks and compare their properties with the spherical one, such as the direction dependency issue. The ideal cloaks are lossless and the cloaking material properties vary in space continuously, which make their practical realization very difficult. The developed FDTD method can also be applied to study the effect of losses in cloaks, evaluate the performance of simplified cloaks, and assist the design and realization of practical cloaks.

## 3.3 FDTD Modeling of Optical Hyperlens

A considerable amount of research effort has been spent on the development of subwavelength imaging devices recently. The concept of a "perfect lens" was originally proposed by Pendry [2], using LHM [1]. The principle of operation of such devices is based on the negative refraction of propagating waves and the amplification of evanescent field components. However, the resonant excitation of surface plasmons that causes evanescent wave amplification is sensitive to losses in LHM and hence limits the maximum thickness of the device [44].

An alternative method of subwavelength imaging has been suggested in literature and termed *canalization* [45]. Its operation is based on the fact that for a certain type of device, the evanescent wave components can be transformed into propagating waves, and therefore the source field can be delivered to its back interface with little or no deterioration. In contrast to the case of LHM, such devices are less sensitive to losses. Typical examples of structures operating in the canalization regime include the wire medium formed by an array of parallel conducting wires [46] and the layered structure consisting of materials with alternating positive and negative permittivity [47]. The thickness of both structures needs to be equal to an integer multiple of a half wavelength at the operating frequency (due to the Fabry-Perot resonance), in order to avoid reflections between the source and the structure. It has been demonstrated numerically and experimentally that such a canalization regime indeed exists, and subwavelength details can be transferred to significant distances at frequencies up to the terahertz (THz) and infrared range [48,49].

Recently, there is a growing interest in the development of devices that are capable of simultaneously imaging and magnifying subwavelength field distributions in the visible frequency range [50–52]. In analogy to the perfect lens, such devices are termed *hyperlenses*. Both the wire medium and the layered structure have been considered in constructing the hyperlens in literature. In the visible frequency range, the hyperlens can be formed by either the layered structure with dielectric-plasmonic materials arranged uniaxially in Cartesian or cylindrical coordinates, or the plasmonic wire medium [53]. In microwave frequencies, the simultaneous enhancement and magnification of evanescent field patterns have been demonstrated experimentally with the help of double cylindrical polariton resonant structures by Alitalo et al. [54], and numerically using a tapered array of metallic wires simulated using a commercial full-wave electromagnetic solver, FEKO [55].

Numerical modeling has always played an important role in the analysis and design of metamaterials. Owning to its simplicity and ability to handle inhomogeneous and anisotropic materials, the FDTD method [6,30] has become one of the most widely used numerical techniques. It has been applied to model the perfect lens formed by LHM [7], the wire medium (using the effective medium theory) [56], plasmonic structures [5], and cloaking devices [3,29]. However, all the above implementations of the FDTD method are performed in the Cartesian coordinates and therefore staircase approximations need to be applied when modeling curved surfaces. In the modeling of a hyperlens, which often possesses cylindrical or spherical symmetry, the most suited approach is to apply the FDTD method in cylindrical or spherical coordinates. In this work, we focus on the numerical modeling of hyperlens structures and propose a spatially dispersive FDTD method in cylindrical coordinates to model the hyperlens formed by a tapered array of metallic wires using the effective medium theory. Not only is the developed method more efficient compared with commercial electromagnetic solvers since the fine details of the hyperlens are not taken into account, but it also offers flexibility in varying material parameters and dimensions of the structure such as its length and magnification ratio.

### 3.3.1 FDTD Modeling of Hyperlens Formed by a Tapered Array of Metallic Wires

The conventional FDTD method deals with frequency-independent materials [30]. However, for the case of metamaterials that possess negative permittivity and/or permeability that can be realized only through material frequency dispersion, the dispersive FDTD method [6] needs to be applied. The ADE dispersive FDTD method is based on Faraday's and Ampere's laws

$$\nabla \times \mathbf{E} = -\frac{\partial \mathbf{B}}{\partial t},$$ (3.23)

$$\nabla \times \mathbf{H} = \frac{\partial \mathbf{D}}{\partial t},$$ (3.24)

as well as the constitutive relations $\mathbf{D} = \varepsilon\mathbf{E}$ and $\mathbf{B} = \mu\mathbf{H}$. To simplify our analysis, in the current work we only consider the 2D hyperlens, which has cylindrical symmetry and only electric properties ($\mu = \mu_0$). The extension to the general 3D case to model the spherical hyperlens is straightforward. Under the 2D assumption and due to the fact that the operation of cylindrical hyperlens requires the electric field component to be aligned along its longitudinal direction, the expanded form of Equations (3.23) and (3.24) in cylindrical coordinates for the 2D transverse electric (TE$_z$) case in the frequency domain is used:

$$j\omega\mu_0 H_z = \frac{1}{r}\frac{\partial E_r}{\partial \phi} - \frac{1}{r}\frac{\partial(rE_\phi)}{\partial r},$$ (3.25)

$$j\omega\varepsilon_r D_r = \frac{1}{r}\frac{\partial H_z}{\partial\phi}, \tag{3.26}$$

$$j\omega\varepsilon_\phi D_\phi = -\frac{\partial H_z}{\partial r}, \tag{3.27}$$

where

$$D_r = \varepsilon_r E_r, \tag{3.28}$$

$$D_\phi = \varepsilon_\phi E_\phi. \tag{3.29}$$

In addition to Maxwell's equations, at the outer boundary of the computational domain, an absorbing boundary condition (ABC) needs to be applied in order to model an unbounded space. The originally proposed PML by Berenger [40] has excellent performance but is only applicable to the Cartesian coordinates [57]. Hence, in this work, we implement the complex coordinate stretching-based PML through the change of variable [58]:

$$\bar{r} = r + \frac{1}{j\omega\varepsilon_0}\int_0^r \sigma(s)ds = r + \frac{\bar{\sigma}}{j\omega\varepsilon_0}, \tag{3.30}$$

where $\sigma(s)$ is the conductivity term, and

$$\bar{\sigma} = \int_0^r \sigma(s)ds. \tag{3.31}$$

By splitting the $H_z$ component [57], Equations (3.25)–(3.27) read

$$j\omega\mu_0\overline{H_{zr}} + \sigma^*\overline{H_{zr}} = -\frac{\partial\overline{E_\phi}}{\partial r}, \tag{3.32}$$

$$j\omega r\mu_0 H_{z\phi} + \overline{\sigma^*}H_{z\phi} = \frac{\partial E_r}{\partial\phi}, \tag{3.33}$$

$$j\omega r D_r + \frac{\bar{\sigma}}{\varepsilon_0}D_r = \frac{\partial H_z}{\partial\phi}, \tag{3.34}$$

$$j\omega D_\phi + \frac{\sigma}{\varepsilon_0}D_\phi = -\frac{\partial H_z}{\partial r}, \tag{3.35}$$

$$\overline{H_{zr}} = \overline{r}H_{zr},$$ (3.36)

$$\overline{E_{\phi}} = \overline{r}E_{\phi},$$ (3.37)

where the chain rule $\partial/\partial r = (\partial/\partial \overline{r})(\partial \overline{r}/\partial r)$ is applied, and $\sigma^*$ and $\overline{\sigma}^*$ are defined as

$$\sigma^* = \sigma\frac{\mu_0}{\varepsilon_0}, \qquad \overline{\sigma}^* = \overline{\sigma}\frac{\mu_0}{\varepsilon_0}.$$

Applying inverse Fourier transform and using $j\omega \rightarrow \partial/\partial t$, Equations (3.32)–(3.37) can be written in the time domain as

$$\mu_0\frac{\partial \overline{H_{zr}}}{\partial t} + \sigma^*\overline{H_{zr}} = -\frac{\partial \overline{E_{\phi}}}{\partial r},$$ (3.38)

$$\mu_0 r\frac{\partial H_{z\phi}}{\partial t} + \overline{\sigma}^*H_{z\phi} = \frac{\partial E_r}{\partial \phi},$$ (3.39)

$$r\frac{\partial D_r}{\partial t} + \frac{\overline{\sigma}}{\varepsilon_0}D_r = \frac{\partial H_z}{\partial \phi},$$ (3.40)

$$\frac{\partial D_{\phi}}{\partial t} + \frac{\sigma}{\varepsilon_0}D_{\phi} = -\frac{\partial H_z}{\partial r},$$ (3.41)

$$r\frac{\partial H_{zr}}{\partial t} + \frac{\overline{\sigma}^*}{\mu_0}H_{zr} = \frac{\partial \overline{H_{zr}}}{\partial t},$$ (3.42)

$$r\frac{\partial E_{\phi}}{\partial t} + \frac{\overline{\sigma}}{\varepsilon_0}E_{\phi} = \frac{\partial \overline{E_{\phi}}}{\partial t}.$$ (3.43)

The FDTD simulation domain is represented by an equally spaced 2D grid with periods $\Delta r$ and $\Delta\phi$ along the $r$-and $\phi$-directions, respectively. For the discretization of Equations (3.38)–(3.43), we use the central finite difference operators in space ($\delta_r$, $\delta_\phi$) and in time ($\delta_t$), and the central average operator in time ($\mu_t$):

$$\frac{\partial}{\partial r} \rightarrow \frac{\delta_r}{\Delta r}, \qquad \frac{\partial}{\partial \phi} \rightarrow \frac{\delta_\phi}{\Delta\phi}, \qquad \frac{\partial}{\partial t} \rightarrow \frac{\delta_t}{\Delta t}, \qquad 1 \rightarrow \mu_t,$$

where the operators $\delta_r$, $\delta_\phi$, $\delta_t$, and $\mu_t$ are defined as [35]

$$\delta_r \mathbf{F}\,\big|_{m_r,m_\phi}^{n} \equiv \mathbf{F}\,\big|_{m_r+\frac{1}{2},m_\phi}^{n} - \mathbf{F}\,\big|_{m_r-\frac{1}{2},m_\phi}^{n},$$

$$\delta_\phi \mathbf{F}\,\big|_{m_r,m_\phi}^{n} \equiv \mathbf{F}\,\big|_{m_r,m_\phi+\frac{1}{2}}^{n} - \mathbf{F}\,\big|_{m_r,m_\phi-\frac{1}{2}}^{n},$$

$$\delta_t \mathbf{F}\,\big|_{m_r,m_\phi}^{n} \equiv \mathbf{F}\,\big|_{m_r,m_\phi}^{n+\frac{1}{2}} - \mathbf{F}\,\big|_{m_r,m_\phi}^{n-\frac{1}{2}},$$

$$\mu_t \mathbf{F}\,\big|_{m_r,m_\phi}^{n} \equiv \frac{\mathbf{F}\,\big|_{m_r,m_\phi}^{n+\frac{1}{2}} + \mathbf{F}\,\big|_{m_r,m_\phi}^{n-\frac{1}{2}}}{2},$$

where $\mathbf{F}$ represents field components and $m_r$, $m_\phi$ are the indices corresponding to a certain discretization point in the FDTD domain. The discretized Equations (3.38)–(3.43) read

$$\mu_0 \frac{\delta_t}{\Delta t}\overline{H_{zr}} + \sigma^* \mu_t \overline{H_{zr}} = -\frac{\delta_t}{\Delta r}\overline{E_\phi}, \tag{3.44}$$

$$\mu_0 r \frac{\delta_t}{\Delta t}H_{z\phi} + \overline{\sigma}^* \mu_t H_{z\phi} = \frac{\delta_\phi}{\Delta \phi}E_r, \tag{3.45}$$

$$r\frac{\delta_t}{\Delta t}D_r + \frac{\overline{\sigma}}{\varepsilon_0}\mu_t D_r = \frac{\delta_\phi}{\Delta\phi}H_z, \tag{3.46}$$

$$\frac{\delta_t}{\Delta t}D_\phi + \frac{\sigma}{\varepsilon_0}\mu_t D_\phi = -\frac{\delta_t}{\Delta r}H_z, \tag{3.47}$$

$$r\frac{\delta_t}{\Delta t}H_{zr} + \frac{\overline{\sigma}^*}{\mu_0}\mu_t H_{zr} = \frac{\delta_t}{\Delta t}\overline{H_{zr}}, \tag{3.48}$$

$$r\frac{\delta_t}{\Delta t}E_\phi + \frac{\overline{\sigma}}{\varepsilon_0}\mu_t E_\phi = \frac{\delta_t}{\Delta t}\overline{E_\phi}. \tag{3.49}$$

Applying the operators and rearranging terms, the following FDTD updating equations can be obtained:

$$\overline{E_\phi}\,\big|_{m_r,m_\phi}^{n+\frac{1}{2}} = \overline{E_\phi}\,\big|_{m_r,m_\phi}^{n-\frac{1}{2}} + \left(r + \frac{\overline{\sigma}\Delta t}{2\varepsilon_0}\right)E_\phi\,\big|_{m_r,m_\phi}^{n+\frac{1}{2}} - \left(r - \frac{\overline{\sigma}\Delta t}{2\varepsilon_0}\right)E_\phi\,\big|_{m_r,m_\phi}^{n-\frac{1}{2}}, \tag{3.50}$$

$$\overline{H_{zr}}\Big|_{m_r,m_\phi}^{n+\frac{1}{2}} = \left(\frac{2\mu_0 - \overset{*}{\sigma}\Delta t}{2\mu_0 + \overset{*}{\sigma}\Delta t}\right)\overline{H_{zr}}\Big|_{m_r,m_\phi}^{n-\frac{1}{2}} - \frac{\overline{E_\phi}\Big|_{m_r+\frac{1}{2},m_\phi}^{n} - \overline{E_\phi}\Big|_{m_r-\frac{1}{2},m_\phi}^{n}}{\Delta r(2\mu_0 + \overset{*}{\sigma}\Delta t)/(2\Delta t)}, \tag{3.51}$$

$$H_{zr}\Big|_{m_r,m_\phi}^{n+\frac{1}{2}} = \left(\frac{2\mu_0 r - \overline{\overset{*}{\sigma}}\Delta t}{2\mu_0 r + \overset{*}{\sigma}\Delta t}\right)H_{zr}\Big|_{m_r,m_\phi}^{n-\frac{1}{2}} + \frac{\overline{H_{zr}}\Big|_{m_r,m_\phi}^{n+\frac{1}{2}} - \overline{H_{zr}}\Big|_{m_r,m_\phi}^{n-\frac{1}{2}}}{\Delta t(2\mu_0 r + \overset{*}{\sigma}\Delta t)/(2\mu_0\Delta t)}, \tag{3.52}$$

$$H_{z\phi}\Big|_{m_r,m_\phi}^{n+\frac{1}{2}} = \left(\frac{2\mu_0 r - \overset{*}{\sigma}\Delta t}{2\mu_0 r + \overset{*}{\sigma}\Delta t}\right)H_{z\phi}\Big|_{m_r,m_\phi}^{n-\frac{1}{2}} + \frac{E_r\Big|_{m_r,m_\phi+\frac{1}{2}}^{n} - E_r\Big|_{m_r,m_\phi-\frac{1}{2}}^{n}}{\Delta\phi(2\mu_0 r + \overset{*}{\sigma}\Delta t)/(2\Delta t)}, \tag{3.53}$$

$$H_z\Big|_{m_r,m_\phi}^{n+\frac{1}{2}} = H_{zr}\Big|_{m_r,m_\phi}^{n+\frac{1}{2}} + H_{z\phi}\Big|_{m_r,m_\phi}^{n+\frac{1}{2}}, \tag{3.54}$$

$$D_r\Big|_{m_r,m_\phi}^{n+\frac{1}{2}} = \left(\frac{2\varepsilon_0 r - \overline{\sigma}\Delta t}{2\varepsilon_0 r + \overset{*}{\sigma}\Delta t}\right)D_r\Big|_{m_r,m_\phi}^{n-\frac{1}{2}} + \frac{H_z\Big|_{m_r,m_\phi+\frac{1}{2}}^{n} - H_z\Big|_{m_r,m_\phi-\frac{1}{2}}^{n}}{\Delta\phi(2\varepsilon_0 r + \overset{*}{\sigma}\Delta t)/(2\varepsilon_0\Delta t)}, \tag{3.55}$$

$$D_\phi\Big|_{m_r,m_\phi}^{n+\frac{1}{2}} = \left(\frac{2\varepsilon_0 - \sigma\Delta t}{2\varepsilon_0 + \sigma\Delta t}\right)D_\phi\Big|_{m_r,m_\phi}^{n-\frac{1}{2}} - \frac{H_z\Big|_{m_r+\frac{1}{2},m_\phi}^{n} - H_z\Big|_{m_r-\frac{1}{2},m_\phi}^{n}}{\Delta r(2\varepsilon_0 + \sigma\Delta t)/(2\varepsilon_0\Delta t)}. \tag{3.56}$$

In addition to the above updating equations, the dielectric properties of hyperlens structures need to be taken into account in FDTD simulations. These additional updating equations can be derived through the discretization of Equations (3.28) and (3.29), which is introduced as follows.

The wire medium has been known for a long time as an artificial dielectric with plasma-like frequency-dependent permittivity [59,60], but only recently has it been shown that this dielectric is nonlocal and possesses strong spatial dispersion even at very low frequencies [61]. Following [61], the wire medium can be described as a uniaxial dielectric (if the lattice period is much smaller than the wavelength) with both frequency and spatially dependent effective permittivity:

$$\overline{\overline{\varepsilon}} = \varepsilon(k,q_x)\mathbf{xx} + \mathbf{yy} + \mathbf{zz}, \qquad \varepsilon(k,q_x) = \varepsilon_0\left(1 - \frac{k_p^2}{k^2 - q_x^2}\right), \tag{3.57}$$

where $k_p = \omega_p/c$ is the wave number corresponding to the plasma frequency $\omega_p$, $k = \omega/c$, is the wave number of free space, $c$ is the speed of light, and $q_x$ is the component of wave vector along the wires. The dependence of permittivity (3.57) on $q_x$ represents the spatial dispersion effect, which is not taken into account in the conventional local uniaxial model of the wire medium [59,60]. The $k_p$ depends on the lattice periods ($a$, $b$) and the radius of wires ($r$) [61]:

$$k_p^2 = \frac{2\pi/(ab)}{\log[\sqrt{ab}/(2\pi r)] + F(a/b)},$$  (3.58)

where

$$F(\xi) = -\frac{1}{2}\log\xi + \sum_{n=1}^{+\infty}\left[\frac{\coth(\pi n\xi) - 1}{n}\right] + \frac{\pi}{6}\xi.$$  (3.59)

For the commonly used case of square grid ($a = b$), $F(1) = 0.5275$.

The material properties of the wire medium specified in (3.57) is in Cartesian coordinates, and the parallel wires are aligned along the $x$-direction. For the construction of a hyperlens using an array of wires, the spacing of wires can be gradually increased from the front to the back interface of the device, as shown in Figure 3.9(b). Thus the hyperlens can be represented in cylindrical coordinates for the 2D case or in spherical/cylindrical coordinates for the 3D case, by assuming the wires to be aligned along the radial direction. In the 2D $TE_z$ case, the permittivity of the hyperlens is given by

$$\varepsilon_r(k, q_r) = \varepsilon_0\left[1 - \frac{k_p^2(r)}{k^2 - q_r^2}\right], \qquad \varepsilon_\phi = \varepsilon_0.$$  (3.60)

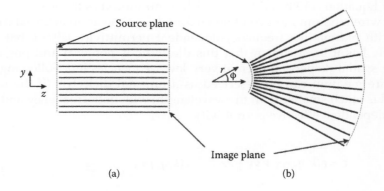

(a)                                           (b)

**FIGURE 3.9**
(a) The 2D view of a parallel wire medium as a transmission device. (b) The 2D view of a hyperlens formed by a tapered array of wires, which can be conveniently represented in cylindrical coordinates.

To correctly model the wire medium by taking into account the spatial dispersion effect, a spatially dispersive FDTD method in Cartesian coordinates has been developed in [56]. For the modeling of hyperlens with its permittivity specified in cylindrical coordinates using the conventional FDTD method, it is necessary to perform a coordinates transformation and convert the permittivity (3.60) to Cartesian coordinates in simulations, as has been done for the modeling of cylindrical cloaking devices in Cartesian FDTD simulations [29]. However, the angles involved in the transformation matrix need to be calculated numerically and staircase approximations are required to represent curved surfaces in a Cartesian grid. These approximations introduce numerical errors that reduce the simulation accuracy. Alternatively, the permittivity of the structure specified in cylindrical coordinates can be directly taken into account in the FDTD method in cylindrical coordinates.

Because the permittivity of the tapered array of wires along the $\phi$-direction is equal to the free space one, the updating equation for the $E_\phi$ component can be directly expressed as

$$E_\phi \left. \right|_{m_r, m_\phi}^{n+\frac{1}{2}} = \frac{1}{\varepsilon_0} D_\phi \left. \right|_{m_r, m_\phi}^{n+\frac{1}{2}}. \tag{3.61}$$

For the $r$-component of electric field, since $D_r(\omega, q_r)$ is related to $E_r(\omega, q_r)$ in the spectral (frequency-wave vector) domain as

$$D_r(\omega) = \varepsilon(\omega, q_r) E_r(\omega), \tag{3.62}$$

one can write that

$$\left(k^2 - q_r^2\right) D_r + \left(q_r^2 - k^2 + k_p^2\right)\varepsilon_0 E_r = 0, \tag{3.63}$$

Using inverse Fourier transformation and the following rules,

$$k^2 \rightarrow -\frac{1}{c^2}\frac{\partial^2}{\partial t^2}, \qquad q_r^2 \rightarrow -\frac{\partial^2}{\partial r^2},$$

the constitutive relation in the time-space domain can be written as

$$\left(\frac{\partial^2}{\partial r^2} - \frac{1}{c^2}\frac{\partial^2}{\partial t^2}\right) D_r + \left(\frac{1}{c^2}\frac{\partial^2}{\partial t^2} - \frac{\partial^2}{\partial r^2} + k_p^2\right)\varepsilon_0 E_r = 0. \tag{3.64}$$

The above equation can be discretized using the following operators [35]:

$$\delta_t^2 \mathbf{F} \big|_{m_r, m_\phi}^n \equiv \mathbf{F} \big|_{m_r, m_\phi}^{n+1} - 2\mathbf{F} \big|_{m_r, m_\phi}^n + \mathbf{F} \big|_{m_r, m_\phi}^{n-1},$$

$$\delta_r^2 \mathbf{F} \big|_{m_r, m_\phi}^n \equiv \mathbf{F} \big|_{m_r+1, m_\phi}^n - 2\mathbf{F} \big|_{m_r, m_\phi}^n + \mathbf{F} \big|_{m_r-1, m_\phi}^n,$$

$$\mu_t^2 \mathbf{F} \big|_{m_r, m_\phi}^n \equiv \frac{\mathbf{F} \big|_{m_r, m_\phi}^{n+1} + 2\mathbf{F} \big|_{m_r, m_\phi}^n + \mathbf{F} \big|_{m_r, m_\phi}^{n-1}}{4},$$

Thus the updating equation for the $E_r$ component can be derived as

$$E_r \big|_{m_r, m_\phi}^{n+1} = \left\{ -\frac{1}{c^2(\Delta t)^2} D_r \big|_{m_r, m_\phi}^{n+1} + \frac{1}{(\Delta r)^2} D_r \big|_{m_r+1, m_\phi}^n + \frac{1}{(\Delta r)^2} D_r \big|_{m_r-1, m_\phi}^n \right.$$

$$-\frac{1}{c^2(\Delta t)^2} D_r \big|_{m_r, m_\phi}^{n-1} + \left[ \frac{2}{c^2(\Delta t)^2} - \frac{2}{(\Delta r)^2} \right] D_r \big|_{m_r, m_\phi}^n - \frac{\varepsilon_0}{(\Delta r)^2} E_r \big|_{m_r+1, m_\phi}^n$$

$$-\left[ \frac{2\varepsilon_0}{c^2(\Delta t)^2} - \frac{2\varepsilon_0}{(\Delta r)^2} - \frac{\varepsilon_0 k_p^2}{2} \right] E_r \big|_{m_r, m_\phi}^n - \frac{\varepsilon_0}{(\Delta r)^2} E_r \big|_{m_r-1, m_\phi}^n$$

$$\left. + \left[ \frac{\varepsilon_0}{c^2(\Delta t)^2} + \frac{\varepsilon_0 k_p^2}{4} \right] E_r \big|_{m_r, m_\phi}^{n-1} \right\} \Big/ \left[ \frac{\varepsilon_0}{c^2(\Delta t)^2} + \frac{\varepsilon_0 k_p^2}{4} \right]. \tag{3.65}$$

Note from Equation (3.58) that $k_p$ depends on the spacing between wires. Hence in the simulation of the hyperlens structure in which the spacing gradually increases from the front interface of the device to its back one, $k_p$ needs to be reduced accordingly.

In Equation (3.65), the central finite difference approximations in time (for the frequency dispersion) for both $D_r$ and $E_r$ are used at position $m_r$, and the central finite difference approximations in space (for the spatial dispersion) are used at the time step $n$ in order to update $E_r^{n+1}$. Therefore the storage of $D_r$ and $E_r$ at two previous time steps are required. At the free space–tapered array interface along the $\gamma$-direction, the updating equation (3.65) involves $D_r$ and $E_r$ in both free space and the tapered array, as indicated by the large arrows in Figure 3.10. Outside the region of the tapered array, the updating equation (3.65) reduces to the equation relating $D_r$ and $E_r$ in the free space, which can be realized by letting the plasma frequency equal zero ($k_p = 0$). Along the $\phi$-direction, the transverse dimension of the hyperlens, $w$, can be equal to either a complete circle of $2\pi$ radian or a finite angular length. For the case of $w = 2\pi$, no material interfaces exist and thus no special treatment is necessary; however, for the modeling of a hyperlens with finite transverse dimension, the boundary of the tapered array needs to be aligned with the $H_z$ component to avoid the averaging of material parameters of the free

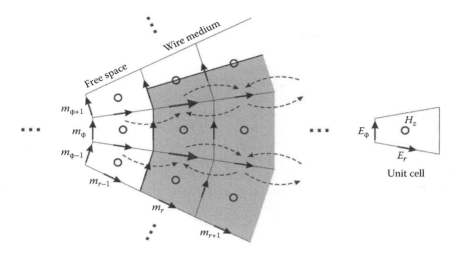

**FIGURE 3.10**
A partial grid layout in cylindrical FDTD domain for the modeling of a hyperlens formed by a tapered array of wires. The dashed arrows indicate the dependency between adjacent field components.

space and the tapered array at their interface [7], as illustrated in Figure 3.10. Equations (3.50)–(3.56), (3.61), and (3.65) form a complete set of updating equations for the modeling of the hyperlens formed by a tapered array of wires.

For the computer programming of the conventional FDTD method in Cartesian coordinates, 2D arrays are often used to store field values conveniently. However, in cylindrical coordinates, a mapping of 2D arrays to FDTD grid needs to be performed. Figure 3.11 shows the mapping of a

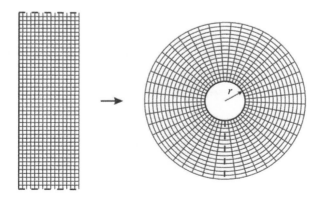

**FIGURE 3.11**
Mapping of a 2D array to the FDTD grid in cylindrical coordinates. The thick solid/dashed lines indicate the same array of elements.

15 × 45-element array to a cylindrical FDTD grid, where the first column of the array is mapped to the inner boundary of the grid, as indicated by the thick solid lines. The top and bottom rows of the 2D array merge to form the same array of radial elements, shown by the thick dashed lines. Due to the merge, the original boundaries where the arrays are located disappear, and the FDTD updating equations need to be modified accordingly. Note that such a way of mapping is different from the conventional implementation of the FDTD method in cylindrical coordinates, which directly assumes the radius of the grid to be zero. This approach allows one to model a perfect electric conductor (PEC) centered at the origin with reduced computer memory since no fields can enter the area enclosed by PEC; thus no FDTD grids need to be assigned in that region. However, for the modeling of free space at the origin, it is necessary to set $r = 0$. Then the issue of numerical singularity at the origin arises. Figure 3.12 shows an example of detailed FDTD grid layout in cylindrical coordinates in the vicinity of the origin for the case of eight elements along the $\phi$-direction (for illustration purposes). In general, either the $E_z$ or $H_z$ component can be assumed to be located at the origin. In our 2D case, the nonzero component is $H_z$, and it is then aligned with the origin of the coordinates and treated separately from the FDTD updating equations to avoid the singularity issue.

Methods for dealing with the numerical singularity at the origin have been proposed in [62–64]. Following the same approach as in [63,64] by evaluating

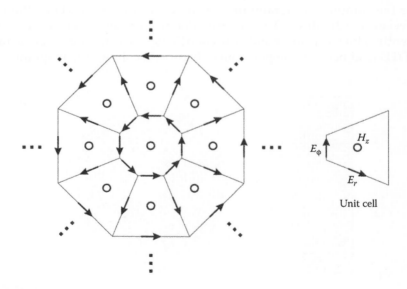

**FIGURE 3.12**
Layout of the FDTD grid in cylindrical coordinates in the vicinity of the origin, where the $H_z$ component is located and treated separately.

the integral form of Faraday's law, the $H_z$ component at the origin can be calculated as

$$H_z|_{0,m_\phi}^{n+\frac{1}{2}} = H_z|_{0,m_\phi}^{n-\frac{1}{2}} - \frac{4\Delta t}{\mu_0 N_\phi \Delta r} \sum_{k=1}^{N_\phi} E_\phi|_{1,k}^n,$$

(3.66)

where $N_\phi$ is the number of elements along the $\phi$-direction.

At the outer boundary of the FDTD domain, in order to model an unbounded space, the above introduced complex coordinates stretching-based PML can be applied by gradually increasing the conductivity term from zero to its maximum value, $\sigma_{max}$, which is defined as

$$\sigma_{max} = -\frac{(m+1)\ln[R(0)]}{2N_{PML}\Delta r \sqrt{\mu_0/\varepsilon_0}},$$

(3.67)

where $R(0)$ is the reflection coefficient at normal incidence between the free space and the PML layer, and $N_{PML}$ is the number of cells in the PML layer. The value of the conductivity term in each PML layer can be calculated as

$$\sigma(n) = \sigma_{max}\left(\frac{n}{N_{PML}}\right)^m,$$

(3.68)

where the layer index, $n$, is an integer and $1 \leq n \leq N_{PML}$. Then the integral factor can be numerically calculated as

$$\bar{\sigma} = \int_0^r \sigma(s)ds = \frac{\sigma_{max}n\Delta r}{m+1}\left(\frac{n}{N_{PML}}\right)^m.$$

(3.69)

In the above equations, $m$ defines the profile of the location-dependent conductivity term within the PML layer and satisfactory results can be obtained by letting $m = 2$. The outmost layer of the PML is defined as the PEC condition by specifying zero values for the tangential electric field component, $E_\phi$.

For stable FDTD simulations, the time step $\Delta t$ needs to be bounded by the following condition [6]:

$$c\Delta t \leq \frac{1}{\sqrt{\frac{1}{(\Delta r)^2} + \frac{1}{(\Delta r\Delta\phi)^2}}}.$$

(3.70)

It is worth mentioning that since the FDTD spatial resolution decreases toward the boundary of the cylindrical domain due to the expansion of the unit cell along the radial direction, for the simulation of large domains, it is necessary to ensure that the largest mesh size is small enough (e.g., $r\Delta\phi < \lambda/20$) to guarantee sufficient numerical accuracy.

The developed cylindrical FDTD method has been implemented to investigate the image transfer and magnification capability of the hyperlens. The simulation results are presented in the next section.

### 3.3.2 Subwavelength Image Magnification by the Hyperlens

The 2D FDTD simulation domain is shown in Figure 3.13, where the outer boundary of the domain is terminated by a 20-layer complex-coordinates stretching-based PML. The FDTD cell size along the $r$-direction is $\Delta r = \lambda/60$, where $\lambda$ is the wavelength at the operating frequency of $f = 1.0$ GHz. The number of FDTD cells along the $\phi$-direction is 800 in all simulations. Three magnetic line sources (located at equal distances to the origin) are excited at a distance of $\lambda/10$ to the front interface of the hyperlens. The distance between adjacent sources is $\lambda/20$ and the central source is excited out of phase with respect to the neighboring ones. The proposed three-source configuration creates a distribution with two strong maxima at the front interface of the hyperlens and the distance between the maxima is about $\lambda/5$.

In order to investigate the imaging and magnification properties of the hyperlens, various lengths, transverse (angular) dimensions of the device,

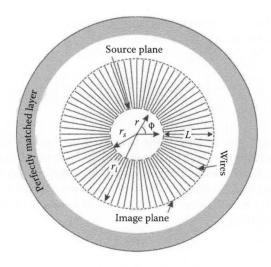

**FIGURE 3.13**
The 2D FDTD simulation domain in cylindrical coordinates for the modeling of hyperlens formed by a tapered array of wires.

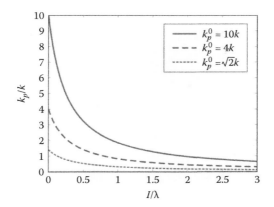

**FIGURE 3.14**
Normalized value of $k_p$ versus the location within the hyperlens (with reference to the inner interface), where $k_p^0$ is the value of $k_p$ at the inner interface.

and $k_p$ are considered in simulations. The value of $k_p$ can be calculated from Equation (3.34) by choosing appropriate values of $a$, $b$, and $r$. Figure 3.14 shows the variation of normalized $k_p$ versus the location $l$ within the hyperlens (with reference to the inner interface) for three different cases: $k_p^0 = 10k$ ($a_0 = b_0 = 11.3$ mm), $k_p^0 = 4k$ ($a_0 = b_0 = 22.3$ mm), and $k_p^0 = \sqrt{2}k$ ($a_0 = b_0 = 52.1$ mm), where $k_p^0$ is the value of $k_p$ at the inner interface, $a_0$ and $b_0$ are the lattice periods at the inner interface, and $r = 1$ mm. It can be seen that for all cases, as the spacing between wires increases, the value of $k_p$ decreases accordingly and becomes less than $k$ beyond a certain location. However, the operation of the wire medium as a transmission device requires $k_p > k$ [56]. This means that there may exist a maximum length for the proper operation of the device. This effect is demonstrated through the following simulation results.

It has been shown in [56] that due to the canalization principle, which allows waves to travel only along the direction of wires, the subwavelength imaging property of the wire medium device is insensitive to its transverse dimension. It is also of practical importance to investigate if the magnification of subwavelength distributions using hyperlens formed by a tapered array of wires is affected by its angular dimension, $w$. Figure 3.15 shows the comparison of magnetic field distributions using two hyperlens structures with $w = 2\pi$ and $w = \pi/4$, respectively. For both devices, $k_p^0 = 10k$, the length is $L = \lambda$, and the locations of the inner and outer interfaces are $r_s = \lambda/2$ and $r_i = 3\lambda/2$, respectively (see Figure 3.13). The magnetic field distributions at the source and image planes of both devices are plotted in Figure 3.16. The nearly identical field distributions within the area of the device between the cases of $w = 2\pi$ and $w = \pi/4$ demonstrate that the image magnification property of the hyperlens is insensitive to its transverse (angular) dimension. In addition, since the angular width increases

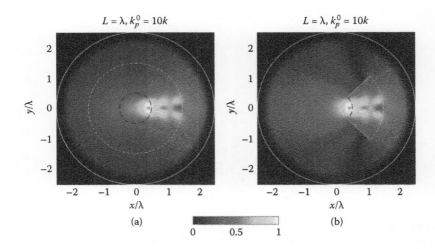

**FIGURE 3.15**
Normalized magnetic field distributions calculated from cylindrical FDTD simulations of two hyperlens structures with angular widths of (a) $w = 2\pi$ and (b) $w = \pi/4$. (See color figure.)

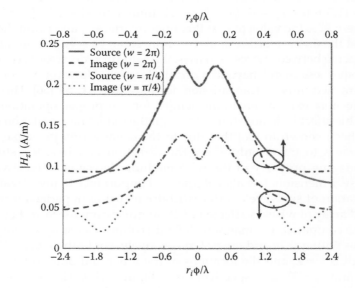

**FIGURE 3.16**
Comparison of magnetic field distributions at the source and image planes of two hyperlens structures with angular widths of $w = 2\pi$ and $w = \pi/4$. In both cases, $k_p^0 = 10k$. (See color figure.)

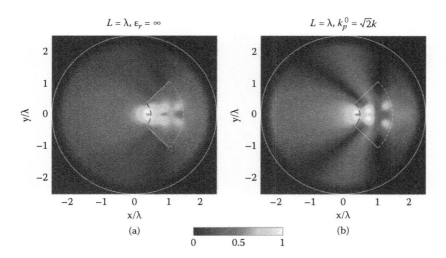

$L = \lambda, \, \varepsilon_r = \infty$          $L = \lambda, \, k_p^0 = \sqrt{2}k$

(a)          (b)

0     0.5     1

**FIGURE 3.17**
Normalized magnetic field distributions for two hyperlens structures with (a) $\varepsilon_r = \infty$ and (b) $k_p^0 = \sqrt{2}k$. In both cases, $L = \lambda$. (See color figure.)

from the inner interface to the outer one by three times, a threefold magnification of subwavelength source distribution is observed, as shown by the two clearly resolved maxima at the image plane in Figure 3.16. In the rest of the simulations, the angular width is always kept as $w = \pi/4$.

To investigate the effect of varying $k_p^0$ on the image magnification capability of the hyperlens, different values of $k_p^0$ are chosen in simulations. Moreover, when the value of $k_p^0$ is large, the device can be approximated to have infinite permittivity along the radial direction, i.e., $E_r = 0$. Hence simulations are performed for two hyperlens devices with $E_r = \infty$ and $k_p^0 = \sqrt{2}k$, respectively. The rest of the parameters remain unchanged. The magnetic field distributions for these structures are shown in Figure 3.17, and the comparison of magnetic field distributions in the source and image planes of both devices is shown in Figure 3.18. It can be seen that Figures 3.17(a) and 3.15(b) have very similar distributions, but Figure 3.17(b) is considerably different. This is due to the small value of $k_p^0$ at the front interface; thus the value of $k_p$ decreases to be less than $k$ beyond a distance of $0.12\lambda$ from the source plane. Similar to the case of subwavelength imaging using the wire medium, the low value of $k_p$ leads to the low imaging resolution of the device [56]. As a consequence, the two maxima at the image plane become much less distinguishable, as shown in Figure 3.18.

If the length of the hyperlens is extended, it is expected that the maxima can be hardly resolved by the device. Figure 3.19 shows magnetic field distributions for two $3\lambda$-long hyperlens structures with $k_p^0 = \sqrt{2}k$ and $k_p^0 = 10k$, respectively. It can be seen that the distributions are dramatically different

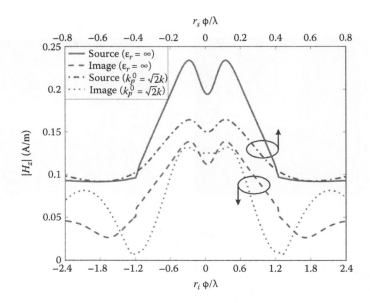

**FIGURE 3.18**
Comparison of magnetic field distributions in the source and image planes of two hyperlens structures with $\varepsilon_r = \infty$ and $k_p^0 = \sqrt{2}k$. In both cases, $L = \lambda$. (See color figure.)

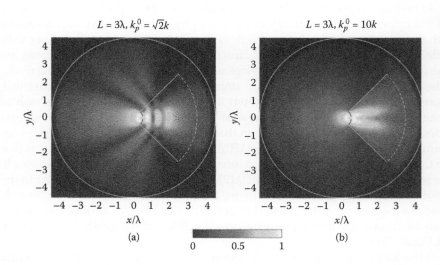

**FIGURE 3.19**
Normalized magnetic field distributions for two hyperlens structures with (a) $k_p^0 = \sqrt{2}k$ and (b) $k_p^0 = 10k$. In both cases, $L = 3\lambda$. (See color figure.)

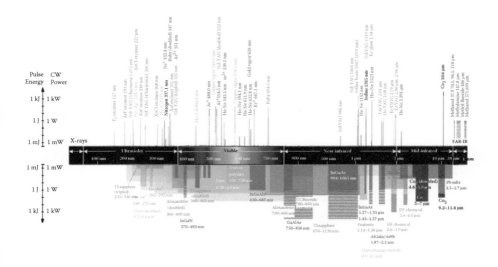

**COLOR FIGURE B.2**
Wavelengths of commercially available lasers. (From M. J. Weber, *Handbook of Laser Wavelengths*. Boca Raton, FL: CRC Press, 1999. ISBN 0-8493-3508-6.)

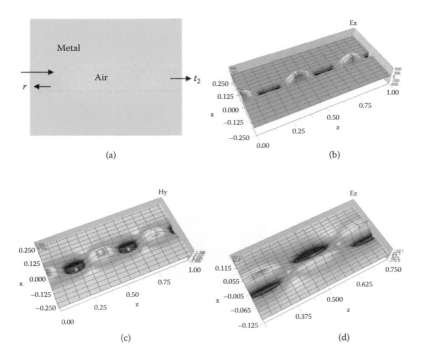

**COLOR FIGURE 1.2**
The propagation field inside a plasmonic slot for the different field components.

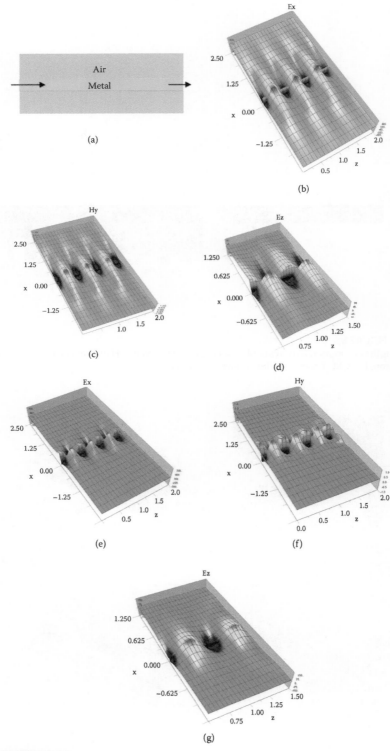

**COLOR FIGURE 1.3**
The propagation field inside an IMI for the symmetric mode (b,c,d) for different field components and for the asymmetric mode (e,f,g).

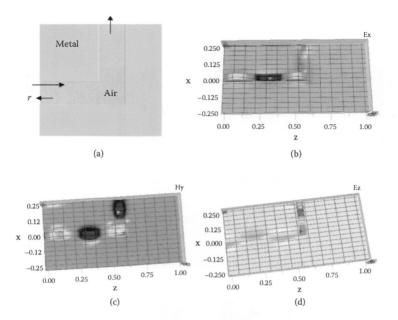

**COLOR FIGURE 1.4**
The field inside an L-shaped sharp bend.

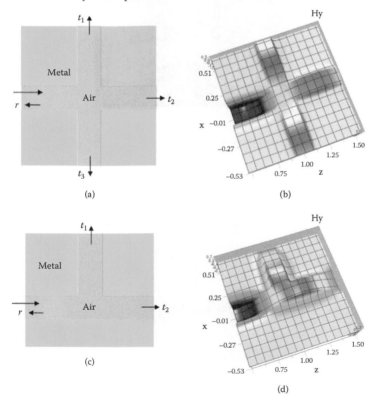

**COLOR FIGURE 1.5**
The field inside X and T junctions.

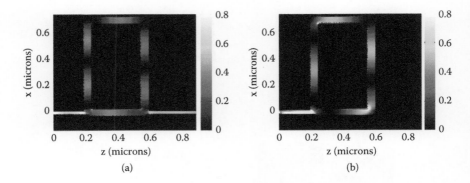

**COLOR FIGURE 1.8**
Field intensity of the asymmetric structure at wavelength (a) 950 nm and (b) 1250 nm.

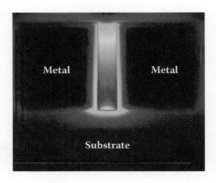

**COLOR FIGURE 1.10**
Schematic diagram of the plasmonic slot.

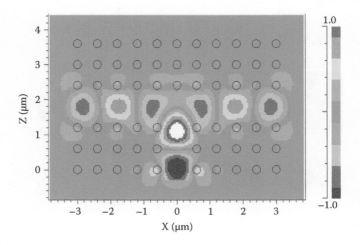

**COLOR FIGURE 1.13**
Field propagation inside the *T*-shaped photonic crystal.

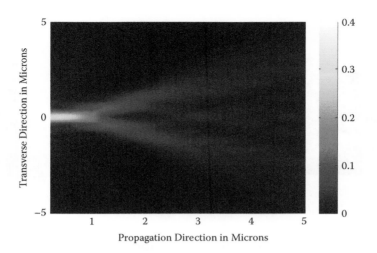

**COLOR FIGURE 1.17**
The field intensity for subwavelength focusing.

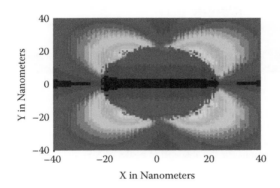

**COLOR FIGURE 1.18**
The electric field localization around Al nanorod of radius 30 nm.

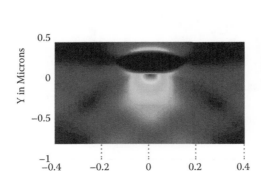

**COLOR FIGURE 1.20**
Enhancement of optical field due to the plasmonic nanoparticles.

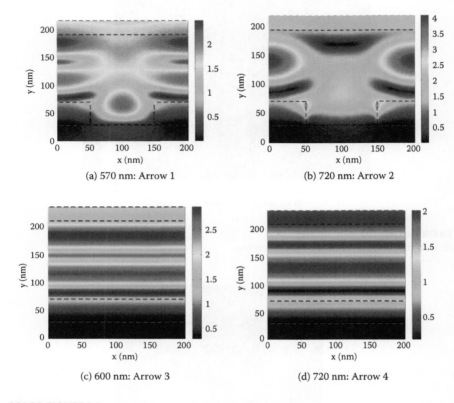

(a) 570 nm: Arrow 1

(b) 720 nm: Arrow 2

(c) 600 nm: Arrow 3

(d) 720 nm: Arrow 4

**COLOR FIGURE 2.7**
The near-field distributions for the absorption peaks of A-Si material denoted in Figure 2.6.

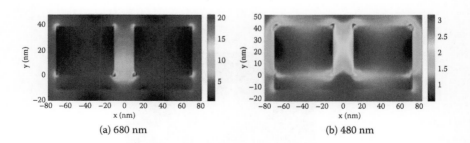

(a) 680 nm

(b) 480 nm

**COLOR FIGURE 2.11**
The near-field distributions of the antenna respectively at the resonance (680 nm) and off the resonance (480 nm).

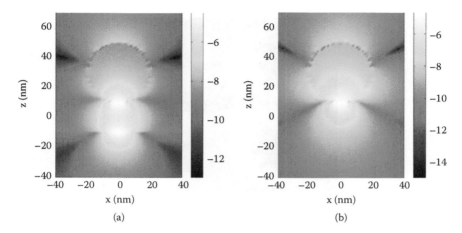

**COLOR FIGURE 2.14**
The scattered near-field distributions $E_z$ of (a) a hybrid plasmonic system and (b) a single Au sphere excited with a $z$-polarized dipole (in the logarithmic scale). The dipole is located at $z = 0$ nm.

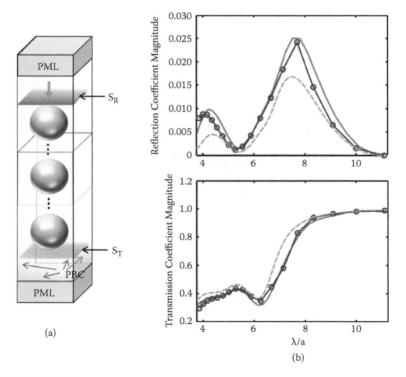

**COLOR FIGURE 2.16**
(a) A schematic pattern for the EMP retrieval. $S_R$ and $S_T$ are the planar surfaces for extracting the reflection and transmission coefficients. PML and PBC denote perfectly matched layer and periodic boundary condition, respectively. (b) The reflection and transmission coefficients. Solid red lines: real model simulation; blue circles: an equivalent slab with the EMPs obtained by the homogenization from FDTD solutions; dashed green lines: an equivalent slab with the EMPs obtained from the Lewin's theory.

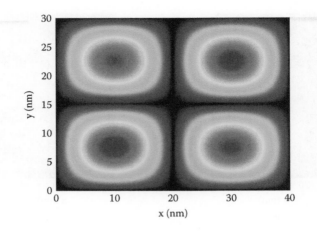

**COLOR FIGURE 2.21**
The eigenstate corresponding to the eigenfrequency $f_{22}$ for a 2D quantum well obtained by the quantum FDTD method and discrete Fourier transform.

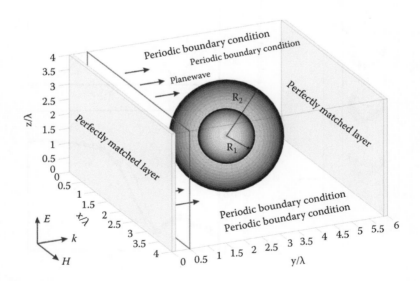

**COLOR FIGURE 3.2**
The 3D parallel dispersive FDTD simulation domain for the case of plane-wave incidence on the cloak. The red rectangle indicates the location of the source plane.

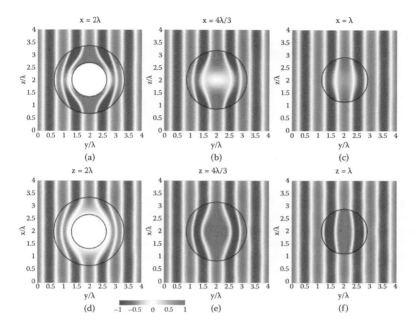

**COLOR FIGURE 3.3**
Normalized field distributions for the $E_z$ component in (a)–(c) $y$-$z$ plane and (d)–(f) $x$-$y$ plane in the steady state of the parallel dispersive FDTD simulations. The cutting planes are (see Figure 3.2) (a) $x = 2\lambda$, (b) $x = 4\lambda/3$, (c) $x = \lambda$, (d) $z = 2\lambda$, (e) $z = 4\lambda/3$, (f) $z = \lambda$. The wave propagation direction is from left to right.

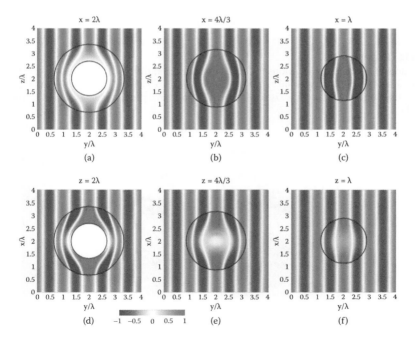

**COLOR FIGURE 3.4**
Normalized field distributions for the $H_x$ component in (a)–(c) $y$-$z$ plane and (d)–(f) $x$-$y$ plane in the steady state of the parallel dispersive FDTD simulations. The cutting planes are (see Figure 3.2) (a) $x = 2\lambda$, (b) $x = 4\lambda/3$, (c) $x = \lambda$, (d) $z = 2\lambda$, (e) $z = 4\lambda/3$, (f) $z = \lambda$. The wave propagation direction is from left to right.

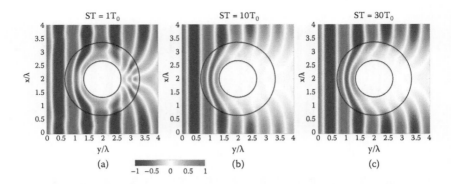

**COLOR FIGURE 3.6**
Comparison of the influence of different switching time (ST) of the sinusoidal source on the simulation results: (a) $ST = T_0$, (b) $ST = 10T_0$, (c) $ST = 30T_0$, where $T_0$ is the period of the sinusoidal wave. The wave propagation direction is from left to right and the normalized field distributions are plotted in the $x$-$y$ plane ($z = 2\lambda$; see Figure 3.2) and at the time step $t = 1320\Delta t$.

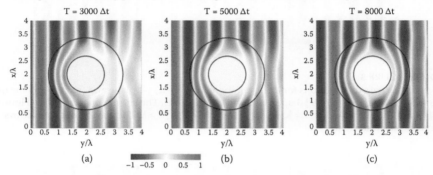

**COLOR FIGURE 3.7**
Snapshots of the field distributions for the $E_z$ component at different time steps in the parallel dispersive FDTD simulations: (a) $t = 3000\Delta t$ (5.77 ns), (b) $t = 5000\Delta t$ (9.62 ns), (c) $t = 8000\Delta t$ (15.40 ns), plotted in the $x$-$y$ plane ($z = 2\lambda$; see Figure 3.2). The wave propagation direction is from left to right.

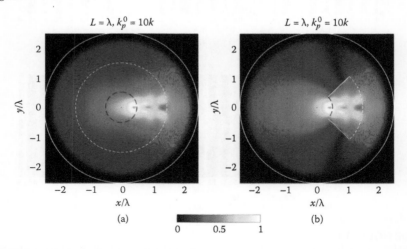

**COLOR FIGURE 3.15**
Normalized magnetic field distributions calculated from cylindrical FDTD simulations of two hyperlens structures with angular widths of (a) $w = 2\pi$ and (b) $w = \pi/4$.

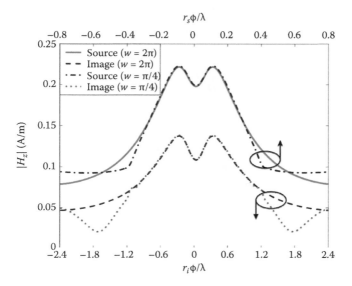

**COLOR FIGURE 3.16**
Comparison of magnetic field distributions at the source and image planes of two hyperlens structures with angular widths of $w = 2\pi$ and $w = \pi/4$. In both cases, $k_p^0 = 10k$.

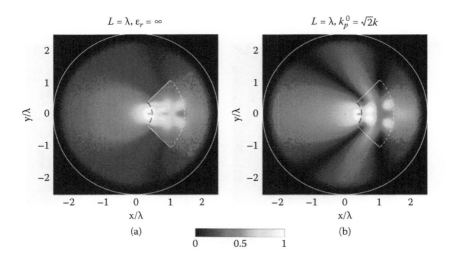

**COLOR FIGURE 3.17**
Normalized magnetic field distributions for two hyperlens structures with (a) $\varepsilon_r = \infty$ and (b) $k_p^0 = \sqrt{2}k$. In both cases, $L = \lambda$.

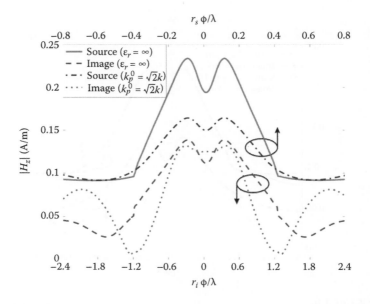

**COLOR FIGURE 3.18**
Comparison of magnetic field distributions in the source and image planes of two hyperlens structures with $\varepsilon_r = \infty$ and $k_p^0 = \sqrt{2}k$. In both cases, $L = \lambda$.

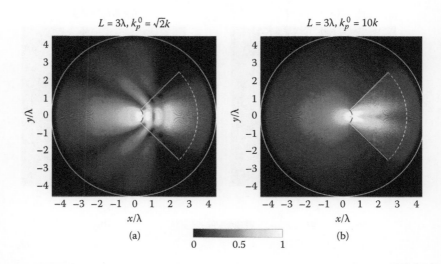

**COLOR FIGURE 3.19**
Normalized magnetic field distributions for two hyperlens structures with (a) $k_p^0 = \sqrt{2}k$ and (b) $k_p^0 = 10k$. In both cases, $L = 3\lambda$.

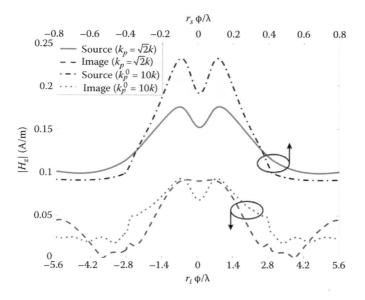

**COLOR FIGURE 3.20**
Comparison of magnetic field distributions in the source and image planes of two hyperlens structures with $k_p^0 = \sqrt{2}k$ and $k_p^0 = 10k$. In both cases, $L = 3\lambda$.

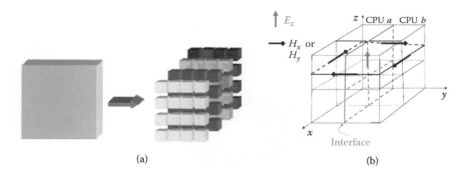

**COLOR FIGURE 4.2**
Schematic of the parallel computation scheme. (a) Basic idea of parallel FDTD processing on a computer cluster, in which the original job is split into small pieces that are assigned to each CPU. (b) Distribution of the electric and magnetic fields near the interface between two CPUS.

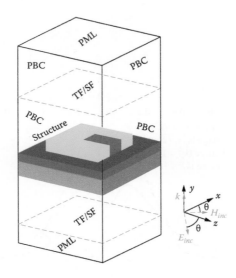

**COLOR FIGURE 4.4**
Schematic illustration of our computation space and the polarization of the incident fields.

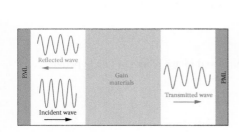

**COLOR FIGURE 4.5**
Schematic of the simulated 1D structure.

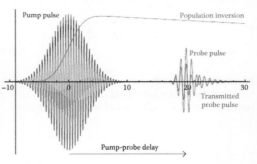

**COLOR FIGURE 4.6**
Schematic illustration of pump-probe experiments.

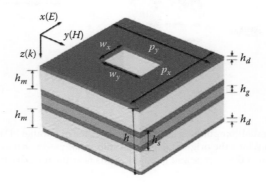

**COLOR FIGURE 4.10**
Unit cell of the perforated fishnet structure with gain embedded between two metal (silver) layers. The geometric parameters are $p_x = p_y = 280$ nm, $w_x = 75$ nm, $w_y = 115$ nm, $h = 170$ nm, $h_m = h_s = 50$ nm, $h_d = 10$ nm and $h_g = 20$ nm. The thicknesses of the silver (yellow) and gain (magenta) layer are $h_m$ and $h_g$, respectively. The dielectric layer (blue) and the gain have a refractive index $n = 1.65$.

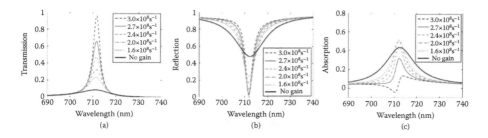

**COLOR FIGURE 4.11**
The transmission (a), reflection (b), and absorption (c) as a function of wavelength for different pumping rates.

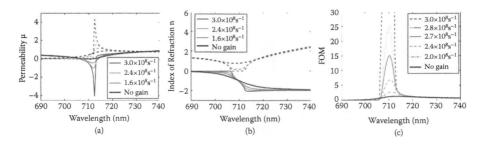

**COLOR FIGURE 4.12**
The retrieved results for the real (solid lines) and the imaginary (dashed lines) parts of (a) the effective permeability, $m$, and (b) the corresponding effective index of refraction, $n$, without and with gain for different pumping rates. (c) The figure-of-merit (FOM) as a function of wavelength for different pumping rates.

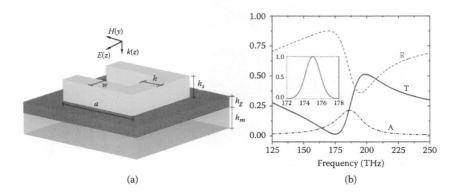

**COLOR FIGURE 4.13**
(a) Schematic of the unit cell for the silver-based SRR structure (yellow) with the electric field polarization parallel to the gap. The dielectric constants $\varepsilon$ for gain (red) and GaAs (light blue) are 9.0 and 11.0, respectively. (b) Calculated spectra for transmittance $T$ (black), reflectance $R$ (red), and absorptance $A$ (blue) for the structure shown in Figure 4.13(a). The inset shows the profile of the probe pulse with a center frequency of 175 THz (FWHM = 2 THz).

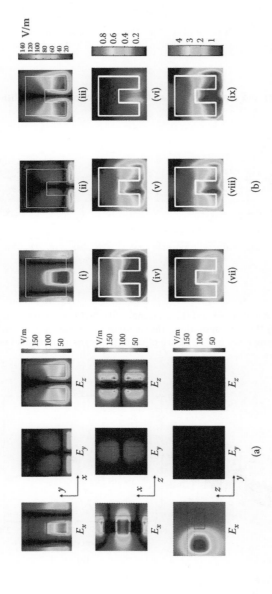

**COLOR FIGURE 4.14**

The spatial distribution of electric field and gain. (a) The electric field in different cut plane without pumping. (b) The first row corresponds to the electric field amplitude distributions at 175 THz with pump ($P_0 = 3 \times 10^9\,\text{s}^{-1}$ and $\Delta t = 5$ ps) in the cross section of the gain layer ($z = 40$ nm from the top of the structure) for different components: (i) $E_x$ (ii) $E_y$ and (iii) $E_z$. The second row corresponds to the near-field differential $\Delta E$ with $P_0 = 3 \times 10^9\,\text{s}^{-1}$ for three different time delays, namely (iv) 5 ps, (v) 45 ps, and (vi) 135 ps, respectively. The third row corresponds to the near-field differential, $\Delta E$ at $\Delta t = 5$ ps, for three different pumping strengths, namely (vii) $P_0 = 6 \times 10^9\,\text{s}^{-1}$, (viii) $P_0 = 9 \times 10^9\,\text{s}^{-1}$, and (ix) $P_0 = 12 \times 10^9\,\text{s}^{-1}$, respectively. Here the $\Delta E$ is defined by taking the difference of the measured total electric field with pumping the active structure minus the same without pumping. The area enclosed by the white line is the projection of the SRRs on the gain layer.

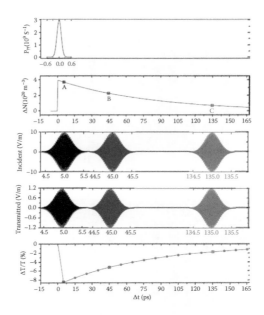

**COLOR FIGURE 4.16**
Schematic of the numerical pump-probe experiments for the case on resonance. From the top to the bottom, each row corresponds to the pump pulse, population inversion, incident signal (with time delays 5, 45, and 135 ps), transmitted signal, and differential transmittance $\Delta T/T$. It should be mentioned here that the incident frequency of the probe pulse is 175 THz with a FWHM of 2 THz and is equal to the SRR resonance frequency.

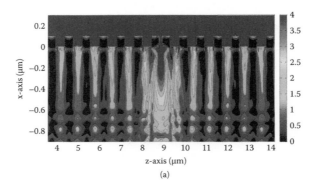

(a)

**COLOR FIGURE 6.7**
Total electric field intensity distribution in the cross section of computational volume behind the nanograting with different profiles or grooves cross section and the following nanograting parameters: nanograting height of 90 nm, unperturbed gold layer thickness of 10 nm, period 810 nm, slit (subwavelength aperture) width of 50 nm. The design parameters used for this computation are the same as those resulting in the maximum predicted (almost 50 times when the nanogratings shape is rectangular) light absorption enhancement due to plasmon-assisted light intensity concentration under the subwavelength apertures. The color scale has been optimized for representing the small weak-field intensity variations. (a) For a rectangular-shaped nanograting profile with the aspect ratio of 1.

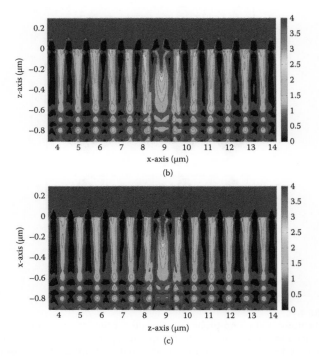

(b)

(c)

**COLOR FIGURE 6.7**
(b) for a trapezoidal-shaped nanograting profile (for this case, the aspect ratio is 0.5), the absorption enhancement is about 11 times; and (c) for a triangular-shaped nanograting profile with the aspect ratio of 0 (zero), the absorption enhancement is about 5 times.

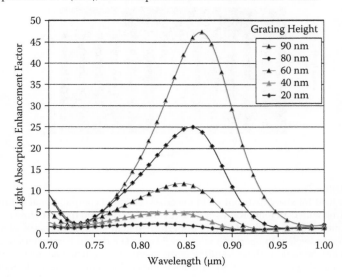

**COLOR FIGURE 6.8**
Light absorption enhancement factor spectra for nanostructured MSM-PDs with several metal nanograting heights. For this simulation, the metal nanograting heights were varied from 20 nm to 90 nm, while keeping the subwavelength aperture width constant at 50 nm; the shapes of all grating corrugations were rectangular type.

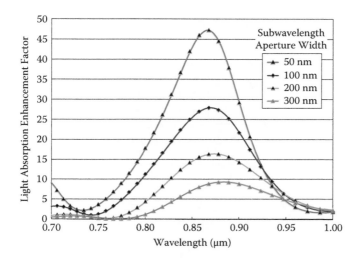

**COLOR FIGURE 6.9**
The light absorption enhancement factor spectra for nanostructured MSM-PDs with different subwavelength aperture widths. The subwavelength aperture widths were varied between 50 to 300 nm, and the groove shapes were all rectangular-type for this result.

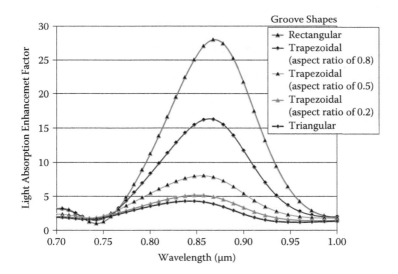

**COLOR FIGURE 6.10**
Light absorption enhancement factor versus the wavelength characteristics for different groove shapes or profiles when the subwavelength aperture width was fixed at 100 nm and the nanograting heights were kept constant at 90 nm.

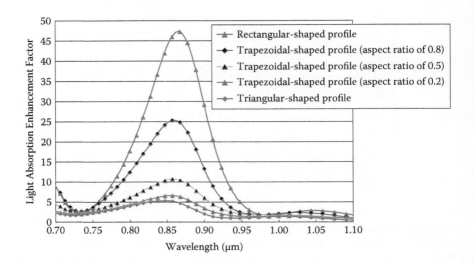

**COLOR FIGURE 6.11**
Light absorption enhancement factor spectra for different groove shape (or profile) geometries of nano-patterned gratings. For the simulation, the subwavelength aperture width was kept constant at 50 nm.

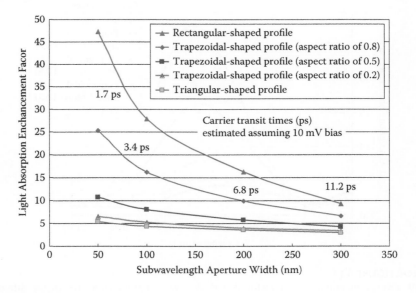

**COLOR FIGURE 6.12**
Maximum light absorption enhancement factor versus the subwavelength aperture width characteristics for different groove geometries (or profiles). The lower limits (drift time-limited) of the carrier transit time at 10 mV bias conditions are shown as a rough-guide order-of-magnitude estimate only.

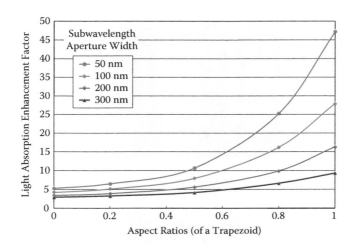

**COLOR FIGURE 6.13**
Maximum light absorption enhancement factor versus the aspect ratios characteristics for different subwavelength aperture widths and groove-shapes (or profiles). Here, the subwavelength aperture widths are varied from 50 nm to 300 nm and the aspect ratios are from 0 (triangular-shaped) to 1 (rectangular-shaped).

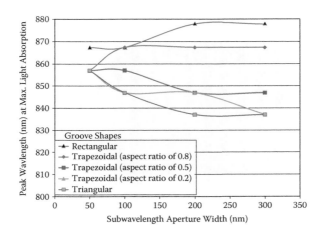

**COLOR FIGURE 6.14**
Light absorption-enhancement peak wavelength versus the subwavelength aperture width characteristics for different groove shapes (or profile) of metal nanogratings of height 90 nm.

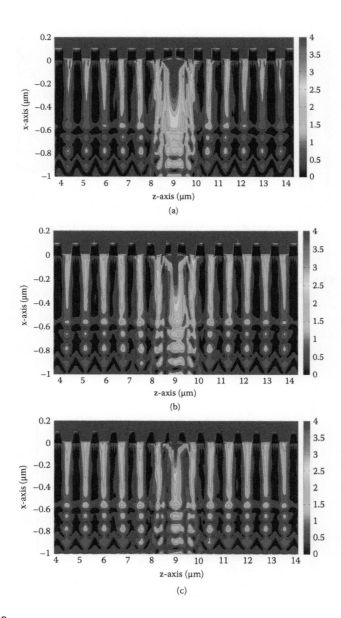

**FIGURE 6.19**
Total electric field intensity distribution in the cross section of computational volume behind the nanograting with different grooves cross section and the following grating parameters: the nanograting height is 90 nm, unperturbed gold (Au) layer thickness is 10 nm, grating period is 810 nm, and the subwavelength aperture width is 100 nm. The color scale has been optimized for representing small weak-field intensity variations. The nanograting profiles are (a) rectangular-shaped (aspect ratio 1) with the nanograting phase shift of 0°, (b) trapezoidal-shaped (with an aspect ratio of 0.8) with a nanograting phase shift of ~45°, and (c) trapezoidal-shaped (with an aspect ratio of 0.5) with a nanograting phase shift of ~90°.

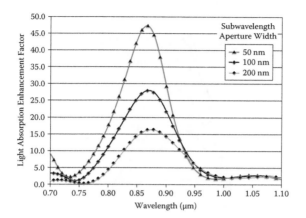

**COLOR FIGURE 6.20**
Light absorption enhancement factor spectra for nanostructured MSM-PDs with plasmon-assisted operation for different subwavelength aperture widths, such as 50 nm, 100 nm, and 200 nm. For this case, the nanograting grooves shape was rectangular type.

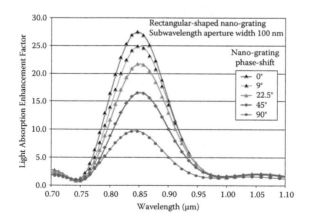

**COLOR FIGURE 6.21**
Light absorption enhancement factor spectra for nanostructured MSM-PDs with plasmon-assisted operation for various nanograting phase shifts. Here, the nanograting phase shift was varied from 0° to 90°. The subwavelength aperture width was kept constant at 100 nm and the nanograting groove shape was rectangular type.

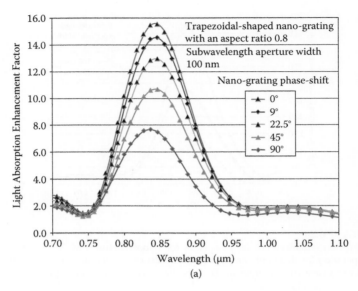

(a)

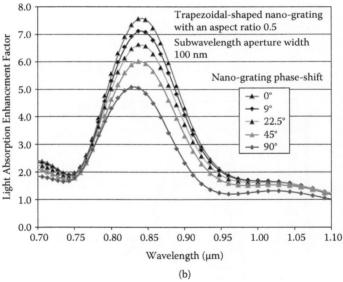

(b)

**COLOR FIGURE 6.22**
Light absorption enhancement factor spectra for trapezoidal-shaped nanograting with the aspect ratios of 0.8 (a) and 0.5 (b) for different nanograting phase shifts. In this case, the subwavelength aperture width was kept constant at 100 nm. (See color figure.)

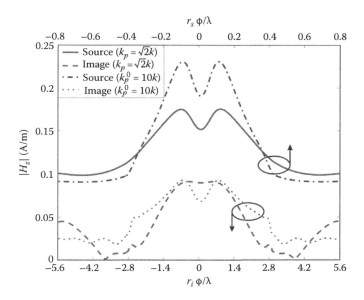

**FIGURE 3.20**
Comparison of magnetic field distributions in the source and image planes of two hyperlens structures with $k_p^0 = \sqrt{2}k$ and $k_p^0 = 10k$. In both cases, $L = 3\lambda$. (See color figure.)

and the hyperlens with $k_p^0 = 10k$ provides a sevenfold image magnification of the subwavelength source distribution, as shown in Figure 3.20. The spacing between these maxima is $1.4\lambda$; thus they can be detected in far-field by conventional scanning devices. On the other hand, the hyperlens with $k_p^0 = \sqrt{2}k$ introduces strong reflections that distort the distributions in both the source and image planes, resulting in two hardly distinguishable maxima. If the length of the device is further extended, the maxima may merge together and the subwavelength details of the source distribution may not be resolved at the image plane. This comparison clearly shows that for a hyperlens formed by a tapered array of metallic wires with a certain value of $k_p^0$, there exists a maximum length below which the device can only be used for image transfer and magnification.

The above numerical results show that the image transfer and magnification capability of such hyperlens structures is insensitive to their transverse (angular) dimensions. A sevenfold magnification of a source distribution with $\lambda/5$ resolution to a distance of $3\lambda$ is demonstrated. However, for the hyperlens device considered in this study, there exists a maximum length for the effective transfer and magnification of subwavelength source distributions to the far-field. Furthermore, using the proposed method, image demagnification properties of hyperlens devices formed by tapered arrays of wires can be analyzed conveniently.

## 3.4 FDTD Modeling of Plasmonic Waveguide

It is well known that photonic crystals (PCs) offer unique opportunities to control the flow of light [65]. The basic idea is to design periodic dielectric structures that have a bandgap for a particular frequency range. Periodic dielectric rods in which one or several rows of elements have been removed can be used as waveguiding devices when operating at bandgap frequencies. A lot of effort has been made to obtain a complete and wider bandgap. It has been shown that a triangular lattice of air holes in a dielectric background has a complete bandgap for TE (transverse electric) mode, while a square lattice of dielectric rods in air has a bandgap for TM (transverse magnetic) mode [66]. The devices operating in the bandgap frequencies are not the only option to guide the flow of light. Another waveguiding mechanism is the total internal reflection (TIR) in one-dimensional (1D) periodic dielectric rods [67]. It is analyzed in [67] that a single row of dielectric rods or air holes supports waveguiding modes and therefore can also be used as a waveguide. In [68], the design of such waveguides consisting of several rows of dielectric rods with various spacing is proposed.

Recently, a new method for guiding electromagnetic waves in structures whose dimensions are below the diffraction limit has been proposed. The structures are termed *plasmonic waveguides*, which have an operation of principle based on near-field interactions between closely spaced noble metal nanoparticles (spacing $\ll \lambda$) that can be efficiently excited at their surface plasmon frequency. The guiding principle relies on coupled plasmon modes set up by near-field dipole interactions that lead to coherent propagation of energy along the array. Analogous structures to waveguides in the microwave regime include periodic metallic cylinders to support propagating waves [69], arrays of flat dipoles that support guided waves [70], and Yagi-Uda antennas [71,72]. However, although these structures can be scaled to optical frequencies with appropriate material properties, their dimensions are limited by the so-called diffraction limit $\lambda/(2n)$. On the other hand, plasmonic waveguides employ the localization of electromagnetic fields near metal surfaces to confine and guide light in regions much smaller than the free-space wavelength and can effectively overcome the diffraction limit. Previous analysis of plasmonic structures includes the plasmon propagation along metal stripes, wires, or grooves in metal [73–78], and the coupling between plasmons on metal particles to guide energy [79,80]. Such subwavelength structures can also find their applications—for example, efficient absorbers and electrically small receiving antennas at microwave frequencies. Recently composite materials containing randomly distributed electrically conductive material and non–electrically conductive material have been designed [81]. They are noted to exhibit plasma-like responses at frequencies well below plasma frequencies of the bulk material.

The FDTD method [6,30] is seen as the most popular numerical technique especially because of its flexibility in handling material dispersion as well as arbitrarily shaped inclusions. In [82], the optical pulse propagation below the diffraction limit is shown using the FDTD method. Also with the FDTD method, the waveguide formed by several rows of silver nanorods arranged in hexagonal is studied [83]. Despite these examples of applying the FDTD method for the plasmonic structures, the accuracy of modeling has not been proven yet. When modeling curved structures, unless using extremely fine meshes, due to the nature of orthogonal and staggered grid of conventional FDTD, often modifications need to be applied in order to improve the numerical accuracy, such as the treatment of interfaces between different materials even for planar structures [84], and the improved conformal algorithms using structured meshes [85] for curved surfaces.

In addition to the modifications at material interfaces, the material frequency dispersion has also to be taken into account in FDTD modeling. However, modeling dispersive materials with curved surfaces remains a challenging topic due to the complexity of the algorithm as well as the introduction of numerical instability. An alternative way to solve this problem is based on the idea of effective permittivities (EPs) [86–88] in the underlying Cartesian coordinate system, and the dispersive FDTD scheme can be therefore modified accordingly without affecting the stability of algorithm. In this work, we first propose a novel conformal dispersive FDTD algorithm combining the EPs together with the ADE method [32], then apply the developed method to the modeling of plasmonic waveguides formed by an array of circular- or elliptical-shaped silver cylinders at optical frequencies. Numerical FDTD simulation results are verified by a frequency domain embedding method [89].

### 3.4.1 Conformal Dispersive FDTD Method Using Effective Permittivities

Conventionally, staircase approximations are often used to model curved electromagnetic structures in an orthogonal FDTD domain. Figure 3.21(a) shows an example layout of an infinitely long cylinder in the free space represented in a 2D orthogonal FDTD domain. The approximated shape introduces spurious numerical resonant modes that do not exist in actual structures. On the other hand, using the concept of filling ratio, which is defined as the ratio of the area of material $\varepsilon_2$ to the area of a particular FDTD cell, the curvature can be properly represented in the FDTD domain as shown in Figure 3.21(b), where different levels of darkness indicate different filling ratios of material $\varepsilon_2$. The accuracy of modeling can be significantly improved in comparison to staircase approximations, as will be shown in a later section.

According to [88], the EP in a general form is given by

$$\varepsilon_{\text{eff}} = \varepsilon_{\parallel}(1 - n^2) + \varepsilon_{\perp} n^2, \tag{3.71}$$

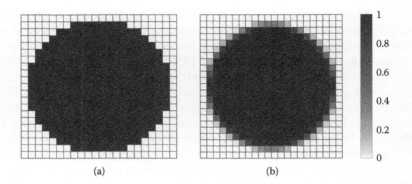

**FIGURE 3.21**
Comparison of the filling ratio for $E_y$ component in FDTD modeling of a circular cylinder using (a) staircase approximations and (b) a conformal scheme. The radius of circular cylinder is 10 cells.

where $n$ is the projection of the unit normal vector **n** along the field component as shown in Figure 3.22, and $\varepsilon_\parallel$ and $\varepsilon_\perp$ are parallel and perpendicular permittivities to the material interface, respectively, and defined as

$$\varepsilon_\parallel = f\varepsilon_2 + (1-f)\varepsilon_1, \tag{3.72}$$

$$\varepsilon_\perp = \left[\frac{f}{\varepsilon_2} + \frac{1-f}{\varepsilon_1}\right]^{-1}, \tag{3.73}$$

where $f$ is the filling ratio of material $\varepsilon_2$ in a certain FDTD cell.

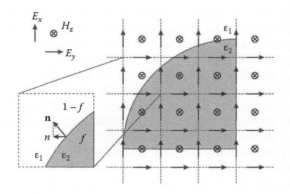

**FIGURE 3.22**
Layout of a quarter circular inclusion in orthogonal FDTD grid for $E_y$ component. The radius of circular cylinder is three cells.

Here we consider the inclusions as silver cylinders, which at optical frequencies can be modeled using the Drude dispersion model

$$\varepsilon_2(\omega) = \varepsilon_0 \left( 1 - \frac{\omega_p^2}{\omega^2 - j\omega\gamma} \right),\tag{3.74}$$

where $\omega_p$ is the plasma frequency and $\gamma$ is the collision frequency. At the frequencies below the plasma frequency, the real part of permittivity is negative. In this work, we assume that the silver cylinders are embedded in the free space ($\varepsilon_1 = \varepsilon_0$).

In order to take into account the frequency dispersion of the material, the electric flux density **D** is introduced into standard FDTD updating equations. At each time step, **D** is updated directly from **H** and **E** can be calculated from **D** through the following steps. Substitute (3.72) and (3.73) into (3.71), and, using the expressions for $\varepsilon_1$ and $\varepsilon_2$ (3.74), the constitutive relation in the frequency domain reads

$$\left\{ \omega^4 - 2\gamma j\omega^3 - \left[ \gamma^2 + (1-f)\omega_p^2 \right]\omega^2 + \gamma(1-f)\omega_p^2 j\omega \right\} \mathbf{D}$$

$$= \left[ \omega^4 - 2\gamma j\omega^3 - \left( \gamma^2 + \omega_p^2 \right)\omega^2 + \gamma\omega_p^2 j\omega + f(1-f)(1-n^2)\omega_p^4 \right] \varepsilon_0 \mathbf{E}.\tag{3.75}$$

Using the inverse Fourier transformation (i.e., $j\omega \to \partial/\partial t$), we obtain the constitutive relation in the time domain as

$$\left\{ \frac{\partial^4}{\partial t^4} + 2\gamma \frac{\partial^3}{\partial t^3} + \left[ \gamma^2 + (1-f)\omega_p^2 \right]\frac{\partial^2}{\partial t^2} + \gamma(1-f)\omega_p^2 \frac{\partial}{\partial t} \right\} \mathbf{D}$$

$$= \left[ \frac{\partial^4}{\partial t^4} + 2\gamma \frac{\partial^3}{\partial t^3} + \left( \gamma^2 + \omega_p^2 \right)\frac{\partial^2}{\partial t^2} + \gamma\omega_p^2 \frac{\partial}{\partial t} + f(1-f)(1-n^2)\omega_p^4 \right] \varepsilon_0 \mathbf{E}.\tag{3.76}$$

The FDTD simulation domain is represented by an equally spaced 3D grid with periods $\Delta x$, $\Delta y$, and $\Delta z$ along $x$-, $y$-, and $z$-directions, respectively. The time step is $\Delta t$. For discretization of (3.76), we use the central finite difference operators in time ($\delta_t$) and the central average operator with respect to time ($\mu_t$):

$$\frac{\partial^4}{\partial t^4} \to \frac{\delta_t^4}{(\Delta t)^4}, \quad \frac{\partial^3}{\partial t^3} \to \frac{\delta_t^3}{(\Delta t)^3}\mu_t, \quad \frac{\partial^2}{\partial t^2} \to \frac{\delta_t^2}{(\Delta t)^2}\mu_t^2, \quad \frac{\partial}{\partial t} \to \frac{\delta_t}{\Delta t}\mu_t^3, \quad 1 \to \mu_t^4,$$

where the operators $\delta_t$ and $\mu_t$ are defined as in [35]:

$$\delta_t F \big|_{m_x,m_y,m_z}^n \equiv F \big|_{m_x,m_y,m_z}^{n+\frac{1}{2}} - F \big|_{m_x,m_y,m_z}^{n-\frac{1}{2}},$$

$$\mu_t F \big|_{m_x,m_y,m_z}^n \equiv \frac{F \big|_{m_x,m_y,m_z}^{n+\frac{1}{2}} + F \big|_{m_x,m_y,m_z}^{n-\frac{1}{2}}}{2}.$$

Here $F$ represents field components and $m_x$, $m_y$, $m_z$ are indices corresponding to a certain discretization point in the FDTD domain. The discretized Equation (3.76) reads

$$\left\{ \frac{\delta_t^4}{(\Delta t)^4} + 2\gamma \frac{\delta_t^3}{(\Delta t)^3} \mu_t + \left[ \gamma^2 + (1-f)\omega_p^2 \right] \frac{\delta_t^2}{(\Delta t)^2} \mu_t^2 + \gamma(1-f)\omega_p^2 \frac{\delta_t}{\Delta t} \mu_t^3 \right\} D$$

$$= \left[ \frac{\delta_t^4}{(\Delta t)^4} + 2\gamma \frac{\delta_t^3}{(\Delta t)^3} \mu_t + \left( \gamma^2 + \omega_p^2 \right) \frac{\delta_t^2}{(\Delta t)^2} \mu_t^2 \right.$$

$$\left. + \gamma\omega_p^2 \frac{\delta_t}{\Delta t} \mu_t^3 + f(1-f)(1-n^2)\omega_p^4 \mu_t^4 \right] \varepsilon_0 E. \tag{3.77}$$

Note that in the above equations we have kept all terms at fourth-order to guarantee numerical stability. Equation (3.77) can be written as

$$\frac{D^{n+1} - 4D^n + 6D^{n-1} - 4D^{n-2} + D^{n-3}}{(\Delta t)^4} + \gamma \frac{D^{n+1} - 2D^n + 2D^{n-2} - D^{n-3}}{(\Delta t)^3}$$

$$+ \left[ \gamma^2 + (1-f)\omega_p^2 \right] \frac{D^{n+1} - 2D^n + D^{n-3}}{4(\Delta t)^2} + \gamma(1-f)\omega_p^2 \frac{D^{n+1} + 2D^n - 2D^{n-2} - D^{n-3}}{8\Delta t}$$

$$= \varepsilon_0 \frac{E^{n+1} - 4E^n + 6E^{n-1} - 4E^{n-2} + E^{n-3}}{(\Delta t)^4} + \varepsilon_0 \gamma \frac{E^{n+1} - 2E^n + 2E^{n-2} - E^{n-3}}{(\Delta t)^3}$$

$$+ \varepsilon_0 \left( \gamma^2 + \omega_p^2 \right) \frac{E^{n+1} - 2E^n + E^{n-3}}{4(\Delta t)^2} + \varepsilon_0 \gamma\omega_p^2 \frac{E^{n+1} + 2E^n - 2E^{n-2} - E^{n-3}}{8\Delta t}$$

$$+ f(1-f)(1-n^2)\omega_p^4 \frac{E^{n+1} + 4E^n + 6E^{n-1} + 4E^{n-2} + E^{n-3}}{16}. \tag{3.78}$$

The indices $m_x$, $m_y$, and $m_z$ are omitted from (3.78) since $\mathbf{E}$ and $\mathbf{D}$ are at the same locations. If we solve for $\mathbf{E}^{n+1}$, then the updating equation for $\mathbf{E}^n$ FDTD iterations reads

$$
\mathbf{E}^{n+1} = \left[ b_0 \mathbf{D}^{n+1} + b_1 \mathbf{D}^n + b_2 \mathbf{D}^{n-1} + b_3 \mathbf{D}^{n-2} + b_4 \mathbf{D}^{n-3} \right.
$$
$$
\left. - \left( a_1 \mathbf{E}^n + a_2 \mathbf{E}^{n-1} + a_3 \mathbf{E}^{n-2} + a_4 \mathbf{E}^{n-3} \right) \right] / a_0 ,
\tag{3.79}
$$

with the coefficients given by

$$
a_0 = \varepsilon_0 \left[ \frac{1}{(\Delta t)^4} + \frac{\gamma}{(\Delta t)^3} + \frac{\gamma^2 + \omega_p^2}{4(\Delta t)^2} + \frac{\gamma \omega_p^2}{8\Delta t} + \frac{f(1-f)(1-n^2)\omega_p^4}{16} \right],
$$

$$
a_1 = \varepsilon_0 \left[ -\frac{4}{(t)^4} - \frac{2\gamma}{(t)^3} + \frac{\gamma \omega_p^2}{4\Delta t} + \frac{f(1-f)(1-n^2)\omega_p^4}{4} \right],
$$

$$
a_2 = \varepsilon_0 \left[ \frac{6}{(t)^4} - \frac{\gamma^2 + \omega_p^2}{2(t)^2} + \frac{3f(1-f)(1-n^2)\omega_p^4}{8} \right],
$$

$$
a_3 = \varepsilon_0 \left[ -\frac{4}{(t)^4} + \frac{2\gamma}{(t)^3} - \frac{\gamma \omega_p^2}{4\Delta t} + \frac{f(1-f)(1-n^2)\omega_p^4}{4} \right],
$$

$$
a_4 = \varepsilon_0 \left[ \frac{1}{(t)^4} - \frac{\gamma}{(t)^3} + \frac{\gamma^2 + \omega_p^2}{4(t)^2} - \frac{\gamma \omega_p^2}{8\Delta t} + \frac{f(1-f)(1-n^2)\omega_p^4}{16} \right],
$$

$$
b_1 = -\frac{4}{(t)^4} - \frac{2\gamma}{(t)^3} + \frac{\gamma(1-f)\omega_p^2}{4\Delta t},
$$

$$
b_2 = \frac{6}{(t)^4} - \frac{\gamma^2 + (1-f)\omega_p^2}{2(t)^2},
$$

$$
b_3 = -\frac{4}{(t)^4} + \frac{2\gamma}{(t)^3} - \frac{\gamma(1-f)\omega_p^2}{4\Delta t},
$$

$$
b_4 = \frac{1}{(t)^4} - \frac{\gamma}{(t)^3} + \frac{\gamma^2 + (1-f)\omega_p^2}{4(t)^2} - \frac{\gamma(1-f)\omega_p^2}{8\Delta t}.
$$

The computations of $\mathbf{E}$ and $\mathbf{D}$ are performed using Yee's standard updating equations in the free space. Note that if the plasma frequency is equal to zero ($\omega_p = 0$), then (3.79) reduces to the updating equation in the free space, i.e., $\mathbf{E} = \mathbf{D}/\varepsilon_0$.

### 3.4.2 FDTD Calculation of Dispersion Diagram

Applying the Bloch's periodic boundary conditions (PBCs) [90–95], the FDTD method can be used to model periodic structures and calculate their dispersion diagrams [96,97]. For any periodic structures, the field at any time should satisfy the Bloch theory, i.e.,

$$\mathbf{E}(\mathbf{d}+\mathbf{a}) = \mathbf{E}(\mathbf{d})e^{j\mathbf{k}\mathbf{a}}, \quad \mathbf{H}(\mathbf{d}+\mathbf{a}) = \mathbf{H}(\mathbf{d})e^{j\mathbf{k}\mathbf{a}}, \tag{3.80}$$

where *d* is the distance vector of any location in the computation domain, *k* is the wave vector, and *a* is the lattice vector along the direction of periodicity. When updating the fields at the boundary of the computation domain using FDTD, the required fields outside the computation domain can be calculated using known field values inside the domain through Equation (3.80). Although instead of using real values in conventional FDTD computations, the calculation of dispersion diagrams requires complex field values, since only one unit cell is modeled, the computation load is not significantly increased.

First we apply the developed conformal dispersive FDTD method to calculate the dispersion diagram for 1D plasmonic waveguides formed by an array of periodic infinitely long (along *z*-direction) circular silver cylinders. As shown in Figure 3.23, the 2D simulation domain (*x-y*) with TE modes (only $E_x$, $E_y$, and $H_z$ are nonzero fields) is truncated using Bloch's PBCs in the *x*-direction and Berenger's PML [40] in the *y*-direction. The Berenger's PML has excellent performance for absorbing propagating waves; however, for evanescent waves, field shows growing behavior inside. Since the waves

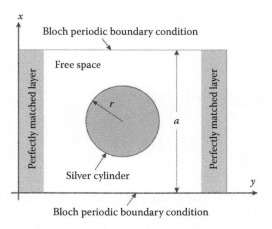

**FIGURE 3.23**
The layout of the 2D FDTD computation domain for calculating dispersion diagram for 1D periodic structures.

radiated by point or line sources consist of both propagating and evanescent components, extra space (typically a quarter wavelength at the frequency of interest) between PMLs and the circular inclusion is added to allow the evanescent waves to decay before reaching the PMLs.

The radius of silver cylinders is $r = 2.5 \times 10^{-8}$ m and the period is $a = 7.5 \times 10^{-8}$ m. The plasma and collision frequencies are $\omega_p = 9.39 \times 10^{15}$ rad/s and $\gamma = 3.14 \times 10^{13}$ Hz, respectively, to closely match the bulk dielectric function of silver [98]. The FDTD cell size is $\Delta x = \Delta y = 2.5 \times 10^{-9}$ m with the time step $\Delta t = \Delta x/(\sqrt{2}c)$ s (where $c$ is the speed of light in the free space) according to the Courant stability criterion [6]. Although the stability condition for high-order FDTD method is typically stricter than the conventional one, since the average operator $\mu_t$ is applied to develop the algorithm, we have not found any instability for a complete time period of more than 40,000 time steps used in all simulations.

A wideband magnetic line source is placed at an arbitrary location in the free space region of the 2D simulation domain to excite all resonant modes of the structure within the frequency range of interest (normalized frequency $\bar{f} = \omega a/(2\pi c) \in [0 \sim 0.5]$):

$$g(t) = e^{-\left(\frac{t-t_0}{\tau}\right)^2} e^{j\omega t}, \qquad (3.81)$$

where $t_0$ is the initial time delay, $\tau$ defines the pulse width, and $\omega$ is the center frequency of the pulse ($\bar{f} = 0.25$). The magnetic fields at 100 random locations in the free-space region are recorded during simulations, transformed into the frequency domain, and combined to extract the individual resonant mode corresponding to each local maximum. For each wave vector, a total number of 40,000 time steps are used in our simulations to obtain enough accurate frequency domain results.

To demonstrate the advantage of EPs and validate the proposed conformal dispersive FDTD method, we have also performed simulations using staircase approximations for the circular cylinder, as shown in Figure 3.21(a). Figure 3.24 shows the comparison of the first resonant frequency (transverse mode) at wave vector $k_x = \pi/a$ of the plasmonic waveguide calculated using the FDTD method with staircase approximations, the FDTD method with EPs, and the frequency domain embedding method [99]. With the same FDTD spatial resolutions, the model using EP shows excellent agreement with the results from the frequency domain embedding method; on the contrary, the staircase approximation not only leads to a shift of the main resonant frequency but also introduces a spurious numerical resonant mode that does not exist in actual structures. The same effect has also been found for nondispersive dielectric cylinders. It is also shown in Figure 3.24 that although one may correct the main resonant frequency using finer meshes, the spurious resonant mode remains.

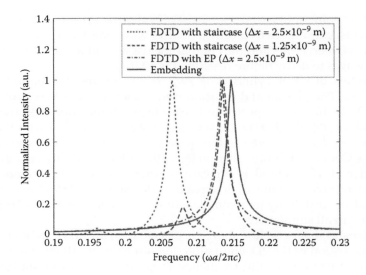

**FIGURE 3.24**
Comparison of the first resonant frequency (transverse mode) at wave vector $k_x = \pi/a$ calculated using the FDTD method with staircase approximations, the FDTD method with EPs, and the frequency domain embedding method.

The problem of frequency shift and spurious modes become more severe when calculating higher guided modes near the "flat band" region (i.e., the region where waves travel at a very low phase velocity). Even with a refined spatial resolution, the staircase approximation fails to provide correct results (not shown). On the other hand, using the proposed conformal dispersive FDTD scheme, all resonant modes are correctly captured in FDTD simulations as demonstrated by the comparison with the embedding method as shown in Figure 3.25.

According to the previous analysis using the frequency embedding method, the fundamental mode in the modeled plasmonic waveguide is transverse mode and the second guided mode is longitudinal [99], which is also shown by the distribution of electric field intensities in Figure 3.26 from our FDTD simulations. The higher guided modes are referred to as plasmon modes. For demonstration of field symmetries and due to the TE mode considered in our simulations, we have plotted the distributions of magnetic field corresponding to different resonant modes at wave number $k_x = \pi/a$ as marked in Figure 3.25, as shown in Figure 3.27. Sinusoidal sources for excitation of certain single modes are used and the sources are placed at different locations corresponding to different symmetries of field patterns. All field patterns are plotted after the steady state is reached in simulations. The modes (a), (c), and (d) in Figure 3.27 are even modes (relative to the direction

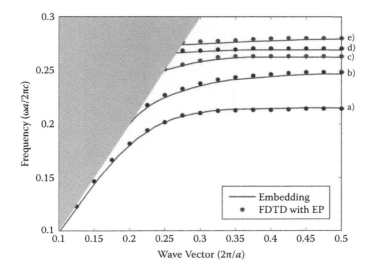

**FIGURE 3.25**
Comparison of dispersion diagrams for an array of infinitely long (along $z$-direction) circular silver cylinders calculated using the FDTD method with EPs and the frequency domain embedding method.

of periodicity of the waveguide, i.e., the $x$-axis), and (b) and (e) are considered as odd modes.

The above comparison of the simulation results calculated using the conformal dispersive FDTD method and the embedding method clearly demonstrates the effectiveness of applying the EPs in FDTD modeling. Furthermore, in contrast to the embedding method, the main advantage of the FDTD method is that arbitrarily shaped geometries can be easily

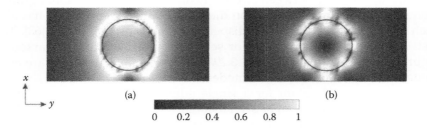

**FIGURE 3.26**
Normalized total electric field intensities corresponding to (a) transverse and (b) longitudinal modes [99] at wave number $k_x = \pi/a$ as marked in Figure 3.25. The structure is infinite along $x$-direction.

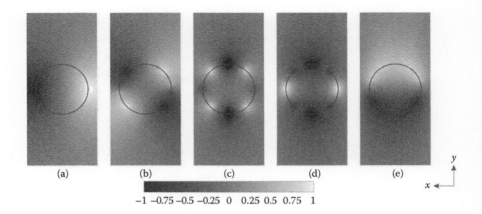

**FIGURE 3.27**
Normalized distributions of magnetic field corresponding to different resonant modes at wave number $k_x = \pi/a$ as marked in Figure 3.25. The structure is infinite along $x$-direction.

modeled. We have applied the conformal dispersive FDTD method to study the effect of different inclusions on the dispersion diagrams of 1D plasmonic waveguides. The geometries considered are two rows of periodic infinitely long (along $z$-direction) circular silver cylinders arranged in a square lattice and a single row of periodic infinitely long (along $z$-direction) elliptical silver cylinders. The elliptical element has a ratio of semimajor-to-semiminor axis 2:1, where the semiminor axis is equal to the radius of the circular element (25.0 nm). For the two rows of circular nanorods, the spacing between two rows (center-to-center distance) is 75.0 nm. The dispersion diagrams for these structures are plotted in Figures 3.28 and 3.30. Comparing the dispersion diagrams for a single circular element in Figure 3.25 and two circular elements in Figure 3.28, we can see that the dispersion diagram has been modified due to the change of inclusion. The strong coupling between two elements introduces additional guided modes to appear in the dispersion diagram. Such a phenomenon has also been studied for dielectric (nondispersive) nanorods previously [68]. The distributions of magnetic field for selected guided modes as marked in Figure 3.28 are plotted in Figure 3.29. The modes (a), (c), and (d) are even modes while (b) and (e) are odd modes.

The dispersion diagram for a single elliptical element as inclusion is shown in Figure 3.30. It can be seen that more guided modes appear, which is caused by the change of the inclusion's geometrical shape from circular to elliptical. The frequency corresponding to the lowest mode has been lowered due to the increase of the inclusion's volume. The distributions of magnetic field for selected guided modes are plotted in Figure 3.31. The modes (a) and (d) are even modes and (b), (c), and (e) are odd modes, respectively.

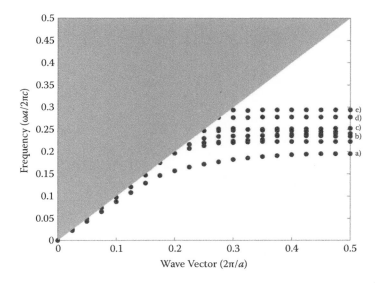

**FIGURE 3.28**
Dispersion diagram for two rows of periodic infinitely long (along *z*-direction) circular silver cylinders arranged in square lattice calculated from conformal dispersive FDTD simulations.

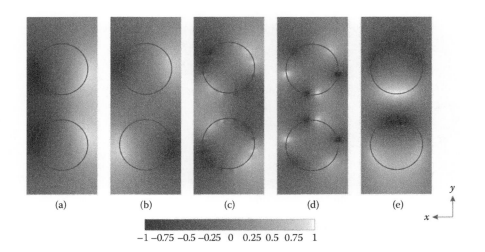

**FIGURE 3.29**
Normalized distributions of magnetic field corresponding to different guided modes as marked in Figure 3.28. The structure is infinite along *x*-direction.

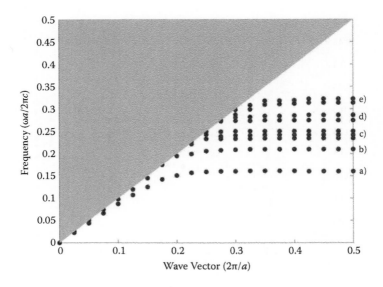

**FIGURE 3.30**
Dispersion diagram for a single row of periodic infinitely long (along *x*-direction) elliptical silver cylinders calculated from conformal dispersive FDTD simulations.

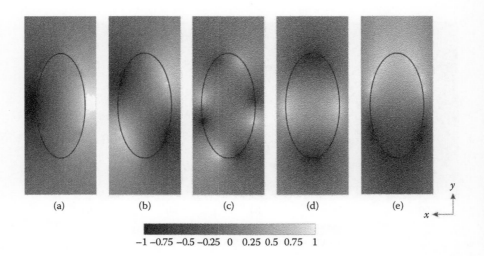

**FIGURE 3.31**
Normalized distributions of magnetic field corresponding to different guided modes as marked in Figure 3.30. The structure is infinite along *x*-direction.

### 3.4.3 Wave Propagation in Plasmonic Waveguides Formed by Finite Number of Elements

In order to study wave propagations in plasmonic waveguides formed by a finite number of silver nanorods, we have replaced PBCs in the $x$-direction with PMLs and added additional cells for the free-space region to the simulation domain. The number of nanorods under study is seven. The spacing (pseudo-period) between adjacent elements remains the same as for infinite structures considered in the previous section. For a single mode excitation, we choose the frequency of the corresponding mode from the dispersion diagram and excite sinusoidal sources at different locations with respect to the symmetry of different guided modes at one end of the waveguides.

For the plasmonic waveguides formed by different types of inclusions, we have chosen certain eigenmodes: mode Figure 3.27(a) for a single row of circular cylinders, mode Figure 3.29(d) for two rows of circular cylinders, and mode Figure 3.31(e) for a single row of elliptical cylinders. The distributions of magnetic field intensity for different waveguides operating in these guided modes are plotted in Figure 3.32. The field plots are taken after the

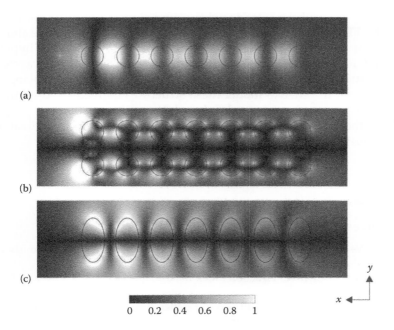

**FIGURE 3.32**
Normalized distributions of magnetic field intensity corresponding to different guided modes for seven-element plasmonic waveguides formed by (a) a single row of circular nanorods (the corresponding eigenmode is shown in Figure 3.27(a)), (b) two rows of circular nanorods arranged in square lattice (the corresponding eigenmode is shown in Figure 3.29(c)), (c) a single row of elliptical nanorods (the corresponding eigenmode is shown in Figure 3.31(e)).

steady state is reached in simulations. It is clearly seen that single guided modes are coupled into these waveguides but the excitation of certain modes highly depends on the symmetry of field patterns. The energy that can be coupled into the waveguides also depends on the matching between source and the plasmonic waveguide.

The above results demonstrate that our developed conformal dispersive FDTD method significantly improves the simulation accuracy. Further work includes the investigation of the effects of different number of elements in plasmonic waveguides on guided modes, and the calculation of group velocity of different modes propagating in these waveguides. Although results presented above have been focused on optical frequencies, with future advances in microwave plasmonic materials, novel applications can be found in the design of small antennas and efficient absorbers.

## 3.5 Conclusion

In this chapter, we have introduced the modeling of optical metamaterials using the FDTD method. The structures considered here include the optical cloaking structures for reducing scattering cross sections, the hyperlens structures for simultaneous imaging and magnification of subwavelength source distributions, and nanoplasmonic waveguides for enhancing light transmissions. In order to model these structures correctly and accurately, new FDTD schemes have been developed. We anticipate that our proposed FDTD methods may assist the analysis, design, and practical realization of novel optical devices.

## References

1. V. G. Veselago, "The electrodynamics of substances with simultaneously negative values of $\varepsilon$ and $\mu$," *Sov. Phys. Usp.* 10, 509–514 (1968).
2. J. Pendry, "Negative refraction index makes a perfect lens," *Phys. Rev. Lett.* 85, 3966–3969 (2000).
3. Y. Zhao and Y. Hao, "Full-wave parallel dispersive finite-difference time-domain modeling of three dimensional electromagnetic cloaking structures," *J. Comput. Phys.* 228, 7300–7312 (2009).
4. Y. Zhao, "Investigation of image magnification property of hyperlens formed by a tapered array of metallic wires using a spatially dispersive finite-difference time-domain method in cylindrical coordinates," *J. Opt.* 14, 035102 (2012).
5. Y. Zhao and Y. Hao, "Finite-difference time-domain study of guided modes in nano-plasmonic waveguides," *IEEE Trans. Antennas Propag.* 55, 3070–3077 (2007).

6. A. Taflove, *Computational Electrodynamics: The Finite-Difference Time-Domain Method*, 2nd ed. Norwood, MA: Artech House, 2000.

7. Y. Zhao, P. A. Belov, and Y. Hao, "Accurate modeling of left-handed metamaterials using finite-difference time-domain method with spatial averaging at the boundaries," *J. Opt. A—Pure Appl. Opt.* 9, 468–475 (2007).

8. J. B. Pendry, D. Schurig, and D. R. Smith, "Controlling electromagnetic fields," *Science* 312, 1780–1782 (2006).

9. S. A. Cummer, B.-I. Popa, D. Schurig, and D. R. Smith, "Full-wave simulations of electromagnetic cloaking structures," *Phys. Rev. E* 74, 036621 (2006).

10. W. Cai, U. K. Chettiar, A. V. Kildishev, and V. M. Shalaev, "Optical cloaking with metamaterials," *Nat. Photonics* 1, 224–227 (2007).

11. M. Yan, Z. Ruan, and M. Qiu, "Scattering characteristics of simplified cylindrical invisibility cloaks," *Opt. Express* 15, 17772–17782 (2007).

12. W. Cai, U. K. Chettiar, A. V. Kildishev, and V. M. Shalaev, "Non-magnetic cloak with minimized scattering," *Appl. Phys. Lett.* 91, 111105 (2007).

13. B. Zhang, H. Chen, and B.-I. Wu, "Limitations of high-order transformation and incident angle on simplified invisibility cloaks," *Opt. Express* 16, 14655–14660 (2008).

14. U. Leonhardt, "Optical conformal mapping," *Science* 312, 1777–1780 (2006).

15. M. Tsang and D. Psaltis, "Magnifying perfect lens and superlens design by coordinate transformation," *Phys. Rev. B* 77, 035122 (2008).

16. H. Chen and C. T. Chan, "Transformation media that rotate electromagnetic fields," *Appl. Phys. Lett.* 90, 241105 (2007).

17. M. Rahm, S. A. Cummer, D. Schurig, J. B. Pendry, and D. R. Smith, "Optical design of reflectionless complex media by finite embedded coordinate transformations," *Phys. Rev. Lett.* 100, 063903 (2008).

18. Y. Luo, J. Zhang, L. Ran, H. Chen, and J. A. Kong, "Controlling the emission of electromagnetic sources by coordinate transformation," *PIERS Online* 4, 795–800 (2008).

19. D. Schurig, J. J. Mock, B. J. Justice, S. A. Cummer, J. B. Pendry, A. F. Starr, and D. R. Smith, "Metamaterial electromagnetic cloak at microwave frequencies," *Science* 314, 977–980 (2006).

20. Y. Huang, Y. Feng, and T. Jiang, "Electromagnetic cloaking by layered structure of homogeneous isotropic materials," *Opt. Express* 15, 11133–11141 (2007).

21. I. I. Smolyaninov, Y. J. Hung, and C. C. Davis, "Electromagnetic cloaking in the visible frequency range," *ArXiv.org:0709.2862v2* (2007).

22. D. Schurig, J. B. Pendry, and D. R. Smith, "Calculation of material properties and ray tracing in transformation media," *Opt. Express* 14, 9794–9804 (2006).

23. Z. Ruan, M. Yan, C. W. Neff, and M. Qiu, "Ideal cylindrical cloak: perfect but sensitive to tiny perturbations," *Phys. Rev. Lett.* 99, 113903 (2007).

24. H. Chen, B.-I. Wu, B. Zhang, and J. A. Kong, "Electromagnetic wave interactions with a metamaterial cloak," *Phys. Rev. Lett.* 99, 063903 (2007).

25. Y. Luo, H. Chen, J. Zhang, L. Ran, and J. A. Kong, "Design and analytical full-wave validation of the invisibility cloaks, concentrators, and field rotators created with a general class of transformations," *Phys. Rev. B* 77, 125127 (2008).

26. B. Vasic, G. Isic, R. Gajic, and K. Hinger, "Coordinate transformation based design of confined metamaterial structures," *Phys. Rev. B* 79, 085103 (2009).

27. R. Weder, "A rigorous time-domain analysis of full-wave electromagnetic cloaking (invisibility)," *ArXiv.org:0704.0248v4* (2007).

28. C. Blanchard, J. Porti, B.-I. Wu, J. A. Morente, A. Salinas, and J. A. Kong, "Time domain simulation of electromagnetic cloaking structures with TLM method," *Opt. Express* 16, 6461–6470 (2008).

29. Y. Zhao, C. Argyropoulos, and Y. Hao, "Full-wave finite-difference time-domain simulation of electromagnetic cloaking structures," *Opt. Express* 16, 6717–6730 (2008).

30. K. S. Yee, "Numerical solution of initial boundary value problems involving Maxwell's equations in isotropic media," *IEEE Trans. Antennas Propag.* 14, 302–307 (1966).

31. R. Luebbers, F. P. Hunsberger, K. Kunz, R. Standler, and M. Schneider, "A frequency-dependent finite-difference time-domain formulation for dispersive materials," *IEEE Trans. Electromagn. Compat.* 32, 222–227 (1990).

32. O. P. Gandhi, B.-Q. Gao, and J.-Y. Chen, "A frequency-dependent finite-difference time-domain formulation for general dispersive media," *IEEE Trans. Microw. Theory Tech.* 41, 658–664 (1993).

33. D. M. Sullivan, "Frequency-dependent FDTD methods using Z transforms," *IEEE Trans. Antennas Propag.* 40, 1223–1230 (1992).

34. A. F. Bower, *Applied Mechanics of Solids*. CRC Press, 2009.

35. F. B. Hildebrand, *Introduction to Numerical Analysis*. New York: McGraw Hill, 1956.

36. J.-Y. Lee and N.-H. Myung, "Locally tensor conformal FDTD method for modeling arbitrary dielectric surfaces," *Microw. Opt. Tech. Lett.* 23, 245–249 (1999).

37. K. C. Chew and V. F. Fusco, "A parallel implementation of the finite-difference time-domain algorithm," *Int. J. Numerical Modeling* 8, 293–299 (1995).

38. S. Gedney, "Finite-difference time-domain analysis of microwave circuit devices on high performance vector/parallel computers," *IEEE Trans. Microwave Theory Tech.* 43, 2510–2514 (1995).

39. http://wwwunix.mcs.anl.gov/mpi/mpich/.

40. J. P. Berenger, "A perfectly matched layer for the absorption of electromagnetic waves," *J. Comput. Phys.* 114, 185 (1994).

41. H. Chen and C. T. Chan, "Time delays and energy transport velocities in three dimensional ideal cloaking devices," *J. Appl. Phys.* 104, 033113 (2008).

42. Z. Liang, P. Yao, X. Sun, and X. Jiang, "The physical picture and the essential elements of the dynamical process for dispersive cloaking structures," *Appl. Phys. Lett.* 92, 131118 (2008).

43. B. Zhang, B.-I. Wu, H. Chen, and J. A. Kong, "Rainbow and blueshift effect of a dispersive spherical invisibility cloak impinged on by a non-monochromatic plane wave," *Phys. Rev. Lett.* 101, 063902 (2008).

44. V. A. Podolskiy and E. E. Narimanov, "Near-sighted superlens," *Opt. Lett.* 30, 75–77 (2005).

45. P. A. Belov, C. R. Simovski, and P. Ikonen, "Canalization of subwavelength images by electromagnetic crystals," *Phys. Rev. B* 71, 193105 (2005).

46. P. A. Belov, Y. Hao, and S. Sudhakaran, "Subwavelength microwave imaging using an array of parallel conducting wires as a lens," *Phys. Rev. B* 73, 033108 (2006).

47. P. A. Belov and Y. Hao, "Subwavelength imaging at optical frequencies using a transmission device formed by a periodic layered metal-dielectric structure operating in the canalization regime," *Phys. Rev. B* 73, 113110 (2006).

48. G. Shvets, S. Trendafilov, J. B. Pendry, and A. Sarychev, "Guiding, focusing, and sensing on the subwavelength scale using metallic wire arrays," *Phys. Rev. Lett.* 99, 053903 (2007).

49. P. A. Belov, C. R. Simovski, P. Ikonen, M. G. Silveirinha, and Y. Hao, "Image transmission with the subwavelength resolution in microwave, terahertz, and optical frequency bands," *J. Commun. Technol. Electron.* 52, 1009 (2007).
50. Z. Jacob, L. V. Alekseyev, and E. Narimanov, "Optical Hyperlens: Far-field imaging beyond the diffraction limit," *Opt. Express* 14, 8247 (2006).
51. Z. Liu, H. Lee, Y. Xiong, C. Sun, and X. Zhang, "Far-field optical hyperlens magnifying subdiffraction-limited objects," *Science* 315, 1686 (2007).
52. I. I. Smolyaninov, Y.-J. Hung, and C. C. Davis, "Magnifying superlens in the visible frequency range," *Science* 315, 1699 (2007).
53. S. Kawata, A. Ono, and P. Verma, "Subwavelength color imaging with a metallic nanolens," *Nature Photon.* 2, 438–442 (2008).
54. P. Alitalo, S. Maslovski, and S. Tretyakov, "Near-field enhancement and imaging in double cylindrical polariton-resonant structures: Enlarging superlens," *Phys. Lett. A* 357, 397 (2006).
55. P. Ikonen, C. Simovski, S. Tretyakov, P. Belov, and Y. Hao, "Magnification of subwavelength field distributions at microwave frequencies using a wire medium slab operating in the canalization regime," *Appl. Phys. Lett.* 91, 104102 (2007).
56. Y. Zhao, P. A. Belov, and Y. Hao, "Spatially dispersive finite-difference time-domain analysis of subwavelength imaging by the wire medium slabs," *Opt. Express* 14, 5154–5167 (2006).
57. F. L. Teixeira and W. C. Chew, "PML-FDTD in cylindrical and spherical grids," *IEEE Microwave Guided Wave Lett.* 7, 285–287 (1997).
58. W. C. Chew and W. Weedon, "A 3D perfectly matched medium from modified Maxwell's equations with stretched coordinates," *Microwave Opt. Tech. Lett.* 7, 599–604 (1994).
59. J. Brown, "Artificial dielectrics," *Progress in Dielectrics* 2, 195–225 (1960).
60. J. B. Pendry, A. J. Holden, W. J. Steward and I. Youngs, "Extremely low frequency plasmons in metallic mesostructures," *Phys. Rev. Lett.* 76, 4773–4776 (1996).
61. P. A. Belov, R. Marques, S. I. Maslovski, I. S. Nefedov, M. Silveirinha, C. R. Simovski, and S. A. Tretyakov, "Strong spatial dispersion in wire media in the very large wavelength limit," *Phys. Rev. B* 67, 113103 (2003).
62. Z. Kancleris, "Handling of singularity in finite-difference time-domain procedure for solving Maxwell's equations in cylindrical coordinate system," *IEEE Trans. Antennas Propag.* 56, 610–613 (2008).
63. N. Dib, T. Weller, M. Scardeletti, and M. Imparato, "Analysis of cylindrical transmission lines with the finite-difference time-domain method," *IEEE Trans. Microw. Theory Tech.* 47, 509–512 (1999).
64. Y. Chen, R. Mittra, and P. Harms, "Finite difference time domain algorithm for solving Maxwell's equations in rotationally symmetric geometries," *IEEE Trans. Microwave Theory Tech.* 44, 832–838 (1996).
65. J. D. Joannopoulos, R. D. Meade, and J. N. Winn, *Photonic Crystals: Molding the Flow of Light*, Princeton, NJ: Princeton U. Press, 1995.
66. G. Qiu, F. Lin, and Y. Li, "Complete two-dimensional bandgap of photonic crystals of a rectangular Bravais lattice," *Opt. Commun.* 219, 285–288 (2003).
67. S. Fan, J. Winn, A. Devenyi, J. C. Chen, R. D. Meade, and J. D. Joannopoulos, "Guided and defect modes in periodic dielectric waveguides," *J. Opt. Soc. Am. B* 12, 1267–1272 (1995).

68. D. Chigrin, A. Lavrinenko, and C. Sotomayor Torres, "Nanopillars photonic crystal waveguides," *Opt. Express* 12, 617–622 (2004).
69. J. Shefer, "Periodic cylinder arrays as transmission lines," *IEEE Trans. Microwave Theory Tech.* 11, 55–61 (1963).
70. B. A. Munk, D. S. Janning, J. B. Pryor, and R. J. Marhefka, "Scattering from surface waves on finite FSS," *IEEE Trans. Antennas Propagat.* 49, 1782–1793 (2001).
71. R. J. Mailloux, "Antenna and wave theories of infinite Yagi-Uda arrays," *IEEE Trans. Antennas Propagat.* 13, 499–506 (1965).
72. A. D. Yaghjian, "Scattering-matrix analysis of linear periodic arrays," *IEEE Trans. Antennas Propagat.* 50, 1050–1064 (2002).
73. J. C. Weeber, A. Dereux, C. Girard, J. R. Krenn, and J. P. Goudonnet, "Plasmon polaritons of metallic nanowires for controlling submicron propagation of light," *Phys. Rev. B* 60, 9061–9068 (1999).
74. B. Lamprecht, J. R. Krenn, G. Schider, H. Ditlbacher, M. Salerno, N. Felidj, A. Leitner, F. R. Aussenegg, and J. C. Weeber, "Surface plasmon propagation in microscale metal stripes," *Appl. Phys. Lett.* 79, 51–53 (2001).
75. T. Yatsui, M. Kourogi, and M. Ohtsu, "Plasmon waveguide for optical far/near-field conversion," *Appl. Phys. Lett.* 79, 4583–4585 (2001).
76. R. Zia, M. D. Selker, P. B. Catrysse, and M. L. Brongersma, "Geometries and materials for subwavelength surface plasmon modes," *J. Opt. Soc. Am. A* 21, 2442–2446 (2004).
77. R. Charbonneau, N. Lahoud, G. Mattiussi, and P. Berini, "Demonstration of integrated optics elements based on long-ranging surface plasmon polaritons," *Opt. Express* 13, 977–984 (2005).
78. D. F. P. Pile, D. K. Gramotnev, "Channel plasmon-polariton in a triangular groove on a metal surface," *Opt. Lett.* 29, 1069–1071 (2004).
79. M. Quinten, A. Leitner, J. R. Krenn, and F. R. Aussenegg, "Electromagnetic energy transport via linear chains of silver nanoparticles," *Opt. Lett.* 23, 1331–1333 (1998).
80. M. L. Brongersma, J. W. Hartman, and H. A. Atwater, "Electromagnetic energy transfer and switching in nanoparticle chain arrays below the diffraction limit," *Phys. Rev. B* 62, 16356–16359 (2000).
81. T. J. Shepherd, C. R. Brewitt-Taylor, P. Dimond, G. Fixter, A. Laight, P. Lederer, P. J. Roberts, P. R. Tapster, and I. J. Youngs, "3D microwave photonic crystals: Novel fabrication and structures," *Electron. Lett.* 34, 787–789 (1998).
82. S. A. Maier, P. G. Kik, and H. A. Atwater, "Optical pulse propagation in metal nanoparticle chain waveguides," *Phys. Rev. B* 67, 205402 (2003).
83. W. M. Saj, "FDTD simulations of 2D plasmon waveguide on silver nanorods in hexagonal lattice," *Opt. Express* 13, 4818–4827 (2005).
84. K.-P. Hwang and A. C. Cangellaris, "Effective permittivities for second-order accurate FDTD equations at dielectric interfaces," *IEEE Microwave Wirel. Compon. Lett.* 11, 158–160 (2001).
85. Y. Hao and C. J. Railton, "Analyzing electromagnetic structures with curved boundaries on Cartesian FDTD meshes," *IEEE Trans. Microwave Theory Tech.* 46, 82–88 (1998).
86. N. Kaneda, B. Houshmand, and T. Itoh, "FDTD analysis of dielectric resonators with curved surfaces," *IEEE Trans. Microwave Theory Tech.* 45, 1645–1649 (1997).
87. J.-Y Lee and N.-H Myung, "Locally tensor conformal FDTD method for modeling arbitrary dielectric surfaces," *Microw. Opt. Tech. Lett.* 23, 245–249 (1999).

88. A. Mohammadi and M. Agio, "Contour-path effective permittivities for the two-dimensional finite-difference time-domain method," *Opt. Express* 13, 10367–10381 (2005).

89. J. E. Inglesfield, "A method of embedding," *J. Phys. C: Solid State Phys.* 14, 3795–3806 (1981).

90. C. T. Chan, Q. L. Yu, and K. M. Ho, "Order-N spectral method for electromagnetic waves," *Phys. Rev. B* 51, 16635–16642 (1995).

91. H. Holter and H. Steyskal, "Infinite phased-array analysis using FDTD periodic boundary conditions-pulse scanning in oblique directions," *IEEE Trans. Antennas Propagat.* 47, 1508–1514 (1999).

92. M. Turner and C. Christodoulou, "FDTD analysis of phased array antennas," *IEEE Trans. Antennas Propagat.* 47, 661–667 (1999).

93. D. T. Prescott and N. V. Shuley, "Extensions to the FDTD method for the analysis of infinitely periodic arrays," *IEEE Microwaves and Guided Waves Letters* 4, 352–354 (1994).

94. J. R. Ren, O. P. Gandhi, L. R. Walker, J. Fraschilla, and C. R. Boerman, "Floquet-based FDTD analysis of two-dimensional phased array antennas," *IEEE Microwave and Guided Wave Letters* 4, 109–111 (1994).

95. J. A. Roden, S. D. Gedney, M. P. Kesler, J. G. Maloney, and P. H. Harms, "Time-domain analysis of periodic structures at oblique incidence: Orthogonal and nonorthogonal FDTD implementations," *IEEE Trans. Microwave Theory and Techniques* 46, 420–427 (1998).

96. S. Fan, P. R. Villeneuve and J. D. Joannopoulos, "Large omnidirectional band gaps in metallodielectric photonic crystals," *Phys. Rev. B* 54, 11245–11251 (1996).

97. M. Qiu and S. He, "A nonorthogonal finite-difference time-domain method for computing the band structure of a two-dimensional photonic crystal with dielectric and metallic inclusions," *J. Appl. Phys.* 87, 8268–8275 (2000).

98. P. B. Johnson and R. W. Christy, "Optical constants of the noble metals," *Phys. Rev. B* 6, 4370–4379 (1972).

99. N. Giannakis, J. Inglesfield, P. Belov, Y. Zhao, and Y. Hao, "Dispersion properties of subwavelength waveguide formed by silver nanorods," *Photon06*, September 3–7, 2006, Manchester, UK.

# 4

# Modeling Optical Metamaterials with and without Gain Materials Using FDTD Method

**Zhixiang Huang, Bo Wu, and Xianliang Wu**

*The Key Laboratory of Intelligent Computing and Signal Processing, Ministry of Education, Anhui University, China*

## CONTENTS

## 4.1 Introduction

Light-matter interaction in micro- and nanostructured photonic media such as metamaterials [1,2], random media [3], and photonic crystals [4] gives rise to a rich variety of physical phenomena. These phenomena are the basis of a wide variety of technologically important devices, such as fiber optics, lasers, perfect lenses [5], invisibility cloaking [6, 7], and enhanced optical nonlinearities [8]. Furthermore, the field of nanostructure has experienced spectacular experimental progress in recent years. Most nanostructures are metal based and eventually suffer from the conductor losses at optical frequencies, which are still orders of magnitude too large for realistic applications. In addition, losses become an increasingly important issue when moving from multiple metal-based layers to the bulk case. Thus, the need for reducing or even compensating for the losses is a key challenge for nanotechnology. One promising way of overcoming the losses is based on introducing the gain material to the structure.

The idea of combination of a metamaterial with an optical gain material has been investigated by several experimental studies [9–11]. Thus a systematic numerical method for simulating of the metallic metamaterials coupled with the gain material, described by a generic four-level atomic system, is of vital importance. To numerically simulate the propagation and location of electromagnetic waves in gain media, we have to deal with time-dependent wave equations in these systems by coupling Maxwell's equations with the rate equations of electron populations described by a multi-level gain system in semiclassical theory [12]. Although simple analytic expressions can be derived from a well-defined system, precise solutions for a realistic system require a numerical solution, particularly in a coupled system. Thus the finite-difference time-domain (FDTD) [13] method has been widely utilized to discretize the coupled systems. The most basic FDTD implication involves incorporating a classical dispersive Lorentzian gain via the auxiliary differential equation (ADE) method. A semiclassical model in which the FDTD method is coupled with the rate equations in a four-level atomic system has been developed in Ref. [3]. Their methodology has been applied to numerous systems such as 2D random media, distributed Bragg reflectors, metamaterials nanostructures [14], and overcoming the loss of fishnet structure [15,16]. Considering the resonant problems like surface modes and finite size effects in metamaterials, this is a very important region of interest. To study these problems, large systems, long time simulation, and high resolutions are necessary, though these may require much computer memory and computational time.

However, the size of memory for each CPU is limited by the operating system. The parallel FDTD algorithm increases the computational efficiency by distributing the burden on a cluster. It also enables one to solve large problems that could be beyond the scope of a single processor because of CPU time limitation [17].

In this chapter, we describe the parallel FDTD method incorporated with a four-level atomic system. Our model incorporates simplified quantized electron energies that provide four energy levels for each of two interacting electrons. Transitions between each energy level are governed by coupled rate equations. The energy exchanges between atoms and fields, electronic pumping, and nonradiative decays are also included in our new model.

## 4.2 Theory Model

### 4.2.1 General Framework

We model the dispersive Lorentz active medium using a generic four-level atomic system, as shown in Figure 4.1. The population density in each level is given by $N_i$ ($i = 0, 1, 2, 3$). The time-dependent Maxwell's equations for isotropic media are given by $\nabla \times \mathbf{E}(\mathbf{r}, t) = -\partial \mathbf{B}(\mathbf{r}, t)/\partial t$ and $\nabla \times \mathbf{H}(\mathbf{r}, t) = \partial \mathbf{D}(\mathbf{r}, t)/\partial t$, where $\mathbf{B}(\mathbf{r}, t) = \mu\mu_o\mathbf{H}(\mathbf{r}, t)$, $\mathbf{D}(\mathbf{r}, t) = \varepsilon\varepsilon_0\mathbf{E}(\mathbf{r}, t) + \mathbf{P}(\mathbf{r}, t)$, and $\mathbf{P}(\mathbf{r}, t) = \Sigma_{i = a,b}\mathbf{P}_i(\mathbf{r}, t)$ is the electric polarization density of the gain material. ($\mathbf{P}_a$ is the induced electric polarization density on the atomic transition between the upper ($N_2$) and lower ($N_1$) lasing levels, and $\mathbf{P}_b$ is between the ground-state level ($N_0$) and the third level ($N_3$).) The induced electric polarizations behave as harmonic oscillators and couple to the local $\mathbf{E}$ field, which is propagated by the Maxwell's equations. The polarization density $\mathbf{P}_i(\mathbf{r}, t)$ locally obeys the following equation of motion:

$$\frac{\partial^2 \mathbf{P}_i(\mathbf{r},t)}{\partial t^2} + \Gamma_i\frac{\partial \mathbf{P}_i(\mathbf{r},t)}{\partial t} + \omega_i^2\mathbf{P}_i(\mathbf{r},t) = -\sigma_i\Delta N_i(\mathbf{r},t)\mathbf{E}(\mathbf{r},t), \quad (i = a,b) \qquad (4.1)$$

where $\Gamma_i$ stands for the linewidth of the atomic transitions at $\omega_i$ and accounts for both the nonradiative energy decay rate and the dephasing

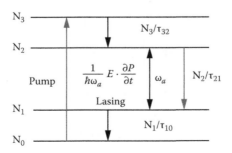

**FIGURE 4.1**
Sketch of the four-level atomic system model.

processes that arise from incoherently driven polarizations. $\sigma_i$ is the coupling strength of $\mathbf{P}_i$ to the external electric field. The factors $\Delta N_a(\mathbf{r}, t) = N_2(\mathbf{r}, t) - N_1(\mathbf{r}, t)$ and $\Delta N_b(\mathbf{r}, t) = N_3(\mathbf{r}, t) - N_0(\mathbf{r}, t)$ are the population inversions that drive the polarizations. Positive inversion is attained when $\Delta N_i(\mathbf{r}, t) > 0$, in which case the medium is amplifying; when $\Delta N_i(\mathbf{r}, t) < 0$, the medium is absorbing.

In the *optical pumping mechanism*, electrons are raised from the ground-state level ($N_0$) to the third level ($N_3$) by an external electromagnetic wave with a pumping frequency $w_b = \frac{E_3 - E_0}{\hbar}$.

Instead of using an external EM wave to optically pump electrons from the ground state level ($N_0$) to the third level ($N_3$), we can simplify this process by pumping electrons either with a *homogeneous pumping* or *Gaussian pumping* rate, which is proportional to the pumping intensity in the experiments. After a short lifetime $\tau_{32}$ electrons experience a nonradiative transfer into metastable second level $N_2$. The second level ($N_2$) and the first level ($N_1$) are called the upper and lower lasing levels. Electrons can be transferred from the upper to the lower lasing level by spontaneous and stimulated emission. At last, electrons undergo a quick and nonradiative transfer from the first level ($N_1$) to the ground-state level ($N_0$). The lifetimes and energies of the upper and lower lasing levels are $\tau_{21}$, $E_2$ and $\tau_{10}$, $E_1$, respectively. The center frequency of the radiation is $\omega_a = \frac{E_2 - E_1}{\hbar}$.

Thus, the atomic population densities obey the following rate equations:

$$\frac{dN_3}{dt} = \frac{1}{\hbar\omega_b}\mathbf{E}\cdot\frac{d\mathbf{P}_b}{dt} - \frac{N_3}{\tau_{32}} \quad \text{(optical pumping)} \tag{4.2a}$$

$$\frac{dN_3}{dt} = P_r N_0 - \frac{N_3}{\tau_{32}} \quad \text{(homogeneous pumping)} \tag{4.2b}$$

$$\frac{dN_3}{dt} = P_g(t)N_0 - \frac{N_3}{\tau_{32}} \quad \text{(Gaussian pumping)} \tag{4.2c}$$

$$\frac{dN_2}{dt} = \frac{1}{\hbar\omega_a}\mathbf{E}\cdot\frac{d\mathbf{P}_a}{dt} + \frac{N_3}{\tau_{32}} - \frac{N_2}{\tau_{21}} \tag{4.3}$$

$$\frac{dN_1}{dt} = -\frac{1}{\hbar\omega_a}\mathbf{E}\cdot\frac{d\mathbf{P}_a}{dt} + \frac{N_2}{\tau_{21}} - \frac{N_1}{\tau_{10}} \tag{4.4}$$

$$\frac{dN_0}{dt} = -\frac{1}{\hbar\omega_b}\mathbf{E}\cdot\frac{d\mathbf{P}_b}{dt} + \frac{N_1}{\tau_{10}} \quad \text{(optical pumping)} \tag{4.5a}$$

$$\frac{dN_0}{dt} = -P_r N_0 + \frac{N_1}{\tau_{10}} \quad \text{(homogeneous pumping)} \tag{4.5b}$$

$$\frac{dN_0}{dt} = -P_g(t)N_0 + \frac{N_1}{\tau_{10}} \quad \text{(Gaussian pumping)} \tag{4.5c}$$

## 4.2.2 Parallel FDTD Simulations of Active Media

To solve the response of the active materials in the electromagnetic fields numerically, the FDTD technique is utilized [13], using an approach similar to the one outlined in Refs. [14,15]. Maxwell's equations are approximated by the second-order center difference scheme so that both three-dimensional (3D) space and time are discretized, leading to spatial and temporal inter-leaving of the electromagnetic fields.

It is important to consider the resonant problems like surface modes and finite size effects in metamaterials, which are very important regions of interest. To study these problems, large systems, long time simulation, and high resolutions are necessary, which may require a lot of memory and computational time.

However, the size of memory for each CPU is limited by the operating system. This problem can be solved with parallel computation, which distributes the computing load into multiple CPUs and hence increases the available computation pace and reduces computational time. As a result, local resources can be effectively used and desired results can be obtained.

Due to its natural capability, parallel computation for FDTD simulations only needs the fields in the neighboring Yee grids. Figure 4.2(a) shows our parallel scheme for FDTD simulations. The FDTD computational domain is split into many blocks and each is assigned to one CPU. Each CPU runs parallel to the same update algorithms inside the block and communicates every time step with its neighboring CPUs to send and receive fields data in the block edges. Technically, for edge data communications, each computational block/domain is extended by a foreign edge buffer. The message passing between different nodes in our code is realized by using the message passing interface (MPI) library as the communication layer. Let us start from

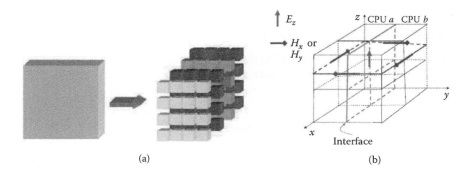

(a)        (b)

**FIGURE 4.2**
Schematic of the parallel computation scheme. (a) Basic idea of parallel FDTD processing on a computer cluster, in which the original job is split into small pieces that are assigned to each CPU. (b) Distribution of the electric and magnetic fields near the interface between two CPUS. (See color figure.)

the $z$-component of Maxwell's equations as an example. In parallel simulations, only the **E** and **H** fields need borrow some information from the adjacent CPUs.

Considering the structure as shown in Figure 4.2(b), the electric field $E_z$ is located on the interface between the CPUs $a$ and $b$. Update of this electric field needs two magnetic fields $H_{x,a}^{n+\frac{1}{2}}(i, j-\frac{1}{2}, k+\frac{1}{2})$, and $H_{x,b}^{n+\frac{1}{2}}(i, j+\frac{1}{2}, k+\frac{1}{2})$, which are located in the CPUs $a$ and $b$, respectively.

We now rewrite the iterative equations of $E_z$ in the CPUs $a$ and $b$ as equations (4.6) and (4.7). The magnetic fields $H_{x,a}^{n+\frac{1}{2}}(i, j-\frac{1}{2}, k+\frac{1}{2})$ in the CPU $a$ and $H_{x,b}^{n+\frac{1}{2}}(i, j+\frac{1}{2}, k+\frac{1}{2})$ in the CPU $b$ are exchanged through the high-performance network at each time step.

$$
\begin{aligned}
& E_{z,a}^{n+1}(i, j, k+\tfrac{1}{2}) \\
& = E_z^{n+1}(i, j, k+\tfrac{1}{2}) + \frac{\Delta t}{\varepsilon_0 \varepsilon_z(i,j,k+\frac{1}{2})} \\
& \times \left( \frac{H_y^{n+\frac{1}{2}}(i+\frac{1}{2},j,k+\frac{1}{2}) - H_y^{n+\frac{1}{2}}(i-\frac{1}{2},j,k+\frac{1}{2})}{\Delta x} - \frac{H_{x,b}^{n+\frac{1}{2}}(i,j+\frac{1}{2},k+\frac{1}{2}) - H_x^{n+\frac{1}{2}}(i,j-\frac{1}{2},k+\frac{1}{2})}{\Delta y} - \frac{P_z^{n+1}(i,j,k+\frac{1}{2}) - P_z^n(i,j,k+\frac{1}{2})}{\Delta t} \right)
\end{aligned}
\tag{4.6}
$$

$$
\begin{aligned}
& E_{z,b}^{n+1}(i, j, k+\tfrac{1}{2}) \\
& = E_z^{n+1}(i, j, k+\tfrac{1}{2}) + \frac{\Delta t}{\varepsilon_0 \varepsilon_z(i,j,k+\frac{1}{2})} \\
& \times \left( \frac{H_y^{n+\frac{1}{2}}(i+\frac{1}{2},j,k+\frac{1}{2}) - H_y^{n+\frac{1}{2}}(i-\frac{1}{2},j,k+\frac{1}{2})}{\Delta x} - \frac{H_x^{n+\frac{1}{2}}(i,j+\frac{1}{2},k+\frac{1}{2}) - H_{x,a}^{n+\frac{1}{2}}(i,j-\frac{1}{2},k+\frac{1}{2})}{\Delta y} - \frac{P_z^{n+1}(i,j,k+\frac{1}{2}) - P_z^n(i,j,k+\frac{1}{2})}{\Delta t} \right)
\end{aligned}
\tag{4.7}
$$

### 4.2.3 The FDTD Update Algorithm

In this section, we will present the detailed 3D FDTD algorithm for vacuum, dispersive Drude metal, gain material, and perfectly matched layer.

#### 4.2.3.1 3D FDTD Algorithm for Vacuum

It's quite easy by just following the Maxwell's equations and the 3D Yell unit cell as shown in Figure 4.3, where the magnetic field dual grid diagonally shifts a vector ($\Delta x/2$, $\Delta y/2$, $\Delta z/2$) from the electric field grid. Here, $\Delta x$, $\Delta y$, and $\Delta z$ are the lattice space increments in the $x$-, $y$-, and $z$-coordinate directions, respectively. Based on this spatial discretization, we show the detailed 3D FDTD update algorithm for a system composed of free space, Drude metals,

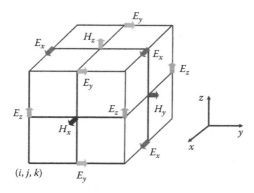

**FIGURE 4.3**
Yee's grid for FDTD algorithm.

and gain materials. To simulate the extension to infinity, the two ends perpendicular to the *y*-direction are terminated by PML absorbing boundary conditions. On the boundaries perpendicular to the *x*- and *z*-directions, periodic boundary conditions are enforced. To apply the total-field and scattered-field (TF/SF) technique to generate a normal incidence plane wave, we choose two interfaces perpendicular to the *y*-direction $j = j_L$ and $j = j_R$ to connect the total and scattered fields, and they are aligned with the electric field components $E_x$ and $E_z$. We denote a space point in a uniform, rectangular lattice as $(i, j, k) = (i\Delta x, j\Delta y, k\Delta z)$. Further, we denote any function $F$ of space and time at a discrete point in the grid and time as $F^n(i, j, k) = F(i\Delta x, j\Delta y, k\Delta z; n\Delta t)$.

Assuming that there are no electric and magnetic current sources, and also that the materials are lossless, we can have the following Maxwell's equations:

$$\frac{\partial H_x}{\partial t} = \frac{1}{\mu}\left(\frac{\partial E_y}{\partial z} - \frac{\partial E_z}{\partial y}\right) \tag{4.8a}$$

$$\frac{\partial H_y}{\partial t} = \frac{1}{\mu}\left(\frac{\partial E_z}{\partial x} - \frac{\partial E_x}{\partial z}\right) \tag{4.8b}$$

$$\frac{\partial H_z}{\partial t} = \frac{1}{\mu}\left(\frac{\partial E_x}{\partial y} - \frac{\partial E_y}{\partial x}\right) \tag{4.8c}$$

$$\frac{\partial E_x}{\partial t} = \frac{1}{\varepsilon}\left(\frac{\partial H_z}{\partial y} - \frac{\partial H_y}{\partial z}\right) \tag{4.8d}$$

$$\frac{\partial E_y}{\partial t} = \frac{1}{\varepsilon}\left(\frac{\partial H_x}{\partial z} - \frac{\partial H_z}{\partial x}\right) \tag{4.8e}$$

$$\frac{\partial E_z}{\partial t} = \frac{1}{\varepsilon}\left(\frac{\partial H_y}{\partial x} - \frac{\partial H_x}{\partial y}\right) \tag{4.8f}$$

Referring to Figure 4.3, consider the typical substitutions of central differences for the time and space derivatives; then we have the 3D FDTD update algorithm for vacuum:

$$H_x^{n+1/2}\left(i-\frac{1}{2},j,k\right) = H_x^{n-1/2}\left(i-\frac{1}{2},j,k\right)$$

$$+ \frac{\Delta t}{\mu\left(i-\frac{1}{2},j,k\right)}\left(\frac{E_y^{n+1}\left(i-\frac{1}{2},j,k+\frac{1}{2}\right) - E_y^{n+1}\left(i-\frac{1}{2},j,k-\frac{1}{2}\right)}{\Delta z}\right.$$

$$\left. - \frac{E_z^n\left(i-\frac{1}{2},j+\frac{1}{2},k\right) - E_z^n\left(i-\frac{1}{2},j-\frac{1}{2},k\right)}{\Delta y}\right) \tag{4.9}$$

$$H_y^{n+1/2}\left(i,j+\frac{1}{2},k\right) = H_y^{n-1/2}\left(i,j+\frac{1}{2},k\right)$$

$$+ \frac{\Delta t}{\mu\left(i,j+\frac{1}{2},k\right)}\left(\frac{E_z^n\left(i+\frac{1}{2},j+\frac{1}{2},k\right) - E_y^{n+1}\left(i-\frac{1}{2},j+\frac{1}{2},k\right)}{\Delta x}\right.$$

$$\left. - \frac{E_x^n\left(i,j+\frac{1}{2},k+\frac{1}{2}\right) - E_z^n\left(i,j+\frac{1}{2},k-\frac{1}{2}\right)}{\Delta z}\right) \tag{4.10}$$

$$H_z^{n+1/2}\left(i,j,k+\frac{1}{2}\right)=H_z^{n-1/2}\left(i,j,k+\frac{1}{2}\right)$$

$$+\frac{\Delta t}{\mu\left(i,j,k+\frac{1}{2}\right)}\left(\frac{E_x^n\left(i,j+\frac{1}{2},k+\frac{1}{2}\right)-E_x^n\left(i,j-\frac{1}{2},k+\frac{1}{2}\right)}{\Delta y}\right.$$

$$\left.-\frac{E_y^n\left(i+\frac{1}{2},j,k+\frac{1}{2}\right)-E_y^n\left(i-\frac{1}{2},j,k+\frac{1}{2}\right)}{\Delta x}\right)$$

$$(4.11)$$

$$E_x^{n+1}\left(i,j+\frac{1}{2},k+\frac{1}{2}\right)=E_x^n\left(i,j+\frac{1}{2},k+\frac{1}{2}\right)$$

$$+\frac{\Delta t}{\varepsilon\left(i,j+\frac{1}{2},k+\frac{1}{2}\right)}\left(\frac{H_z^{n+1/2}\left(i,j+1,k+\frac{1}{2}\right)-H_z^{n+1/2}\left(i,j,k+\frac{1}{2}\right)}{\Delta y}\right.$$

$$\left.-\frac{H_y^{n+1/2}\left(i,j+\frac{1}{2},k+1\right)-H_y^{n+1/2}\left(i,j+\frac{1}{2},k\right)}{\Delta z}\right)$$

$$(4.12)$$

$$E_y^{n+1}\left(i-\frac{1}{2},j,k+\frac{1}{2}\right)=E_y^n\left(i-\frac{1}{2},j,k+\frac{1}{2}\right)$$

$$+\frac{\Delta t}{\varepsilon\left(i-\frac{1}{2},j,k+\frac{1}{2}\right)}\left(\frac{H_x^{n+\frac{1}{2}}\left(i-\frac{1}{2},j,k+1\right)-H_x^{n+\frac{1}{2}}\left(i-\frac{1}{2},j,k\right)}{\Delta z}\right.$$

$$\left.-\frac{H_z^{n+\frac{1}{2}}\left(i,j,k+\frac{1}{2}\right)-H_z^{n+\frac{1}{2}}\left(i-1,j,k+\frac{1}{2}\right)}{\Delta x}\right)$$

$$(4.13)$$

$$E_z^{n+1}\left(i-\frac{1}{2},j+\frac{1}{2},k\right) = E_z^n\left(i-\frac{1}{2},j+\frac{1}{2},k\right)$$

$$+\frac{\Delta t}{\varepsilon\left(i-\frac{1}{2},j+\frac{1}{2},k\right)}\left(\frac{H_y^{n+\frac{1}{2}}\left(i,j+\frac{1}{2},k\right)-H_y^{n+\frac{1}{2}}\left(i-1,j+\frac{1}{2},k\right)}{\Delta x}\right.$$

$$\left.-\frac{H_x^{n+\frac{1}{2}}\left(i-\frac{1}{2},j+1,k\right)-H_x^{n+\frac{1}{2}}\left(i-\frac{1}{2},j,k\right)}{\Delta y}\right)$$

$$(4.14)$$

### 4.2.3.2 3D FDTD Algorithm for Drude Metal

The Drude media are a very important class of material dispersions for FDTD modeling, which are typically characterized by the following dispersive permittivity:

$$\varepsilon(\omega) = \varepsilon_0\left(1-\frac{\omega_{pe}^2}{\omega(\omega-j\Gamma_e)}\right) \tag{4.15}$$

For virtual Drude material, we also have the permeability

$$\mu(\omega) = \mu_0\left(1-\frac{\omega_{pm}^2}{\omega(\omega-j\Gamma_m)}\right)$$

Here we will consider only the Drude metal, i.e., the dispersion of the permittivity $\varepsilon$. Like we did before, we can get the auxiliary differential equation (ADE) for the electric polarization current $\vec{J}_p$ from the dispersive permittivity:

$$\frac{d\vec{J}_p(t)}{dt}+\Gamma_e\vec{J}_p(t) = \varepsilon_0\omega_{pe}^2\vec{E}(t) \tag{4.16}$$

Taking the electric polarization current into account, we have Maxwell's equations as follows:

$$\frac{\partial H_x}{\partial t} = \frac{1}{\mu_0}\left(\frac{\partial E_y}{\partial z} - \frac{\partial E_x}{\partial y}\right) \tag{4.17a}$$

$$\frac{\partial H_y}{\partial t} = \frac{1}{\mu_0}\left(\frac{\partial E_z}{\partial x} - \frac{\partial E_x}{\partial z}\right) \tag{4.17b}$$

$$\frac{\partial Hz}{\partial t} = \frac{1}{\mu_0}\left(\frac{\partial E_x}{\partial y} - \frac{\partial E_y}{\partial x}\right) \tag{4.17c}$$

$$\frac{\partial E_x}{\partial t} = \frac{1}{\varepsilon_0}\left(\frac{\partial H_z}{\partial y} - \frac{\partial H_y}{\partial z} - J_{px}\right) \tag{4.17d}$$

$$\frac{\partial E_y}{\partial t} = \frac{1}{\varepsilon_0}\left(\frac{\partial H_x}{\partial z} - \frac{\partial H_z}{\partial x} - J_{py}\right) \tag{4.17e}$$

$$\frac{\partial E_z}{\partial t} = \frac{1}{\varepsilon_0}\left(\frac{\partial H_y}{\partial x} - \frac{\partial H_x}{\partial y} - J_{pz}\right) \tag{4.17f}$$

Combining the Maxwell's equations with the ADE, we can have the 3D FDTD update algorithm for Drude metal as follows:

$$H_x^{n+1/2}\left(i-\frac{1}{2},j,k\right) = H_x^{n-1/2}\left(i-\frac{1}{2},j,k\right)$$

$$+\frac{\Delta t}{\mu_0}\left(\frac{E_y^n\left(i-\frac{1}{2},j,k+\frac{1}{2}\right) - E_y^n\left(i-\frac{1}{2},j,k-\frac{1}{2}\right)}{\Delta z}\right.$$

$$\left.-\frac{E_z^n\left(i-\frac{1}{2},j+\frac{1}{2},k\right) - E_z^n\left(i-\frac{1}{2},j-\frac{1}{2},k\right)}{\Delta y}\right)$$

$$\tag{4.18}$$

$$H_y^{n+1/2}\left(i, j+\frac{1}{2}, k\right) = H_y^{n-1/2}\left(i, j+\frac{1}{2}, k\right)$$

$$+\frac{\Delta t}{\mu_0}\left(\frac{E_z^n\left(i+\frac{1}{2}, j+\frac{1}{2}, k\right) - E_z^n\left(i-\frac{1}{2}, j+\frac{1}{2}, k\right)}{\Delta z}\right.$$

$$\left.-\frac{E_x^n\left(i, j+\frac{1}{2}, k+\frac{1}{2}\right) - E_x^n\left(i, j+\frac{1}{2}, k-\frac{1}{2}\right)}{\Delta z}\right)$$

$$(4.19)$$

$$H_z^{n+1/2}\left(i, j, k+\frac{1}{2}\right) = H_z^{n-1/2}\left(i, j, k+\frac{1}{2}\right)$$

$$+\frac{\Delta t}{\mu_0}\left(\frac{E_x^n\left(i, j+\frac{1}{2}, k+\frac{1}{2}\right) - E_x^n\left(i, j-\frac{1}{2}, k+\frac{1}{2}\right)}{\Delta y}\right.$$

$$\left.-\frac{E_y^n\left(i+\frac{1}{2}, j, k+\frac{1}{2}\right) - E_y^n\left(i-\frac{1}{2}, j, k+\frac{1}{2}\right)}{\Delta x}\right)$$

$$(4.20)$$

$$J_{px}^{n+1/2}\left(i, j+\frac{1}{2}, k+\frac{1}{2}\right) = \frac{1-0.5\Gamma_{ex}\Delta t}{1+0.5\Gamma_{ex}\Delta t} J_{px}^{n-1/2}\left(i, j+\frac{1}{2}, k+\frac{1}{2}\right)$$

$$+\frac{\varepsilon_0\omega_{pex}^2\Delta t}{1+0.5\Gamma_{ex}\Delta t} E_x^n\left(i, j+\frac{1}{2}, k+\frac{1}{2}\right) \qquad (4.21)$$

$$J_{py}^{n+1/2}\left(i-\frac{1}{2},j,k+\frac{1}{2}\right) = \frac{1-0.5\Gamma_{ey}\Delta t}{1+0.5\Gamma_{ey}\Delta t} J_{py}^{n-1/2}\left(i-\frac{1}{2},j,k+\frac{1}{2}\right)$$

$$+\frac{\varepsilon_0\omega_{pey}^2\Delta t}{1+0.5\Gamma_{ey}\Delta t} E_y^n\left(i-\frac{1}{2},j,k+\frac{1}{2}\right) \tag{4.22}$$

$$J_{pz}^{n+1/2}\left(i-\frac{1}{2},j+\frac{1}{2},k\right) = \frac{1-0.5\Gamma_{ez}\Delta t}{1+0.5\Gamma_{ez}\Delta t} J_{pz}^{n-1/2}\left(i-\frac{1}{2},j+\frac{1}{2},k\right)$$

$$+\frac{\varepsilon_0\omega_{pez}^2\Delta t}{1+0.5\Gamma_{ez}\Delta t} E_z^n\left(i-\frac{1}{2},j+\frac{1}{2},k\right) \tag{4.23}$$

$$E_x^{n+1}\left(i,j+\frac{1}{2},k+\frac{1}{2}\right) = E_x^n\left(i,j+\frac{1}{2},k+\frac{1}{2}\right)$$

$$+\frac{\Delta t}{\varepsilon_0}\left(\frac{H_z^{n+\frac{1}{2}}\left(i,j+1,k+\frac{1}{2}\right)-H_z^{n+\frac{1}{2}}\left(i,j,k+\frac{1}{2}\right)}{\Delta y}\right.$$

$$\left.-\frac{H_y^{n+\frac{1}{2}}\left(i,j+\frac{1}{2},k+1\right)-H_y^{n+\frac{1}{2}}\left(i,j+\frac{1}{2},k\right)}{\Delta z}\right)$$

$$-J_{px}^{n+1/2}\left(i,j+\frac{1}{2},k+\frac{1}{2}\right) \tag{4.24}$$

$$E_y^{n+1}\left(i-\frac{1}{2},j,k+\frac{1}{2}\right) = E_y^n\left(i-\frac{1}{2},j,k+\frac{1}{2}\right)$$

$$+\frac{\Delta t}{\varepsilon_0}\left(\frac{H_x^{n+\frac{1}{2}}\left(i-\frac{1}{2},j,k+1\right)-H_x^{n+\frac{1}{2}}\left(i-\frac{1}{2},j,k\right)}{\Delta z}\right.$$

$$\left.-\frac{H_z^{n+\frac{1}{2}}\left(i,j,k+\frac{1}{2}\right)-H_z^{n+\frac{1}{2}}\left(i-1,j,k+\frac{1}{2}\right)}{\Delta x}\right)$$

$$-J_{py}^{n+1/2}\left(i-\frac{1}{2},j,k+\frac{1}{2}\right) \tag{4.25}$$

$$E_z^{n+1}\left(i-\frac{1}{2}, j+\frac{1}{2}, k\right) = E_z^n\left(i-\frac{1}{2}, j+\frac{1}{2}, k\right)$$

$$+\frac{\Delta t}{\varepsilon_0}\left( \frac{H_y^{n+\frac{1}{2}}\left(i, j+\frac{1}{2}, k\right) - H_y^{n+\frac{1}{2}}\left(i-1, j+\frac{1}{2}, k\right)}{\Delta x} \right.$$

$$\left. -\frac{H_x^{n+\frac{1}{2}}\left(i-\frac{1}{2}, j+1, k\right) - H_x^{n+\frac{1}{2}}\left(i-\frac{1}{2}, j, k\right)}{\Delta y} \right)$$

$$-J_{pz}^{n+1/2}\left(i-\frac{1}{2}, j+\frac{1}{2}, k\right) \tag{4.26}$$

### 4.2.3.3 3D FDTD Algorithm for Gain Material (Four-Level System)

In a four-level atomic system, the polarization density $\mathbf{P}(\mathbf{r}; t)$ obeys the following equation of motion:

$$\frac{\partial^2 \mathbf{P}_i(\mathbf{r},t)}{\partial t^2} + \Gamma_i \frac{\partial \mathbf{P}_i(\mathbf{r},t)}{\partial t} + \omega_i^2 \mathbf{P}_i(\mathbf{r},t) = -\sigma_i \Delta N_i(\mathbf{r},t)\mathbf{E}(\mathbf{r},t), \quad (i = a, b) \tag{4.27}$$

And we have rate equations for the occupation numbers:

$$\frac{dN_3}{dt} = \frac{1}{\hbar\omega_b}\mathbf{E}\cdot\frac{d\mathbf{P}_b}{dt} - \frac{N_3}{\tau_{32}} \quad \textit{(optical pumping)} \tag{4.28a}$$

$$\frac{dN_3}{dt} = P_r N_0 - \frac{N_3}{\tau_{32}} \quad \textit{(homogeneous pumping)} \tag{4.28b}$$

$$\frac{dN_3}{dt} = P_g(t)N_0 - \frac{N_3}{\tau_{32}} \quad \textit{(Gaussian pumping)} \tag{4.28c}$$

$$\frac{dN_2}{dt} = \frac{1}{\hbar\omega_a}\mathbf{E}\cdot\frac{d\mathbf{P}_a}{dt} + \frac{N_3}{\tau_{32}} - \frac{N_2}{\tau_{21}} \tag{4.29}$$

$$\frac{dN_1}{dt} = -\frac{1}{\hbar\omega_a}\mathbf{E}\cdot\frac{d\mathbf{P}_a}{dt} + \frac{N_2}{\tau_{21}} - \frac{N_1}{\tau_{10}} \tag{4.30}$$

$$\frac{dN_0}{dt} = -\frac{1}{\hbar\omega_b}\mathbf{E}\cdot\frac{d\mathbf{P}_b}{dt} + \frac{N_1}{\tau_{10}} \quad (optical\ pumping) \tag{4.31a}$$

$$\frac{dN_0}{dt} = -P_r N_0 + \frac{N_1}{\tau_{10}} \quad (homogeneous\ pumping) \tag{4.31b}$$

$$\frac{dN_0}{dt} = -P_g(t)N_0 + \frac{N_1}{\tau_{10}} \quad (Gaussian\ pumping) \tag{4.31c}$$

Combining the polarization equations, rate equations, and Maxwell's equations, we have the 3D FDTD update algorithm for the gain materials:

$$H_x^{n+1/2}\left(i-\frac{1}{2},j,k\right) = H_x^{n-1/2}\left(i-\frac{1}{2},j,k\right)$$

$$+\frac{\Delta t}{\mu_0}\left(\frac{E_y^n\left(i-\frac{1}{2},j,k+\frac{1}{2}\right)-E_y^n\left(i-\frac{1}{2},j,k-\frac{1}{2}\right)}{\Delta z}\right.$$

$$\left.-\frac{E_z^n\left(i-\frac{1}{2},j+\frac{1}{2},k\right)-E_z^n\left(i-\frac{1}{2},j-\frac{1}{2},k\right)}{\Delta y}\right) \tag{4.32}$$

$$H_y^{n+1/2}\left(i,j+\frac{1}{2},k\right) = H_y^{n-1/2}\left(i,j+\frac{1}{2},k\right)$$

$$+\frac{\Delta t}{\mu_0}\left(\frac{E_z^n\left(i+\frac{1}{2},j+\frac{1}{2},k\right)-E_z^n\left(i-\frac{1}{2},j+\frac{1}{2},k\right)}{\Delta z}\right.$$

$$\left.-\frac{E_x^n\left(i,j+\frac{1}{2},k+\frac{1}{2}\right)-E_x^n\left(i,j+\frac{1}{2},k-\frac{1}{2}\right)}{\Delta z}\right) \tag{4.33}$$

$$H_z^{n+1/2}\left(i,j,k+\frac{1}{2}\right)=H_z^{n-1/2}\left(i,j,k+\frac{1}{2}\right)$$

$$+\frac{\Delta t}{\mu_0}\left(\frac{E_x^n\left(i,j+\frac{1}{2},k+\frac{1}{2}\right)-E_x^n\left(i,j-\frac{1}{2},k+\frac{1}{2}\right)}{\Delta y}\right.$$

$$\left.-\frac{E_y^n\left(i+\frac{1}{2},j,k+\frac{1}{2}\right)-E_y^n\left(i-\frac{1}{2},j,k+\frac{1}{2}\right)}{\Delta x}\right) \qquad (4.34)$$

For the **optical pumping** version:

$$P_{1x}^{n+1}\left(i,j+\frac{1}{2},k+\frac{1}{2}\right)=\frac{2-\omega_a^2\Delta t^2}{1+\Gamma_a\,\Delta t/2}\,P_{1x}^n\left(i,j+\frac{1}{2},k+\frac{1}{2}\right)$$

$$-\frac{1-\Gamma_a\,\Delta t/2}{1+\Gamma_a\,\Delta t/2}\,P_{1x}^{n-1}\left(i,j+\frac{1}{2},k+\frac{1}{2}\right)$$

$$+\frac{\sigma_a\Delta t^2}{1+\Gamma_a\,\Delta t/2}\left[N_{1x}^n\left(i,j+\frac{1}{2},k+\frac{1}{2}\right)-N_{2x}^n\left(i,j+\frac{1}{2},k+\frac{1}{2}\right)\right]$$

$$\cdot E_x^n\left(i,j+\frac{1}{2},k+\frac{1}{2}\right) \qquad (4.35)$$

$$P_{2x}^{n+1}\left(i,j+\frac{1}{2},k+\frac{1}{2}\right)=\frac{2-\omega_b^2\Delta t^2}{1+\Gamma_b\,\Delta t/2}\,P_{2x}^n\left(i,j+\frac{1}{2},k+\frac{1}{2}\right)$$

$$-\frac{1-\Gamma_b\,\Delta t/2}{1+\Gamma_b\,\Delta t/2}\,P_{2x}^{n-1}\left(i,j+\frac{1}{2},k+\frac{1}{2}\right)$$

$$+\frac{\sigma_b\Delta t^2}{1+\Gamma_b\,\Delta t/2}\left[N_{0x}^n\left(i,j+\frac{1}{2},k+\frac{1}{2}\right)-N_{3x}^n\left(i,j+\frac{1}{2},k+\frac{1}{2}\right)\right]$$

$$\cdot E_x^n\left(i,j+\frac{1}{2},k+\frac{1}{2}\right) \qquad (4.36)$$

$$J_{px}^{G\,n+1}\left(i, j+\frac{1}{2}, k+\frac{1}{2}\right) = \frac{1}{\Delta t}\left[P_{1x}^{n+1}\left(i, j+\frac{1}{2}, k+\frac{1}{2}\right) + P_{2x}^{n+1}\left(i, j+\frac{1}{2}, k+\frac{1}{2}\right)\right.$$

$$\left.-P_{1x}^{n}\left(i, j+\frac{1}{2}, k+\frac{1}{2}\right) - P_{2x}^{n}\left(i, j+\frac{1}{2}, k+\frac{1}{2}\right)\right] \quad (4.37)$$

$$P_{1y}^{n+1}\left(i-\frac{1}{2}, j, k+\frac{1}{2}\right) = \frac{2-\omega_a^2\Delta t^2}{1+\Gamma_a\,\Delta t/2}P_{1y}^{n}\left(i-\frac{1}{2}, j, k+\frac{1}{2}\right)$$

$$-\frac{1-\Gamma_a\,\Delta t/2}{1+\Gamma_a\,\Delta t/2}P_{1y}^{n-1}\left(i-\frac{1}{2}, j, k+\frac{1}{2}\right)$$

$$+\frac{\sigma_a\Delta t^2}{1+\Gamma_a\,\Delta t/2}\left[N_{1y}^{n}\left(i-\frac{1}{2}, j, k+\frac{1}{2}\right) - N_{2y}^{n}\left(i-\frac{1}{2}, j, k+\frac{1}{2}\right)\right]$$

$$\cdot E_y^{n}\left(i-\frac{1}{2}, j, k+\frac{1}{2}\right) \quad (4.38)$$

$$P_{2y}^{n+1}\left(i-\frac{1}{2}, j, k+\frac{1}{2}\right) = \frac{2-\omega_b^2\Delta t^2}{1+\Gamma_b\,\Delta t/2}P_{2y}^{n}\left(i-\frac{1}{2}, j, k+\frac{1}{2}\right)$$

$$-\frac{1-\Gamma_b\,\Delta t/2}{1+\Gamma_b\,\Delta t/2}P_{2y}^{n-1}\left(i-\frac{1}{2}, j, k+\frac{1}{2}\right)$$

$$+\frac{\sigma_b\Delta t^2}{1+\Gamma_b\,\Delta t/2}\left[N_{0y}^{n}\left(i-\frac{1}{2}, j, k+\frac{1}{2}\right) - N_{3y}^{n}\left(i-\frac{1}{2}, j, k+\frac{1}{2}\right)\right]$$

$$\cdot E_y^{n}\left(i-\frac{1}{2}, j, k+\frac{1}{2}\right) \quad (4.39)$$

$$J_{py}^{G\,n+1}\left(i-\frac{1}{2}, j, k+\frac{1}{2}\right) = \frac{1}{\Delta t}\left[P_{1y}^{n+1}\left(i-\frac{1}{2}, j, k+\frac{1}{2}\right) + P_{2y}^{n+1}\left(i-\frac{1}{2}, j, k+\frac{1}{2}\right)\right.$$

$$\left.-P_{1y}^{n}\left(i-\frac{1}{2}, j, k+\frac{1}{2}\right) - P_{2y}^{n}\left(i-\frac{1}{2}, j, k+\frac{1}{2}\right)\right] \quad (4.40)$$

$$P_{1z}^{n+1}\left(i-\frac{1}{2},j+\frac{1}{2},k\right)=\frac{2-\omega_a^2\Delta t^2}{1+\Gamma_a\Delta t/2}P_{1z}^n\left(i-\frac{1}{2},j+\frac{1}{2},k\right)$$

$$-\frac{1-\Gamma_a\Delta t/2}{1+\Gamma_a\Delta t/2}P_{1z}^{n-1}\left(i-\frac{1}{2},j+\frac{1}{2},k\right)$$

$$+\frac{\sigma_a\Delta t^2}{1+\Gamma_a\Delta t/2}\left[N_{1z}^n\left(i-\frac{1}{2},j+\frac{1}{2},k\right)-N_{2z}^n\left(i-\frac{1}{2},j+\frac{1}{2},k\right)\right]$$

$$\cdot E_z^n\left(i-\frac{1}{2},j+\frac{1}{2},k\right) \tag{4.41}$$

$$P_{2z}^{n+1}\left(i-\frac{1}{2},j+\frac{1}{2},k\right)=\frac{2-\omega_b^2\Delta t^2}{1+\Gamma_b\Delta t/2}P_{2z}^n\left(i-\frac{1}{2},j+\frac{1}{2},k\right)$$

$$-\frac{1-\Gamma_b\Delta t/2}{1+\Gamma_b\Delta t/2}P_{2z}^{n-1}\left(i-\frac{1}{2},j+\frac{1}{2},k\right)$$

$$+\frac{\sigma_b\Delta t^2}{1+\Gamma_b\Delta t/2}\left[N_{0z}^n\left(i-\frac{1}{2},j+\frac{1}{2},k\right)-N_{3z}^n\left(i-\frac{1}{2},j+\frac{1}{2},k\right)\right]$$

$$\cdot E_z^n\left(i-\frac{1}{2},j+\frac{1}{2},k\right) \tag{4.42}$$

$$J_{pz}^{G\,n+\frac{1}{2}}\left(i-\frac{1}{2},j+\frac{1}{2},k\right)=\frac{1}{\Delta t}\left[P_{1z}^n\left(i-\frac{1}{2},j+\frac{1}{2},k\right)-P_{2z}^n\left(i-\frac{1}{2},j+\frac{1}{2},k\right)\right] \tag{4.43}$$

For the **homogeneous** and **Gaussian pumping** rate versions:

$$P_{1x}^{n+1}\left(i,j+\frac{1}{2},k+\frac{1}{2}\right)=\frac{2-\omega_a^2\Delta t^2}{1+\Gamma_a\Delta t/2}P_{1x}^n\left(i,j+\frac{1}{2},k+\frac{1}{2}\right)$$

$$-\frac{1-\Gamma_a\Delta t/2}{1+\Gamma_a\Delta t/2}P_{1x}^{n-1}\left(i,j+\frac{1}{2},k+\frac{1}{2}\right)$$

$$+\frac{\sigma_a\Delta t^2}{1+\Gamma_a\Delta t/2}\left[N_{1x}^n\left(i,j+\frac{1}{2},k+\frac{1}{2}\right)-N_{2x}^n\left(i,j+\frac{1}{2},k+\frac{1}{2}\right)\right]$$

$$\cdot E_x^n\left(i,j+\frac{1}{2},k+\frac{1}{2}\right)$$

$$\tag{4.44}$$

$$J_{px}^{G^{n+\frac{1}{2}}}\left(i,j+\frac{1}{2},k+\frac{1}{2}\right)=\frac{1}{\Delta t}\left[P_{1x}^{n+1}\left(i,j+\frac{1}{2},k+\frac{1}{2}\right)-P_{1x}^{n}\left(i,j+\frac{1}{2},k+\frac{1}{2}\right)\right] \quad (4.45)$$

$$P_{1y}^{n+1}\left(i-\frac{1}{2},j,k+\frac{1}{2}\right)=\frac{2-\omega_a^2\Delta t^2}{1+\Gamma_a\Delta t/2}P_{1y}^{n}\left(i-\frac{1}{2},j,k+\frac{1}{2}\right)$$

$$-\frac{1-\Gamma_a\Delta t/2}{1+\Gamma_a\Delta t/2}P_{1y}^{n-1}\left(i-\frac{1}{2},j,k+\frac{1}{2}\right)$$

$$+\frac{\sigma_a\Delta t^2}{1+\Gamma_a\Delta t/2}\left[N_{1y}^{n}\left(i-\frac{1}{2},j,k+\frac{1}{2}\right)-N_{2y}^{n}\left(i-\frac{1}{2},j,k+\frac{1}{2}\right)\right]$$

$$\cdot E_y^{n}\left(i-\frac{1}{2},j,k+\frac{1}{2}\right) \quad (4.46)$$

$$J_{py}^{G^{n+\frac{1}{2}}}\left(i-\frac{1}{2},j,k+\frac{1}{2}\right)=\frac{1}{\Delta t}\left[P_{1y}^{n+1}\left(i-\frac{1}{2},j,k+\frac{1}{2}\right)-P_{1y}^{n}\left(i-\frac{1}{2},j,k+\frac{1}{2}\right)\right] \quad (4.47)$$

$$P_{1z}^{n+1}(i-1,j+1,k)=\frac{2-\omega_a^2\Delta t^2}{1+\Gamma_a\Delta t/2}P_{1z}^{n}\left(i-\frac{1}{2},j+\frac{1}{2},k\right)$$

$$-\frac{1-\Gamma_a\Delta t/2}{1+\Gamma_a\Delta t/2}P_{1z}^{n-1}\left(i-\frac{1}{2},j+\frac{1}{2},k\right)$$

$$+\frac{\sigma_a\Delta t^2}{1+\Gamma_a\Delta t/2}\left[N_{1z}^{n}\left(i-\frac{1}{2},j+\frac{1}{2},k\right)-N_{2z}^{n}\left(i-\frac{1}{2},j+\frac{1}{2},k\right)\right]$$

$$\cdot E_z^{n}\left(i-\frac{1}{2},j+\frac{1}{2},k\right) \quad (4.48)$$

$$J_{pz}^{G^{n+\frac{1}{2}}}\left(i-\frac{1}{2},j+\frac{1}{2},k\right)=\frac{1}{\Delta t}\left[P_{1z}^{n+1}\left(i-\frac{1}{2},j+\frac{1}{2},k\right)-P_{1z}^{n}\left(i-\frac{1}{2},j+\frac{1}{2},k\right)\right] \quad (4.49)$$

$$E_x^{n+1}\left(i,j+\frac{1}{2},k+\frac{1}{2}\right)=E_x^n\left(i,j+\frac{1}{2},k+\frac{1}{2}\right)$$

$$+\frac{\Delta t}{\varepsilon}\left(\frac{H_z^{n+1/2}\left(i,j+\frac{1}{2},k+\frac{1}{2}\right)-H_z^{n+1/2}\left(i,j,k+\frac{1}{2}\right)}{\Delta y}\right.$$

$$\left.-\frac{H_y^{n+1/2}\left(i,j+\frac{1}{2},k+1\right)-H_y^{n+1/2}\left(i,j+\frac{1}{2},k\right)}{\Delta z}-J_{px}^{G\,n+\frac{1}{2}}\left(i,j+\frac{1}{2},k+\frac{1}{2}\right)\right)$$

(4.50)

$$E_y^{n+1}\left(i-\frac{1}{2},j,k+\frac{1}{2}\right)=E_y^n\left(i-\frac{1}{2},j,k+\frac{1}{2}\right)$$

$$+\frac{\Delta t}{\varepsilon}\left(\frac{H_x^{n+1/2}\left(i-\frac{1}{2},j,k+1\right)-H_x^{n+1/2}\left(i-\frac{1}{2},j,k\right)}{\Delta z}\right.$$

$$\left.-\frac{H_z^{n+1/2}\left(i,j,k+\frac{1}{2}\right)-H_z^{n+1/2}\left(i-1,j,k+\frac{1}{2}\right)}{\Delta x}-J_{py}^{G\,n+\frac{1}{2}}\left(i-\frac{1}{2},j,k+\frac{1}{2}\right)\right)$$

(4.51)

$$E_z^{n+1}\left(i-1,j+\frac{1}{2},k\right)=E_z^n\left(i-\frac{1}{2},j+\frac{1}{2},k\right)$$

$$+\frac{\Delta t}{\varepsilon}\left(\frac{H_y^{n+1/2}\left(i,j+\frac{1}{2},k\right)-H_y^{n+1/2}\left(i-1,j+\frac{1}{2},k\right)}{\Delta x}\right.$$

$$\left.-\frac{H_x^{n+1/2}\left(i-\frac{1}{2},j+1,k\right)-H_x^{n+1/2}\left(i-\frac{1}{2},j,k\right)}{\Delta y}-J_{pz}^{G\,n+\frac{1}{2}}\left(i-\frac{1}{2},j+\frac{1}{2},k\right)\right)$$

(4.52)

For the *optical pumping* version, the update algorithms for rate equations are as follows.

- Update occupation number $N_3$:

$$N_{3x}^{n+1}\left(i, j+\frac{1}{2}, k+\frac{1}{2}\right) = \frac{1 - \Delta t/2\tau_{32}}{1 + \Delta t/2\tau_{32}} N_{3x}^n\left(i, j+\frac{1}{2}, k+\frac{1}{2}\right)$$

$$+ \frac{1}{2\hbar\omega_b} \frac{1}{1 + \Delta t/2\tau_{32}} \left[ E_x^{n+1}\left(i, j+\frac{1}{2}, k+\frac{1}{2}\right) + E_x^n\left(i, j+\frac{1}{2}, k+\frac{1}{2}\right) \right]$$

$$\cdot \left[ P_{2x}^{n+1}\left(i, j+\frac{1}{2}, k+\frac{1}{2}\right) - P_{2x}^n\left(i, j+\frac{1}{2}, k+\frac{1}{2}\right) \right] \qquad (4.53)$$

$$N_{3y}^{n+1}\left(i-\frac{1}{2}, j, k+\frac{1}{2}\right) = \frac{1 - \Delta t/2\tau_{32}}{1 + \Delta t/2\tau_{32}} N_{3y}^n\left(i-\frac{1}{2}, j, k+\frac{1}{2}\right)$$

$$+ \frac{1}{2\hbar\omega_b} \frac{1}{1 + \Delta t/2\tau_{32}} \left[ E_y^{n+1}\left(i-\frac{1}{2}, j, k+\frac{1}{2}\right) + E_y^n\left(i-\frac{1}{2}, j, k+\frac{1}{2}\right) \right]$$

$$\cdot \left[ P_{2y}^{n+1}\left(i-\frac{1}{2}, j, k+\frac{1}{2}\right) - P_{2y}^n\left(i-\frac{1}{2}, j, k+\frac{1}{2}\right) \right] \qquad (4.54)$$

$$N_{3z}^{n+1}\left(i-\frac{1}{2}, j+\frac{1}{2}, k\right) = \frac{1 - \Delta t/2\tau_{32}}{1 + \Delta t/2\tau_{32}} N_{3z}^n\left(i-\frac{1}{2}, j+\frac{1}{2}, k\right)$$

$$+ \frac{1}{2\hbar\omega_b} \frac{1}{1 + \Delta t/2\tau_{32}} \left[ E_z^{n+1}\left(i-\frac{1}{2}, j+\frac{1}{2}, k\right) + E_z^n\left(i-\frac{1}{2}, j+\frac{1}{2}, k\right) \right]$$

$$\cdot \left[ P_{2z}^{n+1}\left(i-\frac{1}{2}, j+\frac{1}{2}, k\right) - P_{2z}^n\left(i-\frac{1}{2}, j+\frac{1}{2}, k\right) \right] \qquad (4.55)$$

For the **homogeneous pumping** rate version:

$$N_{3x}^{n+1}\left(i, j+\frac{1}{2}, k+\frac{1}{2}\right) = \frac{1-\Delta t/2\tau_{32}}{1+\Delta t/2\tau_{32}} N_{3x}^{n}\left(i, j+\frac{1}{2}, k+\frac{1}{2}\right)$$

$$+\frac{\Delta t P_{rx}}{1+\Delta t/2\tau_{32}} N_{0x}^{n}\left(i, j+\frac{1}{2}, k+\frac{1}{2}\right) \qquad (4.56)$$

$$N_{3y}^{n+1}\left(i-\frac{1}{2}, j, k+\frac{1}{2}\right) = \frac{1-\Delta t/2\tau_{32}}{1+\Delta t/2\tau_{32}} N_{3y}^{n}\left(i-\frac{1}{2}, j, k+\frac{1}{2}\right)$$

$$+\frac{\Delta t P_{ry}}{1+\Delta t/2\tau_{32}} N_{0y}^{n}\left(i-\frac{1}{2}, j, k+\frac{1}{2}\right) \qquad (4.57)$$

$$N_{3z}^{n+1}\left(i-\frac{1}{2}, j+\frac{1}{2}, k\right) = \frac{1-\Delta t/2\tau_{32}}{1+\Delta t/2\tau_{32}} N_{3z}^{n}\left(i-\frac{1}{2}, j+\frac{1}{2}, k\right)$$

$$+\frac{\Delta t P_{rz}}{1+\Delta t/2\tau_{32}} N_{0z}^{n}\left(i-\frac{1}{2}, j+\frac{1}{2}, k\right) \qquad (4.58)$$

For the **Gaussian pumping** rate version:

$$N_{3x}^{n+1}\left(i, j+\frac{1}{2}, k+\frac{1}{2}\right) = \frac{1-\Delta t/2\tau_{32}}{1+\Delta t/2\tau_{32}} N_{3x}^{n}\left(i, j+\frac{1}{2}, k+\frac{1}{2}\right)$$

$$+\frac{\Delta t P_{gx}(n+1)}{1+\Delta t/2\tau_{32}} N_{0x}^{n}\left(i, j+\frac{1}{2}, k+\frac{1}{2}\right) \qquad (4.59)$$

$$N_{3y}^{n+1}\left(i-\frac{1}{2}, j, k+\frac{1}{2}\right) = \frac{1-\Delta t/2\tau_{32}}{1+\Delta t/2\tau_{32}} N_{3y}^{n}\left(i-\frac{1}{2}, j, k+\frac{1}{2}\right)$$

$$+\frac{\Delta t P_{gy}(n+1)}{1+\Delta t/2\tau_{32}} N_{0y}^{n}\left(i-\frac{1}{2}, j, k+\frac{1}{2}\right) \qquad (4.60)$$

$$N_{3z}^{n+1}\left(i-\frac{1}{2}, j+\frac{1}{2}, k\right) = \frac{1-\Delta t/2\tau_{32}}{1+\Delta t/2\tau_{32}} N_{3z}^{n}\left(i-\frac{1}{2}, j+\frac{1}{2}, k\right)$$

$$+\frac{\Delta t P_{gz}(n+1)}{1+\Delta t/2\tau_{32}} N_{0z}^{n}\left(i-\frac{1}{2}, j+\frac{1}{2}, k\right) \qquad (4.61)$$

- Update occupation number $N_2$:

$$N_{2x}^{n+1}\left(i, j+\frac{1}{2}, k+\frac{1}{2}\right) = \frac{1-\Delta t/2\tau_{21}}{1+\Delta t/2\tau_{21}} N_{2x}^n\left(i, j+\frac{1}{2}, k+\frac{1}{2}\right)$$

$$+\frac{1}{2\hbar\omega_a}\frac{1}{1+\Delta t/2\tau_{21}}\left[E_x^{n+1}\left(i, j+\frac{1}{2}, k+\frac{1}{2}\right)+E_x^n\left(i, j+\frac{1}{2}, k+\frac{1}{2}\right)\right]$$

$$\cdot\left[P_{1x}^{n+1}\left(i, j+\frac{1}{2}, k+\frac{1}{2}\right)-P_{1x}^n\left(i, j+\frac{1}{2}, k+\frac{1}{2}\right)\right]$$

$$+\frac{1}{1+\Delta t/2\tau_{21}}\frac{\Delta t}{2\tau_{32}}\left[N_{3x}^{n+1}\left(i, j+\frac{1}{2}, k+\frac{1}{2}\right)+N_{3x}^n\left(i, j+\frac{1}{2}, k+\frac{1}{2}\right)\right]$$

$$(4.62)$$

$$N_{2y}^{n+1}\left(i-\frac{1}{2}, j, k+\frac{1}{2}\right) = \frac{1-\Delta t/2\tau_{21}}{1+\Delta t/2\tau_{21}} N_{2y}^n\left(i-\frac{1}{2}, j, k+\frac{1}{2}\right)$$

$$+\frac{1}{2\hbar\omega_a}\frac{1}{1+\Delta t/2\tau_{21}}\left[E_y^{n+1}\left(i-\frac{1}{2}, j, k+\frac{1}{2}\right)+E_y^n\left(i-\frac{1}{2}, j, k+\frac{1}{2}\right)\right]$$

$$\cdot\left[P_{1y}^{n+1}\left(i-\frac{1}{2}, j, k+\frac{1}{2}\right)-P_{1y}^n\left(i-\frac{1}{2}, j, k+\frac{1}{2}\right)\right]$$

$$+\frac{1}{1+\Delta t/2\tau_{21}}\frac{\Delta t}{2\tau_{32}}\left[N_{3y}^{n+1}\left(i-\frac{1}{2}, j, k+\frac{1}{2}\right)+N_{3y}^n\left(i-\frac{1}{2}, j, k+\frac{1}{2}\right)\right]$$

$$(4.63)$$

$$N_{2z}^{n+1}\left(i-\frac{1}{2}, j+\frac{1}{2}, k\right) = \frac{1-\Delta t/2\tau_{21}}{1+\Delta t/2\tau_{21}} N_{2z}^n\left(i-\frac{1}{2}, j+\frac{1}{2}, k\right)$$

$$+\frac{1}{2\hbar\omega_a}\frac{1}{1+\Delta t/2\tau_{21}}\left[E_z^{n+1}\left(i-\frac{1}{2}, j+\frac{1}{2}, k\right)+E_z^n\left(i-\frac{1}{2}, j+\frac{1}{2}, k\right)\right]$$

$$\cdot\left[P_{1z}^{n+1}\left(i-\frac{1}{2}, j+\frac{1}{2}, k\right)-P_{1z}^n\left(i-\frac{1}{2}, j+\frac{1}{2}, k\right)\right]$$

$$+\frac{1}{1+\Delta t/2\tau_{21}}\frac{\Delta t}{2\tau_{32}}\left[N_{3z}^{n+1}\left(i-\frac{1}{2}, j+\frac{1}{2}, k\right)+N_{3z}^n\left(i-\frac{1}{2}, j+\frac{1}{2}, k\right)\right]$$

$$(4.64)$$

- Update occupation number $N_1$:

$$N_{1x}^{n+1}\left(i, j+\frac{1}{2}, k+\frac{1}{2}\right) = \frac{1-\Delta t/2\tau_{10}}{1+\Delta t/2\tau_{10}} N_{1x}^{n}\left(i, j+\frac{1}{2}, k+\frac{1}{2}\right)$$

$$-\frac{1}{2\hbar\omega_a}\frac{1}{1+\Delta t/2\tau_{10}}\left[E_x^{n+1}\left(i, j+\frac{1}{2}, k+\frac{1}{2}\right)+E_x^{n}\left(i, j+\frac{1}{2}, k+\frac{1}{2}\right)\right]$$

$$\cdot\left[P_{1x}^{n+1}\left(i, j+\frac{1}{2}, k+\frac{1}{2}\right)-P_{1x}^{n}\left(i, j+\frac{1}{2}, k+\frac{1}{2}\right)\right]$$

$$+\frac{1}{1+\Delta t/2\tau_{10}}\frac{\Delta t}{2\tau_{21}}\left[N_{2x}^{n+1}\left(i, j+\frac{1}{2}, k+\frac{1}{2}\right)+N_{2x}^{n}\left(i, j+\frac{1}{2}, k+\frac{1}{2}\right)\right]$$

$$(4.65)$$

$$N_{1y}^{n+1}\left(i-\frac{1}{2}, j, k+\frac{1}{2}\right) = \frac{1-\Delta t/2\tau_{10}}{1+\Delta t/2\tau_{10}} N_{1y}^{n}\left(i-\frac{1}{2}, j, k+\frac{1}{2}\right)$$

$$-\frac{1}{2\hbar\omega_a}\frac{1}{1+\Delta t/2\tau_{10}}\left[E_y^{n+1}\left(i-\frac{1}{2}, j, k+\frac{1}{2}\right)+E_y^{n}\left(i-\frac{1}{2}, j, k+\frac{1}{2}\right)\right]$$

$$\cdot\left[P_{1y}^{n+1}\left(i-\frac{1}{2}, j, k+\frac{1}{2}\right)-P_{1y}^{n}\left(i-\frac{1}{2}, j, k+\frac{1}{2}\right)\right]$$

$$+\frac{1}{1+\Delta t/2\tau_{10}}\frac{\Delta t}{2\tau_{21}}\left[N_{2y}^{n+1}\left(i-\frac{1}{2}, j, k+\frac{1}{2}\right)+N_{2y}^{n}\left(i-\frac{1}{2}, j, k+\frac{1}{2}\right)\right]$$

$$(4.66)$$

$$N_{1z}^{n+1}\left(i-\frac{1}{2}, j+\frac{1}{2}, k\right) = \frac{1-\Delta t/2\tau_{10}}{1+\Delta t/2\tau_{10}} N_{1z}^{n}\left(i-\frac{1}{2}, j+\frac{1}{2}, k\right)$$

$$-\frac{1}{2\hbar\omega_a}\frac{1}{1+\Delta t/2\tau_{10}}\left[E_z^{n+1}\left(i-\frac{1}{2}, j+\frac{1}{2}, k\right)+E_z^{n}\left(i-\frac{1}{2}, j+\frac{1}{2}, k\right)\right]$$

$$\cdot\left[P_{1z}^{n+1}\left(i-\frac{1}{2}, j+\frac{1}{2}, k\right)-P_{1z}^{n}\left(i-\frac{1}{2}, j+\frac{1}{2}, k\right)\right]$$

$$+\frac{1}{1+\Delta t/2\tau_{10}}\frac{\Delta t}{2\tau_{21}}\left[N_{2z}^{n+1}\left(i-\frac{1}{2}, j+\frac{1}{2}, k\right)+N_{2z}^{n}\left(i-\frac{1}{2}, j+\frac{1}{2}, k\right)\right]$$

$$(4.67)$$

- Update occupation number $N_0$.

For the **optical pumping** version:

$$N_{0x}^{n+1}\left(i, j+\frac{1}{2}, k+\frac{1}{2}\right) = N_{0x}^{n}\left(i, j+\frac{1}{2}, k+\frac{1}{2}\right)$$

$$-\frac{1}{2\hbar\omega_b}\left[E_x^{n+1}\left(i, j+\frac{1}{2}, k+\frac{1}{2}\right)+E_x^{n}\left(i, j+\frac{1}{2}, k+\frac{1}{2}\right)\right]$$

$$\cdot\left[P_{2x}^{n+1}\left(i, j+\frac{1}{2}, k+\frac{1}{2}\right)-P_{2x}^{n}\left(i, j+\frac{1}{2}, k+\frac{1}{2}\right)\right]$$

$$+\frac{\Delta t}{2\tau_{10}}\left[N_{1x}^{n+1}\left(i, j+\frac{1}{2}, k+\frac{1}{2}\right)+N_{1x}^{n}\left(i, j+\frac{1}{2}, k+\frac{1}{2}\right)\right]$$

$$\text{(4.68)}$$

$$N_{0y}^{n+1}\left(i-\frac{1}{2}, j, k+\frac{1}{2}\right) = N_{0y}^{n}\left(i-\frac{1}{2}, j, k+\frac{1}{2}\right)$$

$$-\frac{1}{2\hbar\omega_b}\left[E_y^{n+1}\left(i-\frac{1}{2}, j, k+\frac{1}{2}\right)+E_y^{n}\left(i-\frac{1}{2}, j, k+\frac{1}{2}\right)\right]$$

$$\cdot\left[P_{2y}^{n+1}\left(i-\frac{1}{2}, j, k+\frac{1}{2}\right)-P_{2y}^{n}\left(i-\frac{1}{2}, j, k+\frac{1}{2}\right)\right]$$

$$+\frac{\Delta t}{2\tau_{10}}\left[N_{1y}^{n+1}\left(i-\frac{1}{2}, j, k+\frac{1}{2}\right)+N_{1y}^{n}\left(i-\frac{1}{2}, j, k+\frac{1}{2}\right)\right]$$

$$\text{(4.69)}$$

$$N_{0z}^{n+1}\left(i-\frac{1}{2}, j+\frac{1}{2}, k\right) = N_{0z}^{n}\left(i-\frac{1}{2}, j+\frac{1}{2}, k\right)$$

$$-\frac{1}{2\hbar\omega_b}\left[E_z^{n+1}\left(i-\frac{1}{2}, j+\frac{1}{2}, k\right)+E_z^{n}\left(i-\frac{1}{2}, j+\frac{1}{2}, k\right)\right]$$

$$\cdot\left[P_{2z}^{n+1}\left(i-\frac{1}{2}, j+\frac{1}{2}, k\right)-P_{2z}^{n}\left(i-\frac{1}{2}, j+\frac{1}{2}, k\right)\right]$$

$$+\frac{\Delta t}{2\tau_{10}}\left[N_{1z}^{n+1}\left(i-\frac{1}{2}, j+\frac{1}{2}, k\right)+N_{1z}^{n}\left(i-\frac{1}{2}, j+\frac{1}{2}, k\right)\right]$$

$$\text{(4.70)}$$

For the **homogeneous pumping** rate version:

$$N_{0x}^{n+1}\left(i, j+\frac{1}{2}, k+\frac{1}{2}\right) = (1 - P_{rx}\Delta t)N_{0x}^{n}\left(i, j+\frac{1}{2}, k+\frac{1}{2}\right)$$

$$+\frac{\Delta t}{2\tau_{10}}\left[N_{1x}^{n+1}\left(i, j+\frac{1}{2}, k+\frac{1}{2}\right) + N_{1x}^{n}\left(i, j+\frac{1}{2}, k+\frac{1}{2}\right)\right]$$

(4.71)

$$N_{0y}^{n+1}\left(i-\frac{1}{2}, j, k+\frac{1}{2}\right) = (1 - P_{ry}\Delta t)N_{0y}^{n}\left(i-\frac{1}{2}, j, k+\frac{1}{2}\right)$$

$$+\frac{\Delta t}{2\tau_{10}}\left[N_{1y}^{n+1}\left(i-\frac{1}{2}, j, k+\frac{1}{2}\right) + N_{1y}^{n}\left(i-\frac{1}{2}, j, k+\frac{1}{2}\right)\right]$$

(4.72)

$$N_{0z}^{n+1}\left(i-\frac{1}{2}, j+\frac{1}{2}, k\right) = (1 - P_{rz}\Delta t)N_{0z}^{n}\left(i-\frac{1}{2}, j+\frac{1}{2}, k\right)$$

$$+\frac{\Delta t}{2\tau_{10}}\left[N_{1z}^{n+1}\left(i-\frac{1}{2}, j+\frac{1}{2}, k\right) + N_{1z}^{n}\left(i-\frac{1}{2}, j+\frac{1}{2}, k\right)\right]$$

(4.73)

For the **Gaussian pumping** rate version:

$$N_{0x}^{n+1}\left(i, j+\frac{1}{2}, k+\frac{1}{2}\right) = \left[1 - P_{gx}((n+1)\Delta t)\right]N_{0x}^{n}\left(i, j+\frac{1}{2}, k+\frac{1}{2}\right)$$

$$+\frac{\Delta t}{2\tau_{10}}\left[N_{1x}^{n+1}\left(i, j+\frac{1}{2}, k+\frac{1}{2}\right) + N_{1x}^{n}\left(i, j+\frac{1}{2}, k+\frac{1}{2}\right)\right]$$

(4.74)

$$N_{0y}^{n+1}\left(i-\frac{1}{2}, j, k+\frac{1}{2}\right) = \left[1 - P_{gy}((n+1)\Delta t)\right]N_{0y}^{n}\left(i-\frac{1}{2}, j, k+\frac{1}{2}\right)$$

$$+\frac{\Delta t}{2\tau_{10}}\left[N_{1y}^{n+1}\left(i-\frac{1}{2}, j, k+\frac{1}{2}\right) + N_{1y}^{n}\left(i-\frac{1}{2}, j, k+\frac{1}{2}\right)\right]$$

(4.75)

$$N_{0z}^{n+1}\left(i-\frac{1}{2},j+\frac{1}{2},k\right)=\left[1-P_{gz}\left((n+1)\Delta t\right)\right]N_{0z}^{n}\left(i-\frac{1}{2},j+\frac{1}{2},k\right)$$

$$+\frac{\Delta t}{2\tau_{10}}\left[N_{1z}^{n+1}\left(i-\frac{1}{2},j+\frac{1}{2},k\right)+N_{1z}^{n}\left(i-\frac{1}{2},j+\frac{1}{2},k\right)\right]$$

$$(4.76)$$

### 4.2.3.4 3D FDTD Algorithm for Perfectly Matched Layer

In numerical simulations, lots of geometries of interest are defined in "open" regions where the spatial domain of the computed field is unbounded in one or more dimensions. Due to the limitation of time and computation resources, it is impossible to handle such an unbounded region problem directly. So there is a need to introduce an absorbing boundary condition (ABC) at the outer lattice boundary to simulate the extension of the FDTD computation domain to infinity.

There are two different categories for absorbing boundaries. One is derived from differential equations, such as Mur's [13] and Liao's [13] absorbing boundary conditions. Another one is actually not a real "boundary" condition; instead, it terminates the outer boundary of the space lattice by surrounding the computation domain with a lossy, reflectionless material that damps the outgoing fields [13]. In the second category, the perfectly matched layer (PML) technique shows much more accuracy than other ABCs. It's only a few lattices thick, reflectionless to all impinging waves (arbitrary incidence and polarization) over their full frequency spectrum and highly absorbing. Here we present the PML technique proposed by J. P. Berenger [18–20], who derived a novel split-field formulation of Maxwell's equations where each vector field component is split into two orthogonal components.

For the 3D case, the six electric and magnetic field components yield 12 subcomponents, denoted by $E_{xy}$, $E_{xz}$, $E_{yx}$, $E_{yz}$, $E_{zx}$, $E_{zy}$, $H_{xy}$, $H_{xz}$, $H_{yx}$, $H_{yz}$, $H_{zx}$, and $H_{zy}$ in Cartesian coordinates. Using these subcomponents, we have 3D time-domain Maxwell's equations for Berenger's split-fields,

$$\left(\varepsilon\frac{\partial}{\partial t}+\sigma_y\right)E_{xy}=\frac{\partial H_z}{\partial y}$$

$$(4.77a)$$

$$\left(\varepsilon\frac{\partial}{\partial t}+\sigma_z\right)E_{xz}=-\frac{\partial H_y}{\partial z}$$

$$(4.77b)$$

$$\left(\varepsilon\frac{\partial}{\partial t}+\sigma_z\right)E_{yz}=\frac{\partial H_x}{\partial z}$$

$$(4.77c)$$

$$\left(\varepsilon\frac{\partial}{\partial t}+\sigma_x\right)E_{yx}=-\frac{\partial H_z}{\partial x} \tag{4.77d}$$

$$\left(\varepsilon\frac{\partial}{\partial t}+\sigma_x\right)E_{zx}=\frac{\partial H_y}{\partial x} \tag{4.77e}$$

$$\left(\varepsilon\frac{\partial}{\partial t}+\sigma_y\right)E_{zy}=-\frac{\partial H_x}{\partial y} \tag{4.77f}$$

$$\left(\varepsilon\frac{\partial}{\partial t}+\sigma_y^*\right)H_{xy}=-\frac{\partial E_x}{\partial y} \tag{4.77g}$$

$$\left(\varepsilon\frac{\partial}{\partial t}+\sigma_z^*\right)H_{xz}=-\frac{\partial E_y}{\partial z} \tag{4.77h}$$

$$\left(\varepsilon\frac{\partial}{\partial t}+\sigma_z^*\right)H_{yz}=-\frac{\partial E_x}{\partial z} \tag{4.77i}$$

$$\left(\varepsilon\frac{\partial}{\partial t}+\sigma_x^*\right)H_{yx}=\frac{\partial E_z}{\partial x} \tag{4.77j}$$

$$\left(\varepsilon\frac{\partial}{\partial t}+\sigma_x^*\right)H_{zx}=-\frac{\partial E_y}{\partial x} \tag{4.77k}$$

$$\left(\varepsilon\frac{\partial}{\partial t}+\sigma_y^*\right)H_{zy}=\frac{\partial E_x}{\partial y} \tag{4.77l}$$

where we have the following relations:

$$E_x=E_{xy}+E_{xz} \tag{4.78a}$$

$$E_y=E_{yx}+E_{yz} \tag{4.78b}$$

$$E_z=E_{zx}+E_{zy} \tag{4.78c}$$

$$H_x=H_{xy}+H_{xz} \tag{4.78d}$$

$$H_y=H_{yx}+H_{yz} \tag{4.78e}$$

$$H_z=H_{zx}+H_{zy} \tag{4.78f}$$

J. P. Berenger has shown that any outgoing waves from the inner lossless isotropic medium can penetrate without reflection into these unphysical absorbing layers and get highly absorbed, independent of the frequency and the angle of incidence, if the absorbing media satisfy the following matching conditions:

1. At a normal-to-$w$ ($w = x, y, z$) PML interface in the FDTD lattice, the parameter pair $(\sigma_w, \sigma_w^*)$ satisfies $\frac{\sigma_w}{e} = \frac{\sigma_w^*}{\mu}$ and all other $(\sigma_w, \sigma_w^*)$ are zero.

2. In a corner region, the PML is provided with each matched $(\sigma_w, \sigma_w^*)$ pair that is assigned to the overlapping PMLs forming the corner. So, PML media located in dihedral-corner overlapping regions have two nonzero and one zero $(\sigma_w, \sigma_w^*)$ pairs. And PML media located in trihedral-corner overlapping regions have three nonzero $(\sigma_w, \sigma_w^*)$ pairs.

Here $\sigma_w$ can be obtained by polynomial grading. For example, $\sigma_x$ can be obtained as follows:

$$\sigma_x = (x/d)^m \sigma_{x,\max} \tag{4.79}$$

where $\sigma_{x,\max} = -\frac{(m+1)\ln[R(0)]}{2d\eta}$, $\eta$ is the EM impedance in vacuum, and $R(0)$ is the desired reflection error.

Then we can have the update algorithm for 3D PML:

$$E_{xy}^{n+1}\left(i, j+\frac{1}{2}, k+\frac{1}{2}\right) = \frac{1-\sigma_y\,\Delta t/2\varepsilon}{1+\sigma_y\,\Delta t/2\varepsilon} E_{xy}^n\left(i, j+\frac{1}{2}, k+\frac{1}{2}\right)$$

$$+ \frac{\Delta t/\Delta y\varepsilon}{1+\sigma_y\,\Delta t/2\varepsilon} \cdot \left[H_z^{n+1/2}\left(i, j+1, k+\frac{1}{2}\right) - H_z^{n+1/2}\left(i, j, k+\frac{1}{2}\right)\right]$$

$$\tag{4.80}$$

$$E_{xz}^{n+1}\left(i, j+\frac{1}{2}, k+\frac{1}{2}\right) = \frac{1-\sigma_z\,\Delta t/2\varepsilon}{1+\sigma_z\,\Delta t/2\varepsilon} E_{xz}^n\left(i, j+\frac{1}{2}, k+\frac{1}{2}\right)$$

$$- \frac{\Delta t/\Delta z\varepsilon}{1+\sigma_z\,\Delta t/2\varepsilon} \cdot \left[H_y^{n+1/2}\left(i, j+\frac{1}{2}, k+1\right) - H_y^{n+1/2}\left(i, j+\frac{1}{2}, k\right)\right]$$

$$\tag{4.81}$$

$$E_x^{n+1}\left(i, j+\frac{1}{2}, k+\frac{1}{2}\right) = E_{xy}^{n+1}\left(i, j+\frac{1}{2}, k+\frac{1}{2}\right) + E_{xz}^{n+1}\left(i, j+\frac{1}{2}, k+\frac{1}{2}\right) \tag{4.82}$$

$$E_{yz}^{n+1}\left(i-\frac{1}{2},j,k+\frac{1}{2}\right) = \frac{1-\sigma_z\,\Delta t/2\varepsilon}{1+\sigma_z\,\Delta t/2\varepsilon}E_{yz}^n\left(i-\frac{1}{2},j,k+\frac{1}{2}\right)$$

$$-\frac{\Delta t/\Delta z\varepsilon}{1+\sigma_z\,\Delta t/2\varepsilon}\cdot\left[H_x^{n+1/2}\left(i-\frac{1}{2},j,k+\frac{1}{2}\right)+H_x^{n+1/2}\left(i-\frac{1}{2},j,k+\frac{1}{2}\right)\right]$$

(4.83)

$$E_{yx}^{n+1}\left(i-\frac{1}{2},j,k+\frac{1}{2}\right) = \frac{1-\sigma_x\,\Delta t/2\varepsilon}{1+\sigma_x\,\Delta t/2\varepsilon}E_{yx}^n\left(i-\frac{1}{2},j,k+\frac{1}{2}\right)$$

$$-\frac{\Delta t/\Delta x\varepsilon}{1+\sigma_x\,\Delta t/2\varepsilon}\cdot\left[H_z^{n+1/2}\left(i,j,k+\frac{1}{2}\right)-H_z^{n+1/2}\left(i-1,j,k+\frac{1}{2}\right)\right]$$

(4.84)

$$E_y^{n+1}\left(i-\frac{1}{2},j,k+\frac{1}{2}\right) = E_{yx}^{n+1}\left(i-\frac{1}{2},j,k+\frac{1}{2}\right)+E_{yz}^{n+1}\left(i-\frac{1}{2},j,k+\frac{1}{2}\right)$$

(4.85)

$$E_{zx}^{n+1}\left(i-\frac{1}{2},j+\frac{1}{2},k\right) = \frac{1-\sigma_x\,\Delta t/2\varepsilon}{1+\sigma_x\,\Delta t/2\varepsilon}E_{zx}^n\left(i-\frac{1}{2},j+\frac{1}{2},k\right)$$

$$+\frac{\Delta t/\Delta x\varepsilon}{1+\sigma_x\,\Delta t/2\varepsilon}\cdot\left[H_y^{n+1/2}\left(i,j+\frac{1}{2},k\right)-H_y^{n+1/2}\left(i-1,j+\frac{1}{2},k\right)\right]$$

(4.86)

$$E_{zy}^{n+1}\left(i-\frac{1}{2},j+\frac{1}{2},k\right) = \frac{1-\sigma_y\,\Delta t/2\varepsilon}{1+\sigma_y\,\Delta t/2\varepsilon}E_{zy}^n\left(i-\frac{1}{2},j+\frac{1}{2},k\right)$$

$$+\frac{\Delta t/\Delta y\varepsilon}{1+\sigma_y\,\Delta t/2\varepsilon}\cdot\left[H_x^{n+1/2}\left(i-\frac{1}{2},j+1,k\right)-H_x^{n+1/2}\left(i-\frac{1}{2},j,k\right)\right]$$

(4.87)

$$E_z^{n+1}\left(i-\frac{1}{2},j+\frac{1}{2},k\right) = E_{zx}^{n+1}\left(i-\frac{1}{2},j+\frac{1}{2},k\right)+E_{zy}^{n+1}\left(i-\frac{1}{2},j+\frac{1}{2},k\right)$$

(4.88)

$$H_{xy}^{n+1/2}\left(i-\frac{1}{2},j,k\right)=\frac{1-\sigma_y\Delta t/2\varepsilon}{1+\sigma_y\Delta t/2\varepsilon}H_{xy}^{n-1/2}\left(i-\frac{1}{2},j,k\right)-\frac{\Delta t/\Delta y\mu}{1+\sigma_y\Delta t/2\varepsilon}$$

$$\cdot\left[E_z^n\left(i-\frac{1}{2},j+\frac{1}{2},k\right)-E_z^n\left(i-\frac{1}{2},j-\frac{1}{2},k\right)\right] \qquad (4.89)$$

$$H_{xz}^{n+1/2}\left(i-\frac{1}{2},j,k\right)=\frac{1-\sigma_z\Delta t/2\varepsilon}{1+\sigma_z\Delta t/2\varepsilon}H_{xz}^{n-1/2}\left(i-\frac{1}{2},j,k\right)+\frac{\Delta t/\Delta z\mu}{1+\sigma_z\Delta t/2\varepsilon}$$

$$\cdot\left[E_y^n\left(i-\frac{1}{2},j,k+\frac{1}{2}\right)-E_y^n\left(i-\frac{1}{2},j,k-\frac{1}{2}\right)\right] \qquad (4.90)$$

$$H_x^{n+1/2}\left(i-\frac{1}{2},j,k\right)=H_{xy}^{n+1/2}\left(i-\frac{1}{2},j,k\right)+H_{xz}^{n+1/2}\left(i-\frac{1}{2},j,k\right) \qquad (4.91)$$

$$H_{yz}^{n+1/2}\left(i,j+\frac{1}{2},k\right)=\frac{1-\sigma_z\Delta t/2\varepsilon}{1+\sigma_z\Delta t/2\varepsilon}H_{yz}^{n-1/2}\left(i,j+\frac{1}{2},k\right)-\frac{\Delta t/\Delta z\mu}{1+\sigma_z\Delta t/2\varepsilon}$$

$$\cdot\left[E_x^n\left(i,j+\frac{1}{2},k+\frac{1}{2}\right)-E_x^n\left(i,j+\frac{1}{2},k-\frac{1}{2}\right)\right] \qquad (4.92)$$

$$H_{yx}^{n+1/2}\left(i,j+\frac{1}{2},k\right)=\frac{1-\sigma_x\Delta t/2\varepsilon}{1+\sigma_x\Delta t/2\varepsilon}H_{yx}^{n-1/2}\left(i,j+\frac{1}{2},k\right)+\frac{\Delta t/\Delta x\mu}{1+\sigma_x\Delta t/2\varepsilon}$$

$$\cdot\left[E_z^n\left(i+\frac{1}{2},j+\frac{1}{2},k\right)-E_z^n\left(i-\frac{1}{2},j+\frac{1}{2},k\right)\right] \qquad (4.93)$$

$$H_y^{n+1/2}\left(i,j+\frac{1}{2},k\right)=H_{yz}^{n+1/2}\left(i,j+\frac{1}{2},k\right)+H_{yx}^{n+1/2}\left(i,j+\frac{1}{2},k\right) \qquad (4.94)$$

$$H_{zx}^{n+1/2}\left(i,j,k+\frac{1}{2}\right)=\frac{1-\sigma_x\Delta t/2\varepsilon}{1+\sigma_x\Delta t/2\varepsilon}H_{zx}^{n-1/2}\left(i,j,k+\frac{1}{2}\right)-\frac{\Delta t/\Delta x\mu}{1+\sigma_x\Delta t/2\varepsilon}$$

$$\cdot\left[E_y^n\left(i+\frac{1}{2},j,k+\frac{1}{2}\right)-E_y^n\left(i-\frac{1}{2},j,k+\frac{1}{2}\right)\right] \qquad (4.95)$$

$$H_{zy}^{n+1/2}\left(i,j,k+\frac{1}{2}\right)=\frac{1-\sigma_y\Delta t/2\varepsilon}{1+\sigma_y\Delta t/2\varepsilon}H_{zy}^{n-1/2}\left(i,j,k+\frac{1}{2}\right)+\frac{\Delta t/\Delta y\mu}{1+\sigma_y\Delta t/2\varepsilon}$$

$$\cdot\left[E_x^n\left(i,j+\frac{1}{2},k+\frac{1}{2}\right)-E_x^n\left(i,j-\frac{1}{2},k+\frac{1}{2}\right)\right] \qquad (4.96)$$

$$H_z^{n+1/2}\left(i,j,k+\frac{1}{2}\right) = H_{zx}^{n+1/2}\left(i,j,k+\frac{1}{2}\right) + H_{zy}^{n+1/2}\left(i,j,k+\frac{1}{2}\right) \qquad (4.97)$$

### 4.2.3.5 *The Order of Update Algorithms in All Regions Including Vacuum, Gain Material, Drude Metal, and PML*

*4.2.3.5.1 Update $H_x$, $H_y$, $H_z$*

In vacuum, Drude metal and gain material region,

$$H_x^{n+1/2}\left(i-\frac{1}{2},j,k\right) = H_x^{n-1/2}\left(i-\frac{1}{2},j,k\right)$$
$$+\frac{\Delta t}{u\left(i-\frac{1}{2},j,k\right)}\left(\frac{E_y^n\left(i-\frac{1}{2},j,k+\frac{1}{2}\right)-E_y^n\left(i-\frac{1}{2},j,k-\frac{1}{2}\right)}{\Delta z}\right.$$
$$\left.-\frac{E_z^n\left(i-\frac{1}{2},j+\frac{1}{2},k\right)-E_z^n\left(i-\frac{1}{2},j-\frac{1}{2},k\right)}{\Delta y}\right) \qquad (4.98)$$

$$H_y^{n+1/2}\left(i,j+\frac{1}{2},k\right) = H_y^{n-1/2}\left(i,j+\frac{1}{2},k\right)$$
$$+\frac{\Delta t}{u\left(i,j+\frac{1}{2},k\right)}\left(\frac{E_z^n\left(i+\frac{1}{2},j+\frac{1}{2},k\right)-E_z^n\left(i-\frac{1}{2},j+\frac{1}{2},k\right)}{\Delta z}\right.$$
$$\left.-\frac{E_x^n\left(i,j+\frac{1}{2},k+\frac{1}{2}\right)-E_x^n\left(i,j+\frac{1}{2},k-\frac{1}{2}\right)}{\Delta z}\right) \qquad (4.99)$$

$$H_z^{n+1/2}\left(i,j,k+\frac{1}{2}\right) = H_z^{n-1/2}\left(i,j,k+\frac{1}{2}\right)$$

$$+\frac{\Delta t}{u\left(i,j,k+\frac{1}{2}\right)}\left(\frac{E_x^n\left(i,j+\frac{1}{2},k+\frac{1}{2}\right)-E_x^n\left(i,j-\frac{1}{2},k+\frac{1}{2}\right)}{\Delta x}\right.$$

$$\left.-\frac{E_y^n\left(i+\frac{1}{2},j,k+\frac{1}{2}\right)-E_y^n\left(i-\frac{1}{2},j,k+\frac{1}{2}\right)}{\Delta x}\right)$$

(4.100)

In PML region,

$$H_{xy}^{n+1/2}\left(i-\frac{1}{2},j,k\right) = \frac{1-\sigma_y\Delta t/2\varepsilon}{1+\sigma_y\Delta t/2\varepsilon}H_{xy}^{n-1/2}\left(i-\frac{1}{2},j,k\right) - \frac{\Delta t/\Delta y\mu_0}{1+\sigma_y\Delta t/2\varepsilon}$$

$$\cdot\left[E_z^n\left(i-\frac{1}{2},j+\frac{1}{2},k\right)-E_z^n\left(i-\frac{1}{2},j-\frac{1}{2},k\right)\right] \quad (4.101)$$

$$H_{xz}^{n+1/2}\left(i-\frac{1}{2},j,k\right) = \frac{1-\sigma_z\Delta t/2\varepsilon}{1+\sigma_z\Delta t/2\varepsilon}H_{xz}^{n-1/2}\left(i-\frac{1}{2},j,k\right) - \frac{\Delta t/\Delta z\mu_0}{1+\sigma_z\Delta t/2\varepsilon}$$

$$\cdot\left[E_y^n\left(i-\frac{1}{2},j,k+\frac{1}{2}\right)-E_y^n\left(i-\frac{1}{2},j,k-\frac{1}{2}\right)\right] \quad (4.102)$$

$$H_x^{n+1/2}\left(i-\frac{1}{2},j,k\right) = H_{xy}^{n+1/2}\left(i-\frac{1}{2},j,k\right) + H_{xz}^{n+1/2}\left(i-\frac{1}{2},j,k\right) \quad (4.103)$$

$$H_{yz}^{n+1/2}\left(i,j+\frac{1}{2},k\right) = \frac{1-\sigma_z\Delta t/2\varepsilon}{1+\sigma_z\Delta t/2\varepsilon}H_{yz}^{n-1/2}\left(i,j+\frac{1}{2},k\right) - \frac{\Delta t/\Delta z\mu_0}{1+\sigma_z\Delta t/2\varepsilon}$$

$$\cdot\left[E_x^n\left(i,j+\frac{1}{2},k+\frac{1}{2}\right)-E_x^n\left(i,j+\frac{1}{2},k-\frac{1}{2}\right)\right] \quad (4.104)$$

$$H_{yx}^{n+1/2}\left(i, j+\frac{1}{2}, k\right) = \frac{1-\sigma_x \Delta t/2\varepsilon}{1+\sigma_x \Delta t/2\varepsilon} H_{yx}^{n-1/2}\left(i, j+\frac{1}{2}, k\right) - \frac{\Delta t/\Delta x \mu_0}{1+\sigma_x \Delta t/2\varepsilon}$$

$$\cdot \left[E_z^n\left(i+\frac{1}{2}, j+\frac{1}{2}, k\right) - E_z^n\left(i-\frac{1}{2}, j+\frac{1}{2}, k\right)\right] \quad (4.105)$$

$$H_y^{n+1/2}\left(i, j+\frac{1}{2}, k\right) = H_{yz}^{n+1/2}\left(i, j+\frac{1}{2}, k\right) + H_{yx}^{n+1/2}\left(i, j+\frac{1}{2}, k\right) \quad (4.106)$$

$$H_{zx}^{n+1/2}\left(i, j, k+\frac{1}{2}\right) = \frac{1-\sigma_x \Delta t/2\varepsilon}{1+\sigma_x \Delta t/2\varepsilon} H_{zx}^{n-1/2}\left(i, j, k+\frac{1}{2}\right) - \frac{\Delta t/\Delta x \mu_0}{1+\sigma_x \Delta t/2\varepsilon}$$

$$\cdot \left[E_y^n\left(i+\frac{1}{2}, j, k+\frac{1}{2}\right) - E_y^n\left(i-\frac{1}{2}, j, k+\frac{1}{2}\right)\right] \quad (4.107)$$

$$H_{zy}^{n+1/2}\left(i, j, k+\frac{1}{2}\right) = \frac{1-\sigma_y \Delta t/2\varepsilon}{1+\sigma_y \Delta t/2\varepsilon} H_{zy}^{n-1/2}\left(i, j, k+\frac{1}{2}\right) - \frac{\Delta t/\Delta y \mu_0}{1+\sigma_y \Delta t/2\varepsilon}$$

$$\cdot \left[E_x^n\left(i, j+\frac{1}{2}, k+\frac{1}{2}\right) - E_x^n\left(i, j-\frac{1}{2}, k+\frac{1}{2}\right)\right] \quad (4.108)$$

$$H_z^{n+1/2}\left(i, j, k+\frac{1}{2}\right) = H_{zx}^{n+1/2}\left(i, j, k+\frac{1}{2}\right) + H_{zy}^{n+1/2}\left(i, j, k+\frac{1}{2}\right) \quad (4.109)$$

*4.2.3.5.2 Make Corrections for $H_x$ and $H_z$*

First we define $E_{inc}$ as being oriented with an angle $\theta$ ($0^0 < \theta < 180^0$) relative to + z-axis as shown in Figure 4.4.

$$E_{z,inc}^n\left(i-\frac{1}{2}, j_L-\frac{1}{2}, k\right) = E_{inc}^n\left(j_L-\frac{1}{2}\right)\cos\theta$$

$$E_{z,inc}^n\left(i-\frac{1}{2}, j_R+\frac{1}{2}, k\right) = E_{inc}^n\left(j_R+\frac{1}{2}\right)\cos\theta$$

$$(4.110)$$

$$E_{x,inc}^n\left(i, j_L-\frac{1}{2}, k+\frac{1}{2}\right) = -E_{inc}^n\left(j_L-\frac{1}{2}\right)\sin\theta$$

$$E_{x,inc}^n\left(i, j_R+\frac{1}{2}, k+\frac{1}{2}\right) = -E_{inc}^n\left(j_R+\frac{1}{2}\right)\sin\theta$$

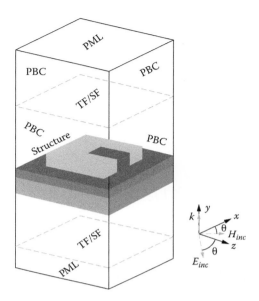

**FIGURE 4.4**
Schematic illustration of our computation space and the polarization of the incident fields. (See color figure.)

- Make $H_x$ corrections.

  At the left interface $j = j_L$,

$$
H_x^{n+1/2}\left(i - \frac{1}{2}, j_L, k\right) = \left\{ H_x^{n+1/2}\left(i - \frac{1}{2}, j_L, k\right) \right\} + \frac{\Delta t}{\mu_0 \Delta y} E_{z,inc}^n\left(i - \frac{1}{2}, j_L - \frac{1}{2}, k\right)
$$

(4.111)

  At the right interface $j = j_R$,

$$
H_x^{n+1/2}\left(i - \frac{1}{2}, j_R, k\right) = \left\{ H_x^{n+1/2}\left(i - \frac{1}{2}, j_R, k\right) \right\} - \frac{\Delta t}{\mu_0 \Delta y} E_{z,inc}^n\left(i - \frac{1}{2}, j_R + \frac{1}{2}, k\right)
$$

(4.112)

- Make $H_z$ corrections.

At the left interface $j = j_L$,

$$H_z^{n+1/2}\left(i, j_L, k+\frac{1}{2}\right) = \left\{H_z^{n+1/2}\left(i, j_L, k+\frac{1}{2}\right)\right\} - \frac{\Delta t}{\mu_0 \Delta y} E_{x,inc}^n\left(i, j_L - \frac{1}{2}, k+\frac{1}{2}\right) \quad (4.113)$$

At the right interface $j = j_R$,

$$H_z^{n+1/2}\left(i, j_R, k+\frac{1}{2}\right) = \left\{H_z^{n+1/2}\left(i, j_R, k+\frac{1}{2}\right)\right\} + \frac{\Delta t}{\mu_0 \Delta y} E_{x,inc}^n\left(i, j_R + \frac{1}{2}, k+\frac{1}{2}\right)$$

$$(4.114)$$

### 4.2.3.5.3 *Update $H_{inc}$ in the Source Grid Generating Lookup Table*

$$H_{inc}^{n+1/2}(j) = \frac{1-\sigma\Delta t / 2\varepsilon_0}{1+\sigma\Delta t / 2\varepsilon_0} H_{inc}^{n-1/2}(j)$$

$$-\frac{\Delta t}{\mu_0 \Delta y} \frac{1}{1+\sigma\Delta t / 2\varepsilon_0}\left[E_{inc}^n\left(j+\frac{1}{2}\right) - E_{inc}^n\left(j-\frac{1}{2}\right)\right] \quad (4.115)$$

### 4.2.3.5.4 *Update the Polarizations $\mathbf{P}$ and Current $\mathbf{J}$*

In the gain material region, we take the homogeneous pumping rate version as an example.

For the **homogeneous pumping** rate version:

$$P_{1x}^{n+1}\left(i, j+\frac{1}{2}, k+\frac{1}{2}\right) = \frac{2-\omega_a^2\Delta t^2}{1+\Gamma_a\Delta t/2} P_{1x}^n\left(i, j+\frac{1}{2}, k+\frac{1}{2}\right)$$

$$-\frac{1-\Gamma_a\Delta t/2}{1+\Gamma_a\Delta t/2} P_{1x}^{n-1}\left(i, j+\frac{1}{2}, k+\frac{1}{2}\right)$$

$$+\frac{\sigma_a\Delta t^2}{1+\Gamma_a\Delta t/2}\left[N_{1x}^n\left(i, j+\frac{1}{2}, k+\frac{1}{2}\right) - N_{2x}^n\left(i, j+\frac{1}{2}, k+\frac{1}{2}\right)\right]$$

$$\cdot E_x^n\left(i, j+\frac{1}{2}, k+\frac{1}{2}\right) \quad (4.116)$$

$$J_{px}^{G^{n+\frac{1}{2}}}\left(i,j+\frac{1}{2},k+\frac{1}{2}\right)=\frac{1}{\Delta t}\left[P_{1x}^{n+1}\left(i,j+\frac{1}{2},k+\frac{1}{2}\right)-P_{1x}^{n}\left(i,j+\frac{1}{2},k+\frac{1}{2}\right)\right] \quad (4.117)$$

$$P_{1y}^{n+1}\left(i-\frac{1}{2},j,k+\frac{1}{2}\right)=\frac{2-\omega_a^2\Delta t^2}{1+\Gamma_a\Delta t/2}P_{1y}^{n}\left(i-\frac{1}{2},j,k+\frac{1}{2}\right)$$

$$-\frac{1-\Gamma_a\Delta t/2}{1+\Gamma_a\Delta t/2}P_{1y}^{n-1}\left(i-\frac{1}{2},j,k+\frac{1}{2}\right)$$

$$+\frac{\sigma_a\Delta t^2}{1+\Gamma_a\Delta t/2}\left[N_{1y}^{n}\left(i-\frac{1}{2},j,k+\frac{1}{2}\right)-N_{2y}^{n}\left(i-\frac{1}{2},j,k+\frac{1}{2}\right)\right]$$

$$\cdot E_y^{n}\left(i-\frac{1}{2},j,k+\frac{1}{2}\right)$$

$$(4.118)$$

$$J_{py}^{G^{n+\frac{1}{2}}}\left(i-\frac{1}{2},j,k+\frac{1}{2}\right)=\frac{1}{\Delta t}\left[P_{1y}^{n+1}\left(i-\frac{1}{2},j,k+\frac{1}{2}\right)-P_{1y}^{n}\left(i-\frac{1}{2},j,k+\frac{1}{2}\right)\right] \quad (4.119)$$

$$P_{1z}^{n+1}\left(i-\frac{1}{2},j+\frac{1}{2},k\right)=\frac{2-\omega_a^2\Delta t^2}{1+\Gamma_a\Delta t/2}P_{1z}^{n}\left(i-\frac{1}{2},j+\frac{1}{2},k\right)$$

$$-\frac{1-\Gamma_a\Delta t/2}{1+\Gamma_a\Delta t/2}P_{1z}^{n-1}\left(i-\frac{1}{2},j+\frac{1}{2},k\right)$$

$$+\frac{\sigma_a\Delta t^2}{1+\Gamma_a\Delta t/2}\left[N_{1z}^{n}\left(i-\frac{1}{2},j+\frac{1}{2},k\right)-N_{2z}^{n}\left(i-\frac{1}{2},j+\frac{1}{2},k\right)\right]$$

$$\cdot E_z^{n}\left(i-\frac{1}{2},j+\frac{1}{2},k\right) \quad (4.120)$$

$$J_{pz}^{G^{n+\frac{1}{2}}}\left(i-\frac{1}{2},j+\frac{1}{2},k\right)=\frac{1}{\Delta t}\left[P_{1z}^{n+1}\left(i-\frac{1}{2},j+\frac{1}{2},k\right)-P_{1z}^{n}\left(i-\frac{1}{2},j+\frac{1}{2},k\right)\right] \quad (4.121)$$

In the Drude metal region,

$$J_{px}^{n+\frac{1}{2}}\left(i,j+\frac{1}{2},k+\frac{1}{2}\right)=\frac{1-0.5\Gamma_{ex}\Delta t}{1+0.5\Gamma_{ex}\Delta t}J_{px}^{n-\frac{1}{2}}\left(i,j+\frac{1}{2},k+\frac{1}{2}\right)$$

$$+\frac{\varepsilon_0\omega_{pex}^2\Delta t}{1+0.5\Gamma_{ex}\Delta t}E_x^n\left(i,j+\frac{1}{2},k+\frac{1}{2}\right) \qquad (4.122)$$

$$J_{py}^{n+\frac{1}{2}}\left(i-\frac{1}{2},j,k+\frac{1}{2}\right)=\frac{1-0.5\Gamma_{ey}\Delta t}{1+0.5\Gamma_{ey}\Delta t}J_{py}^{n-\frac{1}{2}}\left(i-\frac{1}{2},j,k+\frac{1}{2}\right)$$

$$+\frac{\varepsilon_0\omega_{pey}^2\Delta t}{1+0.5\Gamma_{ey}\Delta t}E_y^n\left(i-\frac{1}{2},j,k+\frac{1}{2}\right) \qquad (4.123)$$

$$J_{pz}^{n+\frac{1}{2}}\left(i-\frac{1}{2},j+\frac{1}{2},k\right)=\frac{1-0.5\Gamma_{ez}\Delta t}{1+0.5\Gamma_{ez}\Delta t}J_{pz}^{n-\frac{1}{2}}\left(i-\frac{1}{2},j+\frac{1}{2},k\right)$$

$$+\frac{\varepsilon_0\omega_{pez}^2\Delta t}{1+0.5\Gamma_{ez}\Delta t}E_z^n\left(i-\frac{1}{2},j+\frac{1}{2},k\right) \qquad (4.124)$$

### 4.2.3.5.5 *Update Electric Field $E_x$, $E_y$, $E_z$*
In vacuum, gain material, and Drude metal regions,

$$E_x^{n+1}\left(i,j+\frac{1}{2},k+\frac{1}{2}\right)=E_x^n\left(i,j+\frac{1}{2},k+\frac{1}{2}\right)$$

$$+\frac{\Delta t}{\varepsilon}\left(\frac{H_z^{n+1/2}\left(i,j+1,k+\frac{1}{2}\right)-H_z^{n+1/2}\left(i,j,k+\frac{1}{2}\right)}{\Delta y}\right.$$

$$-\frac{H_y^{n+1/2}\left(i,j+\frac{1}{2},k+1\right)-H_y^{n+1/2}\left(i,j+\frac{1}{2},k\right)}{\Delta z}$$

$$\left.-J_{px}^{n+\frac{1}{2}}\left(i,j+\frac{1}{2},k+\frac{1}{2}\right)\right) \qquad (4.125)$$

$$E_y^{n+1}\left(i-\frac{1}{2},j,k+\frac{1}{2}\right)=E_y^n\left(i-\frac{1}{2},j,k+\frac{1}{2}\right)$$

$$+\frac{\Delta t}{\varepsilon}\left(\frac{H_x^{n+1/2}\left(i-\frac{1}{2},j,k+1\right)-H_x^{n+1/2}\left(i-\frac{1}{2},j,k\right)}{\Delta z}\right.$$

$$-\frac{H_z^{n+1/2}\left(i,j,k+\frac{1}{2}\right)-H_z^{n+1/2}\left(i-1,j,k+\frac{1}{2}\right)}{\Delta x}$$

$$\left.-J_{py}^{n+\frac{1}{2}}\left(i-\frac{1}{2},j,k+\frac{1}{2}\right)\right) \tag{4.126}$$

$$E_z^{n+1}\left(i-\frac{1}{2},j+\frac{1}{2},k\right)=E_z^n\left(i-\frac{1}{2},j+\frac{1}{2},k\right)$$

$$+\frac{\Delta t}{\varepsilon}\left(\frac{H_y^{n+1/2}\left(i,j+\frac{1}{2},k\right)-H_y^{n+1/2}\left(i-1,j+\frac{1}{2},k\right)}{\Delta x}\right.$$

$$-\frac{H_x^{n+1/2}\left(i-\frac{1}{2},j+1,k\right)-H_x^{n+1/2}\left(i-\frac{1}{2},j,k\right)}{\Delta y}$$

$$\left.-J_{pz}^{n+\frac{1}{2}}\left(i-\frac{1}{2},j+\frac{1}{2},k\right)\right) \tag{4.127}$$

In the PML region,

$$E_{xy}^{n+1}\left(i,j+\frac{1}{2},k+\frac{1}{2}\right)=\frac{1-\sigma_y\Delta t/2\varepsilon}{1+\sigma_y\Delta t/2\varepsilon}E_{xy}^n\left(i,j+\frac{1}{2},k+\frac{1}{2}\right)+\frac{\Delta t/\Delta y\varepsilon}{1+\sigma_y\Delta t/2\varepsilon}$$

$$\cdot\left[H_z^{n+1/2}\left(i,j+1,k+\frac{1}{2}\right)-H_z^{n+1/2}\left(i,j,k+\frac{1}{2}\right)\right] \tag{4.128}$$

$$E_{xz}^{n+1}\left(i,j+\frac{1}{2},k+\frac{1}{2}\right) = \frac{1-\sigma_z\Delta t/2\varepsilon}{1+\sigma_z\Delta t/2\varepsilon} E_{xz}^n\left(i,j+\frac{1}{2},k+\frac{1}{2}\right) - \frac{\Delta t/\Delta z\varepsilon}{1+\sigma_z\Delta t/2\varepsilon}$$

$$\cdot\left[H_y^{n+1/2}\left(i,j+\frac{1}{2},k+1\right) - H_y^{n+1/2}\left(i,j+\frac{1}{2},k\right)\right]$$

(4.129)

$$E_x^{n+1}\left(i,j+\frac{1}{2},k+\frac{1}{2}\right) = E_{xy}^{n+1}\left(i,j+\frac{1}{2},k+\frac{1}{2}\right) + E_{xz}^{n+1}\left(i,j+\frac{1}{2},k+\frac{1}{2}\right) \qquad (4.130)$$

$$E_{yz}^{n+1}\left(i-\frac{1}{2},j,k+\frac{1}{2}\right) = \frac{1-\sigma_z\Delta t/2\varepsilon}{1+\sigma_z\Delta t/2\varepsilon} E_{yz}^n\left(i-\frac{1}{2},j,k+\frac{1}{2}\right) + \frac{\Delta t/\Delta z\varepsilon}{1+\sigma_z\Delta t/2\varepsilon}$$

$$\cdot\left[H_x^{n+1/2}\left(i-\frac{1}{2},j,k+1\right) - H_x^{n+1/2}\left(i-\frac{1}{2},j,k\right)\right]$$

(4.131)

$$E_{yx}^{n+1}\left(i-\frac{1}{2},j,k+\frac{1}{2}\right) = \frac{1-\sigma_x\Delta t/2\varepsilon}{1+\sigma_x\Delta t/2\varepsilon} E_{yx}^n\left(i-\frac{1}{2},j,k+\frac{1}{2}\right) - \frac{\Delta t/\Delta x\varepsilon}{1+\sigma_x\Delta t/2\varepsilon}$$

$$\cdot\left[H_z^{n+1/2}\left(i,j,k+\frac{1}{2}\right) - H_z^{n+1/2}\left(i-1,j,k+\frac{1}{2}\right)\right]$$

(4.132)

$$E_y^{n+1}\left(i-\frac{1}{2},j,k+\frac{1}{2}\right) = E_{yz}^{n+1}\left(i-\frac{1}{2},j,k+\frac{1}{2}\right) + E_{yx}^{n+1}\left(i-\frac{1}{2},j,k+\frac{1}{2}\right) \qquad (4.133)$$

$$E_{zx}^{n+1}\left(i-\frac{1}{2},j+\frac{1}{2},k\right) = \frac{1-\sigma_x\Delta t/2\varepsilon}{1+\sigma_x\Delta t/2\varepsilon} E_{zx}^n\left(i-\frac{1}{2},j+\frac{1}{2},k\right) + \frac{\Delta t/\Delta x\varepsilon}{1+\sigma_x\Delta t/2\varepsilon}$$

$$\cdot\left[H_y^{n+\frac{1}{2}}\left(i,j+\frac{1}{2},k\right) - H_y^{n+\frac{1}{2}}\left(i-1,j+\frac{1}{2},k\right)\right]$$

(4.134)

$$E_{zy}^{n+1}\left(i-\frac{1}{2},j+\frac{1}{2},k\right)=\frac{1-\sigma_y\Delta t/2\varepsilon}{1+\sigma_y\Delta t/2\varepsilon}E_{zy}^{n}\left(i-\frac{1}{2},j+\frac{1}{2},k\right)+\frac{\Delta t/\Delta y\varepsilon}{1+\sigma_y\Delta t/2\varepsilon}$$

$$\cdot\left[H_x^{n+\frac{1}{2}}\left(i-\frac{1}{2},j+1,k\right)-H_x^{n+\frac{1}{2}}\left(i-\frac{1}{2},j+\frac{1}{2},k\right)\right]$$

$$(4.135)$$

$$E_z^{n+1}\left(i-\frac{1}{2},j+\frac{1}{2},k\right)=E_{zx}^{n+1}\left(i-\frac{1}{2},j+\frac{1}{2},k\right)+E_{zy}^{n+1}\left(i-\frac{1}{2},j+\frac{1}{2},k\right) \qquad (4.136)$$

#### 4.2.3.5.6 Make $E_x$ and $E_z$ Corrections over the Connecting Interface (TF/SF Method)

$$H_{z,inc}^{n+1/2}\left(i,j_L,k+\frac{1}{2}\right)=H_{inc}^{n+1/2}(j_L)\sin\theta$$

$$H_{z,inc}^{n+1/2}\left(i,j_R,k+\frac{1}{2}\right)=H_{inc}^{n+1/2}(j_R)\sin\theta$$

$$(4.137)$$

$$H_{x,inc}^{n+1/2}\left(i-\frac{1}{2},j_L,k\right)=H_{inc}^{n+1/2}(j_L)\cos\theta$$

$$H_{x,inc}^{n+1/2}\left(i-\frac{1}{2},j_R,k\right)=H_{inc}^{n+1/2}(j_R)\cos\theta$$

- Make $E_x$ corrections.

  At the left interface $j=j_L$,

$$E_x^{n+1}\left(i,j_L-\frac{1}{2},k+\frac{1}{2}\right)=\left\{E_x^{n+1}\left(i,j_L-\frac{1}{2},k+\frac{1}{2}\right)\right\}-\frac{\Delta t}{\Delta y\varepsilon_0}H_{z,inc}^{n+1/2}\left(i,j_L,k+\frac{1}{2}\right)$$

$$(4.138)$$

  At the left interface $j=j_R$,

$$E_x^{n+1}\left(i,j_R+\frac{1}{2},k+\frac{1}{2}\right)=\left\{E_x^{n+1}\left(i,j_R+\frac{1}{2},k+\frac{1}{2}\right)\right\}+\frac{\Delta t}{\Delta y\varepsilon_0}H_{z,inc}^{n+1/2}\left(i,j_R,k+\frac{1}{2}\right)$$

$$(4.139)$$

- Make $E_z$ corrections.

  At the left interface $j = j_L$,

$$E_z^{n+1}\left(i-\frac{1}{2}, j_L-\frac{1}{2}, k\right) = \left\{E_z^{n+1}\left(i-\frac{1}{2}, j_L-\frac{1}{2}, k\right)\right\} + \frac{\Delta t}{\Delta y \varepsilon_0} H_x^{n+1/2}\left(i-\frac{1}{2}, j_L, k\right)$$

(4.140)

At the left interface $j = j_R$,

$$E_z^{n+1}\left(i-\frac{1}{2}, j_L-\frac{1}{2}, k\right) = \left\{E_z^{n+1}\left(i-\frac{1}{2}, j_L-\frac{1}{2}, k\right)\right\} + \frac{\Delta t}{\Delta y \varepsilon_0} H_x^{n+1/2}\left(i-\frac{1}{2}, j_L, k\right)$$

(4.141)

### 4.2.3.5.7  *Update $E_{inc}$ in the Source Grid Generating Lookup Table*

$$E_{inc}^{n+1}(j+\tfrac{1}{2}) = \frac{1-\sigma\Delta t/2\varepsilon_0}{1+\sigma\Delta t/2\varepsilon_0} E_{inc}^n(j+\tfrac{1}{2}) - \frac{\Delta t}{\Delta y \varepsilon_0} \frac{1}{1+\sigma\Delta t/2\varepsilon_0}\left[H_{inc}^{n+\frac{1}{2}}(j+1) - H_{inc}^{n+\frac{1}{2}}(j)\right]$$

(4.142)

### 4.2.3.5.8  *Update Occupation Numbers $N_0$, $N_1$, $N_2$, $N_3$*
Update occupation number $N_3$:

$$N_{3x}^{n+1}\left(i, j+\frac{1}{2}, k+\frac{1}{2}\right) = \frac{1-\Delta t/2\tau_{32}}{1+\Delta t/2\tau_{32}} N_{3x}^n\left(i, j+\frac{1}{2}, k+\frac{1}{2}\right)$$

$$+ \frac{\Delta t P_{rx}}{1+\Delta t/2\tau_{32}} N_{0x}^n\left(i, j+\frac{1}{2}, k+\frac{1}{2}\right)$$

(4.143)

$$N_{3y}^{n+1}\left(i-\frac{1}{2}, j, k+\frac{1}{2}\right) = \frac{1-\Delta t/2\tau_{32}}{1+\Delta t/2\tau_{32}} N_{3y}^n\left(i-\frac{1}{2}, j, k+\frac{1}{2}\right)$$

$$+ \frac{\Delta t P_{ry}}{1+\Delta t/2\tau_{32}} N_{0y}^n\left(i-\frac{1}{2}, j, k+\frac{1}{2}\right)$$

(4.144)

$$N_{3z}^{n+1}\left(i-\frac{1}{2},j+\frac{1}{2},k\right)=\frac{1-\Delta t/2\tau_{32}}{1+\Delta t/2\tau_{32}}N_{3z}^{n}\left(i-\frac{1}{2},j+\frac{1}{2},k\right)$$

$$+\frac{\Delta t P_{rz}}{1+\Delta t/2\tau_{32}}N_{0z}^{n}\left(i-\frac{1}{2},j+\frac{1}{2},k\right)\qquad(4.145)$$

Update occupation number $N_2$:

$$N_{2x}^{n+1}\left(i,j+\frac{1}{2},k+\frac{1}{2}\right)=\frac{1-\Delta t/2\tau_{21}}{1+\Delta t/2\tau_{21}}N_{2x}^{n}\left(i,j+\frac{1}{2},k+\frac{1}{2}\right)$$

$$+\frac{1}{2\hbar\omega_a}\frac{1}{1+\Delta t/2\tau_{21}}\left[E_x^{n+1}\left(i,j+\frac{1}{2},k+\frac{1}{2}\right)+E_x^{n}\left(i,j+\frac{1}{2},k+\frac{1}{2}\right)\right]$$

$$\cdot\left[P_{1x}^{n+1}\left(i,j+\frac{1}{2},k+\frac{1}{2}\right)-P_{1x}^{n}\left(i,j+\frac{1}{2},k+\frac{1}{2}\right)\right]$$

$$+\frac{1}{1+\Delta t/2\tau_{21}}\frac{\Delta t}{2\tau_{32}}\left[N_{3x}^{n+1}\left(i,j+\frac{1}{2},k+\frac{1}{2}\right)+N_{3x}^{n}\left(i,j+\frac{1}{2},k+\frac{1}{2}\right)\right]$$

$$(4.146)$$

$$N_{2y}^{n+1}\left(i-\frac{1}{2},j,k+\frac{1}{2}\right)=\frac{1-\Delta t/2\tau_{21}}{1+\Delta t/2\tau_{21}}N_{2y}^{n}\left(i-\frac{1}{2},j,k+\frac{1}{2}\right)$$

$$+\frac{1}{2\hbar\omega_a}\frac{1}{1+\Delta t/2\tau_{21}}\left[E_y^{n+1}\left(i-\frac{1}{2},j,k+\frac{1}{2}\right)+E_y^{n}\left(i-\frac{1}{2},j,k+\frac{1}{2}\right)\right]$$

$$\cdot\left[P_{1y}^{n+1}\left(i-\frac{1}{2},j,k+\frac{1}{2}\right)-P_{1y}^{n}\left(i-\frac{1}{2},j,k+\frac{1}{2}\right)\right]$$

$$+\frac{1}{1+\Delta t/2\tau_{21}}\frac{\Delta t}{2\tau_{32}}\left[N_{3y}^{n+1}\left(i-\frac{1}{2},j,k+\frac{1}{2}\right)+N_{3y}^{n}\left(i-\frac{1}{2},j,k+\frac{1}{2}\right)\right]$$

$$(4.147)$$

$$N_{2z}^{n+1}\left(i-\frac{1}{2},j+\frac{1}{2},k\right)=\frac{1-\Delta t/2\tau_{21}}{1+\Delta t/2\tau_{21}}N_{2z}^{n}\left(i-\frac{1}{2},j+\frac{1}{2},k\right)$$

$$+\frac{1}{2\hbar\omega_a}\frac{1}{1+\Delta t/2\tau_{21}}\left[E_z^{n+1}\left(i-\frac{1}{2},j+\frac{1}{2},k\right)+E_z^{n}\left(i-\frac{1}{2},j+\frac{1}{2},k\right)\right]$$

$$\cdot\left[P_{1z}^{n+1}\left(i-\frac{1}{2},j+\frac{1}{2},k\right)-P_{1z}^{n}\left(i-\frac{1}{2},j+\frac{1}{2},k\right)\right]$$

$$+\frac{1}{1+\Delta t/2\tau_{21}}\frac{\Delta t}{2\tau_{32}}\left[N_{3z}^{n+1}\left(i-\frac{1}{2},j+\frac{1}{2},k\right)+N_{3z}^{n}\left(i-\frac{1}{2},j+\frac{1}{2},k\right)\right]$$

$$(4.148)$$

Update occupation number $N_1$:

$$N_{1x}^{n+1}\left(i,j+\frac{1}{2},k+\frac{1}{2}\right)=\frac{1-\Delta t/2\tau_{10}}{1+\Delta t/2\tau_{10}}N_{1x}^{n}\left(i,j+\frac{1}{2},k+\frac{1}{2}\right)$$

$$-\frac{1}{2\hbar\omega_a}\frac{1}{1+\Delta t/2\tau_{10}}\left[E_x^{n+1}\left(i,j+\frac{1}{2},k+\frac{1}{2}\right)+E_x^{n}\left(i,j+\frac{1}{2},k+\frac{1}{2}\right)\right]$$

$$\cdot\left[P_{1x}^{n+1}\left(i,j+\frac{1}{2},k+\frac{1}{2}\right)-P_{1x}^{n}\left(i,j+\frac{1}{2},k+\frac{1}{2}\right)\right]$$

$$+\frac{1}{1+\Delta t/2\tau_{10}}\frac{\Delta t}{2\tau_{21}}\left[N_{2x}^{n+1}\left(i,j+\frac{1}{2},k+\frac{1}{2}\right)+N_{2x}^{n}\left(i,j+\frac{1}{2},k+\frac{1}{2}\right)\right]$$

$$(4.149)$$

$$N_{1y}^{n+1}\left(i-\frac{1}{2},j,k+\frac{1}{2}\right)=\frac{1-\Delta t/2\tau_{10}}{1+\Delta t/2\tau_{10}}N_{1y}^{n}\left(i-\frac{1}{2},j,k+\frac{1}{2}\right)$$

$$-\frac{1}{2\hbar\omega_a}\frac{1}{1+\Delta t/2\tau_{10}}\left[E_y^{n+1}\left(i-\frac{1}{2},j,k+\frac{1}{2}\right)+E_y^{n}\left(i-\frac{1}{2},j,k+\frac{1}{2}\right)\right]$$

$$\cdot\left[P_{1y}^{n+1}\left(i-\frac{1}{2},j,k+\frac{1}{2}\right)-P_{1y}^{n}\left(i-\frac{1}{2},j,k+\frac{1}{2}\right)\right]$$

$$+\frac{1}{1+\Delta t/2\tau_{10}}\frac{\Delta t}{2\tau_{21}}\left[N_{2y}^{n+1}\left(i-\frac{1}{2},j,k+\frac{1}{2}\right)+N_{2y}^{n}\left(i-\frac{1}{2},j,k+\frac{1}{2}\right)\right]$$

$$(4.150)$$

$$N_{1z}^{n+1}\left(i-\frac{1}{2},j+\frac{1}{2},k\right) = \frac{1-\Delta t/2\tau_{10}}{1+\Delta t/2\tau_{10}} N_{1z}^{n}\left(i-\frac{1}{2},j+\frac{1}{2},k\right)$$

$$-\frac{1}{2\hbar\omega_a}\frac{1}{1+\Delta t/2\tau_{10}}\left[E_z^{n+1}\left(i-\frac{1}{2},j+\frac{1}{2},k\right)+E_z^{n}\left(i-\frac{1}{2},j+\frac{1}{2},k\right)\right]$$

$$\cdot\left[P_{1z}^{n+1}\left(i-\frac{1}{2},j+\frac{1}{2},k\right)-P_{1z}^{n}\left(i-\frac{1}{2},j+\frac{1}{2},k\right)\right]$$

$$+\frac{1}{1+\Delta t/2\tau_{10}}\frac{\Delta t}{2\tau_{21}}\left[N_{2z}^{n+1}\left(i-\frac{1}{2},j+\frac{1}{2},k\right)+N_{2z}^{n}\left(i-\frac{1}{2},j+\frac{1}{2},k\right)\right]$$

$$(4.151)$$

Update occupation number $N_0$:

$$N_{0x}^{n+1}\left(i,j+\frac{1}{2},k+\frac{1}{2}\right) = (1-P_{rx}\Delta t)N_{0x}^{n}\left(i,j+\frac{1}{2},k+\frac{1}{2}\right)$$

$$+\frac{\Delta t}{2\tau_{10}}\left[N_{1x}^{n+1}\left(i,j+\frac{1}{2},k+\frac{1}{2}\right)+N_{1x}^{n}\left(i,j+\frac{1}{2},k+\frac{1}{2}\right)\right]$$

$$(4.152)$$

$$N_{0y}^{n+1}\left(i-\frac{1}{2},j,k+\frac{1}{2}\right) = (1-P_{ry}\Delta t)N_{0y}^{n}\left(i-\frac{1}{2},j,k+\frac{1}{2}\right)$$

$$+\frac{\Delta t}{2\tau_{10}}\left[N_{1y}^{n+1}\left(i-\frac{1}{2},j,k+\frac{1}{2}\right)+N_{1y}^{n}\left(i-\frac{1}{2},j,k+\frac{1}{2}\right)\right]$$

$$(4.153)$$

$$N_{0z}^{n+1}\left(i-\frac{1}{2},j+\frac{1}{2},k\right) = (1-P_{rz}\Delta t)N_{0z}^{n}\left(i-\frac{1}{2},j+\frac{1}{2},k\right)$$

$$+\frac{\Delta t}{2\tau_{10}}\left[N_{1z}^{n+1}\left(i-\frac{1}{2},j+\frac{1}{2},k\right)+N_{1z}^{n}\left(i-\frac{1}{2},j+\frac{1}{2},k\right)\right]$$

$$(4.154)$$

## 4.3 Numerical Results

### 4.3.1 One-Dimensional Slab with Gain Materials

The following numerical experiments are based on transmission/reflection calculations. The incident pulses dynamically interact with the spatially resolved features of the structure, resulting in parts of the electromagnetic field being reflected, absorbed, and transmitted. We record time-series of the reflected and transmitted fields at each point of the front- and back-plane of the simulation domain, perform Fourier transforms, and determine the complex transmission $t(\omega)$ and reflection $r(\omega)$ coefficients (zero-order diffraction mode). From these we also calculate the spectral energy fluxes expressed in the transmission $T(\omega) = |t(\omega)|^2$, reflection $R(\omega) = |r(\omega)|^2$, and absorption $A(\omega) = 1 - T(\omega) - R(\omega)$.

To demonstrate the validity of our method, we first simulate a one-dimensional (1D) homogenous slab ($d = 200$ nm) embedded in gain material with dielectric constant $\varepsilon = 1$, as shown in Figure 4.5.

In the FDTD calculations, the discrete time step and space step are chosen to be $\Delta t = 1.5 \times 10^{-17}$ s and $\Delta x = \Delta y = \Delta z = 10.0 \times 10^{-9}$ m. The parameters for gain material are $\Gamma_a = 2\pi \times 10$ THz, $\sigma_a = 10^{-4}$ C$^2$/kg, $\omega_a = 2\pi \times 100$ THz. The parameters $\tau_{32}$, $\tau_{21}$, and $\tau_{10}$ are chosen to be $5 \times 10^{-14}$ s, $6.8 \times 10^{-11}$ s, and $5 \times 10^{-14}$ s, respectively. The initial electron density, $N_0(r, t = 0) = N_t = 5.0 \times 10^{23}$/m$^3$. The initial condition is that all the electrons are in the ground state, so there is no field, no polarization, and no spontaneous emission. Then the electrons are pumped from $N_0$ to $N_3$ (then relaxing to $N_2$) with a Gaussian pumping strength $P_g(t)$. The whole system begins to evolve according to the coupled equations. In this numerical simulation, we choose $P_g(t) = P_0 e^{-[(t-t_p)/t_p]^2}$, with $t_p = 3$ ps, $\tau_p = 0.6$ ps, where pumping amplitude $P_0$ is a variable according to the experiments. All the following numerical simulations are done with 16 CPUs parallel cluster.

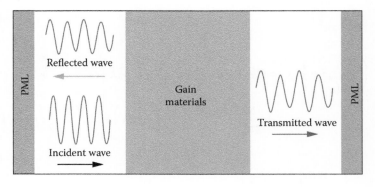

**FIGURE 4.5**
Schematic of the simulated 1D structure. (See color figure.)

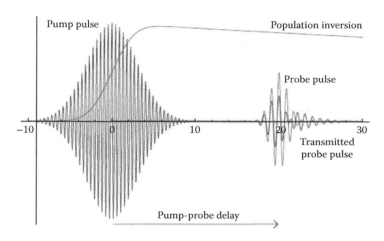

**FIGURE 4.6**
Schematic illustration of pump-probe experiments. (See color figure.)

We conduct numerical pump-probe experiments by first pumping the system at the pumping strength $P_g(t)$, and then we send a Gaussian pulse into the system with a suitable time delay (see Figure 4.6).

Due to the impedance matching, the transmission ($T$) should be one, and the reflection ($R$) and the absorption ($A$) should be zero in the frequency domain for the case without pumping (i.e., $P_0 = 0$ 1/s). This is true as shown in Figure 4.7(a), where we plot the $T$, $R$, and $A$ as a function of frequency in the propagation direction. We can see from the figure that the reflection and absorption are zeros. With the increasing of the pumping $P_0$, we can clearly see the behaviors of the transmission, reflection, and absorption as shown in Figure 4.7(b–d).

If we further increase the pumping amplitude to $P_0 = 3 \times 10^{10}$ 1/s, the absorption can reach the value of $-0.5$ and the transmission can acquire a value as high as 1.48 around the emission frequency of the gain material. Therefore, our model can be used as an effective way to overcome the loss of the metamaterials. To further study the time evolution of the electron population densities in all four levels, the nonlinear response can be observed as shown in Figure 4.8 and Figure 4.9 with $P_0 = 3 \times 10^9$ 1/s and $P_0 = 3 \times 10^{10}$ 1/s, respectively. Strong nonlinear behaviors can be obtained in all population densities, during the procession of Gaussian pumping. Starting with all the electrons in level 0, larger fractions can be excited with increasing pumping amplitudes from $3 \times 10^9$ 1/s to $3 \times 10^{10}$ 1/s. Also the nonlinear behaviors are disappeared as the pumping pulse exits the structure.

### 4.3.2 Loss Compensated Negative Index Material at Optical Wavelengths

The unit cell of the fishnet structure in simulations is shown in Figure 4.10. The structure is perforated and the dielectric in the hole is set as vacuum.

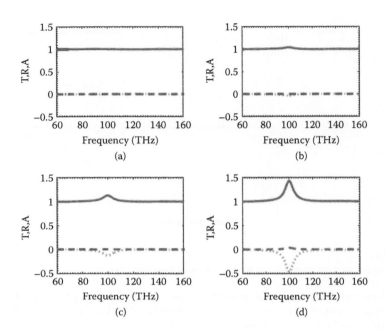

**FIGURE 4.7**
Simulated results for $T$ (solid), $R$ (dashed), and $A$ (dotted). (a) $P_0 = 0$ 1/s. (b) $P_0 = 3 \times 10^9$ 1/s. (c) $P_0 = 1 \times 10^{10}$ 1/s. (d) $P_0 = 3 \times 10^{10}$ 1/s.

The size of the unit cell along the propagation direction is $h = 2h_m + h_s + 2h_d$, where $h_m$, $h_s$, and $h_d$ are the thicknesses of the silver layer, the spacer, and the dielectric layer on the top/bottom surface, respectively. The permittivity of the silver is modeled by a Drude response with $\omega_{pe} = 1.37 \times 10^{16}$ rad/s and $\Gamma_e = 2.73 \times 10^{13}$ rad/s.

Consider two configurations, one without gain and one with gain using a homogeneous pumping rate. In the FDTD calculations, the discrete time and space steps are chosen to be $\Delta t = 8.0 \times 10^{-18}$ s and $\Delta x = 5 \times 10^{-9}$ m. The parameters for gain material are $\Gamma_a = 2\pi \times 10$ THz, $\sigma_a = 5 \times 10^{-7}$ C²/kg, $\lambda_a = 710$ nm. The parameters $\tau_{32}$, $\tau_{21}$, and $\tau_{10}$ are chosen to be $1 \times 10^{-13}$ s, $5 \times 10^{-10}$ s, and $1 \times 10^{-13}$ s, respectively. The initial electron density is $N_0(r, t = 0) = N_t = 6.0 \times 10^{24}$/m³. For the configuration with gain, two thin dielectric layers of thickness, $(h_s - h_g)/2$, are introduced to separate the gain from the metal layers so that the quenching effect can be avoided. The thickness of the gain layer is $h_g$. The emission wavelength of the gain is 710 nm. The full width at half maximum (FWHM) of the gain is 26.8 nm. Notice that the propagation direction is perpendicular to the fishnet plane with the electric and magnetic fields along the x- and y-directions, respectively. All the dimensions are chosen such that the magnetic resonance of the fishnet structure has a resonance wavelength of 710 nm, which overlaps with the peak of the emission of the gain material.

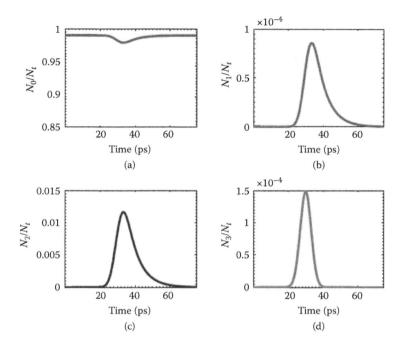

**FIGURE 4.8**
Normalized electron population densities in all four levels with $P_0 = 3 \times 10^9$ 1/s. (a) $N_0/N_t$. (b) $N_1/N_t$. (c) $N_2/N_t$. (d) $N_3/N_t$.

We start the calculations without gain, then increase the pumping rate, $P_r$ from 0 to $3.0 \times 10^8$ 1/s. In Figure 4.11, we plot (a) the transmission, $T$; (b) the reflection, $R$; and (c) the absorption, $A$, as a function of wavelength for different pumping rates. Notice that the wavelength dependence of $T$ and $R$ for different pumping rates away from the resonance wavelength is the same. Around the resonance wavelength, the transmission, $T$, has a peak and increases with the pumping rate. The reflection, $R$, shows a dip around the resonance wavelength, which gets deeper and narrower as the pumping rate increases. Notice in Figure 4.11(a), $T$ without gain is very weak. However, with the introduction of gain, the transmission gets much larger and clearly shows a resonant behavior. The absorption, $A$, is plotted in Figure 4.11(c) as a function of wavelength for different pumping rates. We notice that the absorption peak gets narrower as the pumping rate increases because the gain undamps the resonance. The absorption first increases at the resonance wavelength with the pumping rate because the gain changes the impedance of the fishnet metamaterials and more electromagnetic energy goes through the sample. As the gain further increases, the absorption decreases and finally at the pumping rate of $3.0 \times 10^8$ 1/s, it becomes negative, i.e., the gain

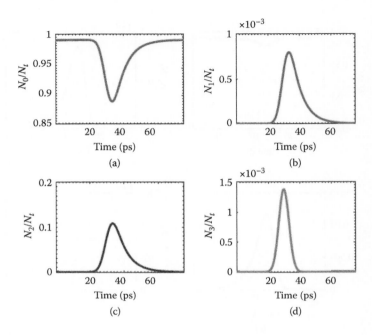

**FIGURE 4.9**
Normalized electron population densities in all four levels with $P_0 = 3 \times 10^{10}$ 1/s. (a) $N_0/N_t$. (b) $N_1/N_t$. (c) $N_2/N_t$. (d) $N_3/N_t$.

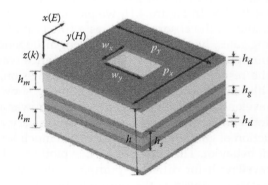

**FIGURE 4.10**
Unit cell of the perforated fishnet structure with gain embedded between two metal (silver) layers. The geometric parameters are $p_x = p_y = 280$ nm, $w_x = 75$ nm, $w_y = 115$ nm, $h = 170$ nm, $h_m = h_s = 50$ nm, $h_d = 10$ nm, and $h_g = 20$ nm. The thicknesses of the silver (yellow) and gain (magenta) layer are $h_m$ and $h_g$, respectively. The dielectric layer (blue) and the gain have a refractive index $n = 1.65$. (See color figure.)

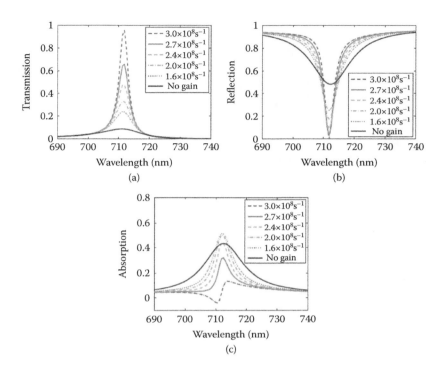

**FIGURE 4.11**
The transmission (a), reflection (b), and absorption (c) as a function of wavelength for different pumping rates. (See color figure.)

overcompensates the losses. So the gain can compensate the losses of the fishnet structure.

In addition, we plot the retrieved results of the real and imaginary parts of the magnetic permeability, μ, with and without gain in Figure 4.12(a) by inverting the scattering amplitudes [21,22]. Without gain, μ is very flat and we almost cannot see the existence of a resonance because the large loss damps the magnetic resonance. As gain increases, the Re(μ) becomes sharper around the resonance wavelength and the Im(μ) becomes much narrower since the gain compensates the losses and hence the resonance is undamped. The negative permittivity mainly comes from the continuous metallic wires. However, the dipole interaction between the corresponding electric dipole and gain material is dominated by the propagation field $O(\omega \ln |kr|)$ and is very weak, such that both the real and imaginary parts of the effective permittivity almost do not change with the introduction of gain. In Figure 4.12(b), we plot the retrieved results for the effective index of refraction, $n$, without and with gain for different pumping rates. The Re(μ) becomes more negative after the gain is introduced and the Im(μ) also drops

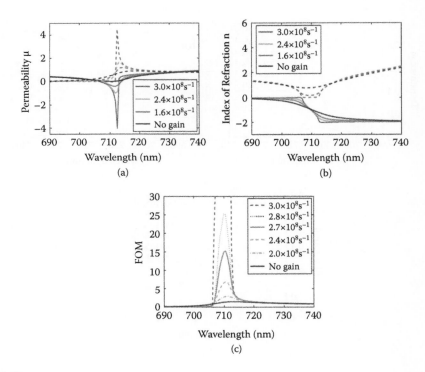

**FIGURE 4.12**
The retrieved results for the real (solid lines) and the imaginary (dashed lines) parts of (a) the effective permeability, μ, and (b) the corresponding effective index of refraction, *n*, without and with gain for different pumping rates. (c) The figure-of-merit (FOM) as a function of wavelength for different pumping rates. (See color figure.)

significantly near the resonance. At a wavelength, $\lambda = 712$ nm, slightly higher than the resonance wavelength, the Re($n$) reduces from −1.19 to −1.40 with a pumping rate, $P_r = 1.6 \times 10^8$ 1/s, and the Im($n$) drops from 0.83 to 0.26. Comparing these two Im($n$), we can find the effective extinction coefficient $\alpha = (\omega/c)\mathrm{Im}(n) \cong 7.35 \times 10^4$ cm$^{-1}$ without gain and $\alpha \cong 2.3 \times 10^4$ cm$^{-1}$ with a pumping rate of $1.6 \times 10^8$ 1/s. Hence, we can obtain an effective amplification coefficient of $\alpha \cong 5.05 \times 10^4$ cm$^{-1}$ for the combined system. This is much larger (on the order of 20) than the expected amplification of $\alpha \cong 2.5 \times 10^3$ cm$^{-1}$ for the bulk gain at the pumping rate $P_r = 1.6 \times 10^8$ 1/s. This difference can be explained by the strong local-field enhancement inside the fishnet structure. In Figure 4.12(c), we plot the figure-of-merit, FOM $= |\ \mathrm{Re}(n)/\mathrm{Im}(n)|$, versus the wavelength for different pumping rates. We find that the FOM can reach a large value on the order of 100 in a certain wavelength range with the introduction of gain. It shows that the gain compensates the losses of the fishnet metamaterials in this range.

### 4.3.3 Pump-Probe Experiments for Metallic Metamaterials Coupled to a Gain Medium

The object of this study is to present pump-probe simulations on arrays of silver split-ring resonators (SRR) coupled to single quantum wells [11,23,24]. The structure considered is a U-shaped SRR fabricated on a gain-GaAs substrate with a square periodicity of $p = 250$ nm (see Figure 4.13[a]). The SRR is made of silver with its permittivity modeled by a Drude response. The incident wave propagates perpendicular to the SRR plane and has the electric field polarization parallel to the gap (see Figure 4.13[a]). The corresponding geometrical parameters are $a = 150$ nm, $h_d = 40$ nm, $h_g = 20$ nm, $h_s = 30$ nm, $w = 50$ nm, and $h = 75$ nm. In the FDTD calculations, the discrete time and space steps are chosen to be $\Delta t = 1.5 \times 10^{-17}$ s and $\Delta x = 10 \times 10^{-9}$ m. The parameters for gain material are $\Gamma_a = 2\pi \times 20$ THz, $\sigma_a = 1 \times 10^{-4}$ C$^2$/kg, $\omega_a = 175$ THz. The parameters $\tau_{32}$, $\tau_{21}$, and $\tau_{10}$ are chosen to be 0.05 ps, 80 ps, and 0.05 ps, respectively. The initial electron density is $N_0(r, t = 0) = N_t = 5.0 \times 10^{23}/$m$^3$. Here we use the Gaussian pumping rate with $P_0 = 3 \times 10^9$ s$^{-1}$, $t_p = 6$ ps, and $\tau_p = 0.15$ ps.

Figure 4.13(b) shows the calculated spectrum (without pump) of transmittance $T$, reflectance $R$, and absorptance $A$ for the structure shown in Figure 4.13(a). The resonant frequency is around 175 THz, and we refer to the resonant frequency according to the dip of the transmittance. In our analysis, we first pump the active structure (see Figure 4.13[a]) with a short intensive Gaussian pump pulse $P_g(t)$ (see Figure 4.3, top panel). After a suitable time delay (i.e., the pump-probe delay), we probe the structure with a weak Gaussian probe pulse with a center frequency close to the SRR resonance frequency of 175 THz.

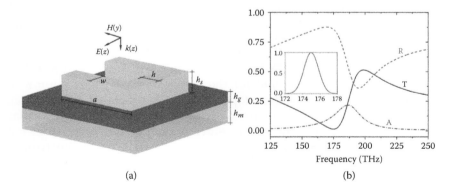

(a)                  (b)

**FIGURE 4.13**

(a) Schematic of the unit cell for the silver-based SRR structure (yellow) with the electric field polarization parallel to the gap. The dielectric constants ε for gain (red) and GaAs (light blue) are 9.0 and 11.0, respectively. (b) Calculated spectra for transmittance $T$ (black), reflectance $R$ (red), and absorptance $A$ (blue) for the structure shown in Figure 4.13(a). The inset shows the profile of the probe pulse with a center frequency of 175 THz (FWHM = 2 THz). (See color figure.)

Typical examples for the spatial distribution of electric field and gain are shown in Figure 4.14. In Figure 4.14(a), we have plotted the spatial distribution of the local electric field components, $E_x$, $E_y$, and $E_z$, with pump ($P_0 = 3 \times 10^9$ 1/s) near the resonance of the SRR (*i–iii*). $E_x$ is very strong inside the SRR gap; also, the normal component $E_z$ is reasonably strong near the ends of the open arms of the SRR. These are the locations where most of the gain originates. We plotted also the spatial distribution of the change in the electric field between pumped and unpumped transmission of the probe pulse, $\Delta E = |E_{\text{pump}}| - |E_{\text{withoutpump}}|$ where $|E| = |E_x^2 + E_y^2 + E_z^2|^{1/2}$, as a function of the pump-probe delay (*iv–vi*) and the strength of the pump pulse (*vii–ix*). As the pump-probe delay increases, the strength of $\Delta E$ becomes smaller; as the pumping rate increases, the amplitude of $\Delta E$ becomes bigger. This is not unexpected and simply is a consequence of the changing population inversion, i.e., the change in available local gain.

The incident electric field amplitude of the probe pulse is 10 V/m, which is well inside the linear response regime (see Figure 4.15). Then, we can Fourier transform the time-dependent transmitted electric field and divide by the Fourier transform of the incident probe pulse to obtain the spectral transmittance of the system as seen by the probe pulse. Additionally, we obtain the total pulse transmittance by dividing the energy in the transmitted pulse by the energy in the incident pulse, integrated in the time domain. We define the differential transmittance $\Delta T/T$ by taking the difference of the measured total plus transmittance with pumping the active structure minus the same without pumping and dividing it by the total plus transmittance without pumping. This differential transmittance is a function of the pump-probe delay.

The bottom panel in Figure 4.16 gives a differential transmittance $\Delta T/T$ that is negative. This result was not expected, and we need to understand this behavior, which agrees with the experiments [11, 23, 24].

Figure 4.17 gives an overview of the results obtained for the case of the SRRs on-resonance, i.e., $f_r = 175$ THz. Data for the structure in Figure 4.13(a) (left column in Figure 4.17) and for the bare gain case (right column in Figure 4.17) without the SRR on top is shown. For parallel polarization, the light does couple to the fundamental SRR resonance; for perpendicular polarization, it does not. The probe center frequency decreases from top (179 THz) to bottom (169 THz). Note that the width of the probe spectrum is 2 THz (see the inset in Figure 4.13[a]). Hence, the data have been taken with 2 THz spectral separation. Inspection of the left column shows a rather different behavior for the SRRs with gain compared to the bare gain case. While the bare gain always delivers positive $\Delta T/T$ signals below +0.16% (right column) over the whole probe spectrum, the sign and magnitude of the signals change for the case SRR with gain.

Under some conditions, $\Delta T/T$ reaches values as negative as −8.50% around $f_{\text{probe}} = 175$ THz. Additionally, we may also get positive $\Delta T/T$ at the very edges of the probe range (see the left column in Figure 4.17). If we turn to the case

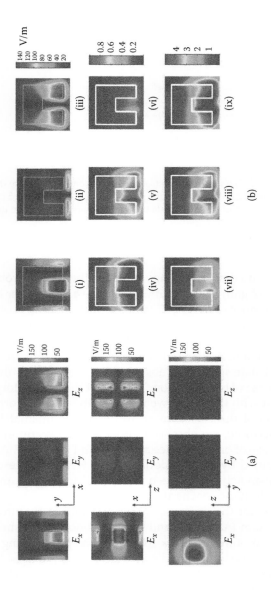

**FIGURE 4.14**

The spatial distribution of electric field and gain. (a) The electric field in different cut plane without pumping. (b) The first row corresponds to the electric field amplitude distributions at 175 THz with pump ($P_0 = 3 \times 10^9\,\text{s}^{-1}$ and $\Delta t = 5$ ps) in the cross section of the gain layer ($z = 40$ nm from the top of the structure) for different components: (i) $E_x$, (ii) $E_y$, and (iii) $E_z$. The second row corresponds to the near-field differential $\Delta E$ with $P_0 = 3 \times 10^9\,\text{s}^{-1}$ for three different time delays, namely (iv) 5 ps, (v) 45 ps, and (vi) 135 ps, respectively. The third row corresponds to the near-field differential, $\Delta E$ at $\Delta t = 5$ ps, for three different pumping strengths, namely (vii) $P_0 = 6 \times 10^9\,\text{s}^{-1}$, (viii) $P_0 = 9 \times 10^9\,\text{s}^{-1}$, and (ix) $P_0 = 12 \times 10^9\,\text{s}^{-1}$, respectively. Here the $\Delta E$ is defined by taking the difference of the measured total electric field with pumping the active structure minus the same without pumping. The area enclosed by the white line is the projection of the SRRs on the gain layer. (See color figure.)

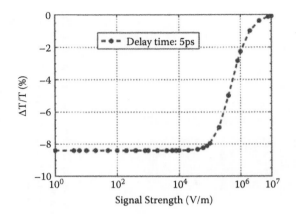

**FIGURE 4.15**

Results for differential transmission ($\Delta T/T$) with different strength of incident signal. Here the case is on-resonance case with FWHM of 2 THz for the probe pulse, and the time delay is 5 ps. For incident fields stronger than $10^4$ V/m the metamaterials behave nonlinearly. Here we use strength of 10 V/m, so we are in the deep linear region.

of perpendicular polarization, no distinct change between the pump-probe results on the SRRs and the bare gain (right column in Figure 4.17), neither in the magnitude nor in the dynamics of the $\Delta T/T$, can be detected.

We argue that the distinct behavior can be attributed to the strong coupling between the resonances of the SRRs and the gain medium. The negative $\Delta T/T$ is not as we expected at first glance: The pump lifts electrons from the ground state to an excited state so that the absorption of the probe pulse is reduced, leading to an increase of transmission. This is not the whole story. The reason lies in the fact that with the pump we not only affect the absorption but disturb the reflection of the structure, resulting in the mismatching of the impedance. Furthermore, we observed either an increasing or a decreasing tendency for the case of on-resonance as shown in Figure 4.17. All those behaviors can be explained by the competing of the weak gain resonance and the impedance mismatching between pump and without pump cases. We will explore the underlying mechanism below.

Figure 4.18 shows the results for the difference in absorptance ($\Delta A$), difference in reflectance ($\Delta R$), their sum ($\Delta A + \Delta R$), and the difference in transmittance ($\Delta T = -[\Delta A + \Delta R]$) between pump ($P_0 = 3 \times 10^9$ s$^{-1}$) and no pump using a wide probe (FWHM = 54 THz) pulse with a fixed pump-probe delay of 5 ps. As expected, we may observe a positive differential transmittance, $\Delta T/T > 0$, when we pump the gain, $\Delta A < 0$, and if $\Delta R$ (impedance match) remains unchanged.

The results of Figure 4.18 are obtained for pump-probe experiments with the probe frequency equal to the resonance frequency of the SRR (175 THz) at a pump-probe delay of 5 ps. Notice that $\Delta R$ is positive, $\Delta A$ is negative, and

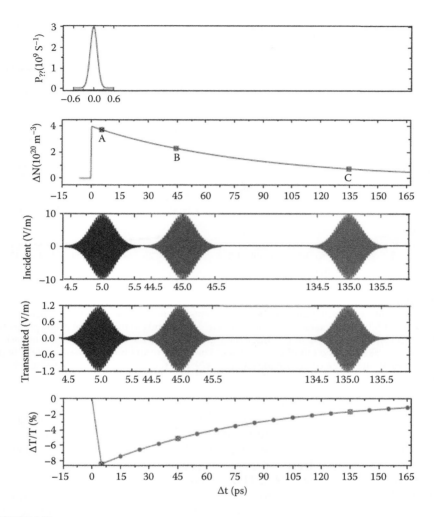

**FIGURE 4.16**
Schematic of the numerical pump-probe experiments for the case on-resonance. From the top to the bottom, each row corresponds to the pump pulse, population inversion, incident signal (with time delays 5, 45, and 135 ps), transmitted signal, and differential transmittance $\Delta T/T$. It should be mentioned here that the incident frequency of the probe pulse is 175 THz with a FWHM of 2 THz and is equal to the SRR resonance frequency. (See color figure.)

$\Delta T$ is also negative very close to the resonance frequency. If the probe center frequency moves away from the SRR resonance frequency, the negative $\Delta T/T$ decreases in magnitude, and finally $\Delta T/T$ becomes positive. These results are shown in Figure 4.19 and agree with experiments [11,23,24]. Results for longer pump-probe delays are shown in Figure 4.20.

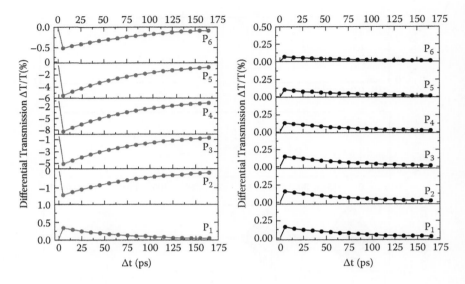

**FIGURE 4.17**

Time domain numerical pump-probe experiments results for the SRR that is nearly on-resonant with the gain material. The left column corresponds to the parallel probe polarization with respect to the gap of the SRRs; the right column is the case for bare gain material, i.e., without SRRs on the top of the substrate. The width of the probe signal is 2 THz with decreasing in the probe center frequency from 179 THz for the top panel to 169 THz for the bottom panel.

In Figure 4.20 we present some additional results for the change in transmittance, reflectance, and absorptance between the pumped and unpumped cases augmenting Figure 4.18. We consistently observe a negative differential transmittance for probe pulses centered at the resonance frequency of the SRR (2 THz probe pulse bandwidth, on-resonance case). We show the results for three different pump-probe delays ranging from 5 ps to 135 ps. The 5 ps case (gray solid line) corresponds to the case shown in Figure 4.18. The differences between the curves shown in Figure 4.20 result from the exponential decay of the population inversion created by the pump (available gain) with increasing pump-probe delay due to nonradiative decay in the rate equations. Thus, the longer we wait, the smaller the effective gain. We see that the absorptance change, $\Delta A$, becomes less negative and the reflectance change, $\Delta R$, and the transmittance change, $\Delta T$, become smaller with increasing pump-probe delay. All three cases maintain the negative differential transmittance. However, in all cases the increase in reflectance (due to increased impedance mismatch near the undamped resonance) is the dominant contribution to the differential transmittance, rendering it negative despite the negative sign in the differential absorptance.

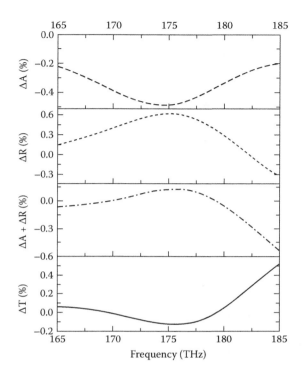

**FIGURE 4.18**
Frequency domain numerical pump-probe experiments results for the on-resonance case. Simulations results for the differences in transmittance ($\Delta T$), reflectance ($\Delta R$), and absorptance ($\Delta A$) versus frequency.

If we can increase the magnitude of the Gaussian pump pulse $P_0$ to $5 \times 10^{10}$ s$^{-1}$ and we repeat the pump-probe experiments, $\Delta T/T \cong -100\%$ at resonance frequency (see Figure 4.21[a]), 175 THz. If we increase the pump amplitude further to $10^{11}$ s$^{-1}$, we can compensate for the losses (see Figure 4.21[b]).

However, such pump intensities are unrealistic experimentally. Here, we have introduced a new approach for pump-probe simulations of metallic metamaterials coupled to gain materials. We study the coupling between the U-shaped SRRs and the gain material described by a four-level gain model. Using pump-probe simulations, we find a distinct behavior for the differential transmittance $\Delta T/T$ of the probe pulse with and without SRRs in both magnitude and sign (negative, unexpected, and/or positive). Our new approach has verified that the coupling between the metamaterial resonance and the gain medium is dominated by near-field interactions. Our model can be used to design new pump-probe experiments to compensate for the losses of metamaterials.

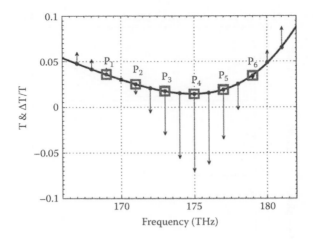

**FIGURE 4.19**
The transmittance $T$ (without pump, solid line) and the on-resonance differential transmittance $\Delta T/T$ results (vector arrow). The direction and the length of the arrow stand for the sign and the amplitude of $\Delta T/T$, respectively. The squares from $P_1$ to $P_6$ correspond to the frequency of probe pulse ranging from 169 to 179 THz with uniform step of 2 THz.

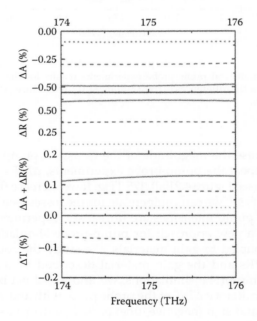

**FIGURE 4.20**
Frequency domain numerical pump-probe experiments results for the on-resonance case (see Figure 4.5) for a pump-probe delay of 5 ps (solid curve), 45 ps (dashed curve) and 135 ps (dotted curve), respectively. Here $\Delta T$, $\Delta R$, and $\Delta A$ are the change in transmittance, reflectance, and absorptance, respectively, between the pumped and unpumped case.

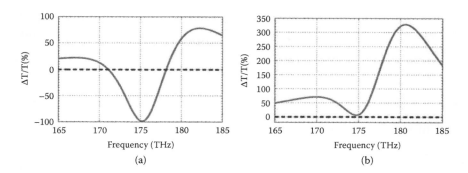

**FIGURE 4.21**
Corresponding frequency domain results of the differential transmission $\Delta T/T$ for the case with pump rate (a) $P_0 = 5.0 \times 10^{10}\,s^{-1}$ and (b) $P_0 = 10.0 \times 10^{10}\,s^{-1}$.

## 4.4 Discussions and Conclusions

In conclusion, we have proposed a parallel FDTD model incorporating a four-level atomic system. The detailed implication and efficiency of the method are also included. A 1D gain slab is used to check the validity of the method. Furthermore, 3D fishnet and SRR structures with gain material underneath are simulated to demonstrate the strong interaction between structures and gain materials.

Our model can be used as an instruction for real pump-probe experiments in metamaterials and provide an insight into the dynamic interaction between nanostructure and gain materials. A significant advantage of our method is the complexity of the FDTD, which scales only linearly with time and space. This ability has been demonstrated by many numerical simulations in photonic crystals [25]. However, the FDTD method is very demanding in terms of memory and speed of the available computer hardware when applied to practical 3D problems such as analysis of transmission spectra of photonic crystals waveguide in layered structures with 2D patterning. Thus, one can turn to the parallel of FDTD method. In addition, there has been less effort to apply the FDTD method in studying dynamical interaction of light and matter in a photonic crystal semiconductor optical amplifier [26]. One can use the self-consistent method presented here, to investigate the amplification and lasing in photonic crystals and other random configurations of nanostructures. Through this 3D nonlinear gain FDTD method, low-threshold laser can be found, which cannot be obtained from conventional travel rate equation approach. The inclusion of both spontaneous emission and performing the pump-probe experiment in active photonic crystals waveguide in order to achieve a more realistic picture of the complex nonlinear dynamics in these devices is currently under investigation.

## Acknowledgments

The authors acknowledge the support of the grant NSFC (Nos. 60931002, 61101064, 51277001, 61201122), Distinguished Natural Science Foundation (No. 1108085J01), Universities Natural Science Foundation of Anhui Province (Nos. KJ2011A002, KJ2011A242, KJ2012A013), DFMEC (No. 20123401110009), and NCET (No. NCET-12-0596) of China.

## References

1. V. M. Shalaev, "Optical negative-index metamaterials," *Nat. Photon.* 1(1), 41–48 (2007).
2. C. M. Soukoulis, S. Linden, and M. Wegener, "Negative refractive index at optical wavelengths," *Science* 315(5808), 47–49 (2007).
3. X. Jiang and C. M. Soukoulis, "Time dependent theory of random lasers," *Phys. Rev. Lett.* 85(1), 70–73 (2000).
4. P. Bermel, E. Lidorikis, Y. Fink, and J. D. Joannopoulos, "Active materials embedded in photonic crystals and coupled to electromagnetic radiation," *Phys. Rev. B* 73(16), 165125(1–8) (2006).
5. J. B. Pendry, "Negative refraction makes a perfect lens," *Phys. Rev. Lett.* 85(18), 3966–3969 (2000).
6. D. Schurig, J. J. Mock, B. J. Justine, S. A. Cummer, J. B. Pendry, A. F. Starr, and D. R. Smith, "Metamaterial electromagnetic cloak at microwave frequencies," *Science* 314(5801), 977–980 (2006).
7. U. Leonhardt, "Optical conformal mapping," *Science*, vol. 312, no. 5781, pp. 1777–1780, 2006.
8. A. A. Zharov, I. A. Shadrivov, and Y. A. Kivshar, "Nonlinear properties of left-handed metamaterials," *Phys. Rev. Lett.* 91(3), 037401(1–4) (2003).
9. S. Xiao, V. P. Drachev, A. V. Kildishev, X. Ni, U. K. Chettiar, H.-K.Yuan, and V. M. Shalaev, "Loss-free and active optical negative-index metamaterials," *Nature* 466(7307), 735–738 (2010).
10. E. Plum, V. A. Fedotov, P. Kuo, D. P. Tsai, and N. I. Zheludev, "Toward the lasing spaser: Controlling metamaterial optical response with semiconductor quantum dots," *Opt. Exp.* 17(10), 8548–8551 (2009).
11. N. Meinzer, M. Ruther, S. Linden, C. M. Soukoulis, G. Khitrova, J.Hendrickson, J. D. Olitzky, H. M. Gibbs, and M. Wegener, "Arrays of Ag split-ring resonators coupled to InGaAs single-quantum-well gain," *Opt. Exp.* 18(23), 24140–24151 (2010).

12. A. E. Siegman, *Lasers*. Hill Valley, CA: Univ. Science Books, 1986.
13. A. Taflove, *Computational Electrodynamics: The Finite Difference Time Domain Method*. Norwood, MA: Artech House, 1995.
14. A. Fang, T. Koschny, and C. M. Soukoulis, "Lasing in metamaterials nanostructures," *J. Opt.* 12(2), 024013(1–13) (2010).
15. A. Fang, T. Koschny, and C. M. Soukoulis, "Self-consistent calculation of metamaterials with gain," *Phys. Rev. B* 79(24), 241104(1–4) (2009).
16. S. Wuestner, A. Pusch, K. L. Tsakmakidis, J. M. Hamm, and O. Hess, "Gain and plasmon dynamics in negative-index metamaterials," *Phil. Trans. Royal Soc. A* 369(1950), 3525–3550 (2011).
17. Z. Huang, Th. Koschny, and C. M. Soukoulis, "Theory of pump-probe experiments of metallic metamaterials coupled to the gain medium," *Phys. Rev. Lett.* 108, 187402(1–5) (2012).
18. J. P. Berenger, "A perfectly matched layer for the absorption of electromagnetic waves," *J. Comput. Phys.* 114(2), 185–200 (1994).
19. J. P. Berenger, "Perfectly matched layer for the FDTD solution of wave-structure interaction problems," *IEEE Trans. Antennas Propag.* 44(1), 110–117 (1996).
20. J. P. Berenger, "Improved PML for the FDTD solution of wave-structure interaction problems," *IEEE Trans. Antennas Propag.* 45(2), 466–473 (1997).
21. D. R. Smith, S. Schultz, P. Markos, C. M. Soukoulis, "Determination of permittivity and permeability of metamaterials from scattering data," *Phys. Rev. B.* 65, 195104(1–5) (2002).
22. T. Koschny, M. Kafesaki, E. N. Economou, and C. M. Soukoulis, "Effective medium theory of left-handed materials," *Phys. Rev. Lett.* 93, 107402(1–4) (2004).
23. N. Meinzer, M. Ruther, S. Linden, C. M. Soukoulis, G. Khitrova, J. Hendrickson, J. D. Olitzky, H. M. Gibbs, and M. Wegener, "Plasmonic metamaterials coupled to single InGaAs-quantum-well gain," *Lasers and Electro-Optics (CLEO)*, 2011 Conference, pp. 1–2.
24. N. Meinzer, M. Konig, M. Ruther, S. Linden, G. Khitrova, H. M. Gibbs, K. Busch, and M. Wegener, "Distance-dependence of the coupling between split-ring resonators and single-quantum-well gain," *Appl. Phys. Lett.* 99(11), 111104(1–3) (2011).
25. S. Obayya, *Computational Photonics*. New York: Wiley, 2010.
26. B. Redding, S. Shi, T. Creazzo, and D. W. Prather, "Electromagnetic modeling of active silicon nanocrystal waveguides," *Opt. Exp.* 16(12), 8792–8799 (2008).

# Appendix

```
//Note:part of our FDTD code is included for reference
void fdtd_update(struct fdtd *fdtd)
{

    //FDTD FIELD UPDATES:


double c1,c2,c3;
//1. update magnetic field Hx, Hy, Hz
```

```
TIMER_START;

for (k = 0;k<MAX_Z;k++)
  for (j = 0;j<MAX_Y;j++)
    for (i = 0;i<MAX_X;i++)

    {

  switch(type_Hx[L(i,j,k)])
    {
      //Update Hx
      case UPDATE_DIELECTRIC:
      case UPDATE_GAIN:
      case UPDATE_DRUDE:
        Hx[L(i,j,k)] = Hx[L(i,j,k)]+dt/(muz*dz)*(Ey[time%2][L(i,j,k+1)]-
        Ey[time%2][L(i,j,k)])-
            dt/(muz*dy)*(Ez[time%2][L(i,j+1,k)]-Ez[time%2][L(i,j,k)]);

        break;


      case UPDATE_PML:

      //Update Hxy

        c1 = (1-sigma_y_Hx[L(i,j,k)]*dt/(2.0*epsz))/(1+sigma_y_Hx[L(i,j,k)]*dt/
            (2.0*epsz));
        c2 = dt/(dy*muz*(1.0+sigma_y_Hx[L(i,j,k)]*dt/(2.0*epsz)));
        Hxy[L(i,j,k)] = c1*Hxy[L(i,j,k)]-c2*(Ez[time%2][L(i,j+1,k)]-Ez[time%2]
        [L(i,j,k)]);


      //Update Hxz

        c1 = (1-sigma_z_Hx[L(i,j,k)]*dt/(2.0*epsz))/(1+sigma_z_Hx[L(i,j,k)]*dt/
            (2.0*epsz));
        c2 = dt/(dz*muz*(1.0+sigma_z_Hx[L(i,j,k)]*dt/(2.0*epsz)));
        Hxz[L(i,j,k)] = c1*Hxz[L(i,j,k)]+c2*(Ey[time%2][L(i,j,k+1)]-Ey[time%2]
        [L(i,j,k)]);

        Hx[L(i,j,k)] = Hxy[L(i,j,k)]+Hxz[L(i,j,k)];


        break;


    }

        switch(type_Hy[L(i,j,k)])

    {
        case UPDATE_DIELECTRIC:
        case UPDATE_GAIN:
        case UPDATE_DRUDE:
        //Update Hy
          Hy[L(i,j,k)] = Hy[L(i,j,k)]+dt/(muz*dx)*(Ez[time%2][L(i+1,j,k)]-
          Ez[time%2][L(i,j,k)])-
              dt/(muz*dz)*(Ex[time%2][L(i,j,k+1)]-Ex[time%2][L(i,j,k)]);


          break;
        case UPDATE_PML:
        //Update Hyz
```

```
            c1 = (1-sigma_z_Hy[L(i,j,k)]*dt/(2.0*epsz))/(1+sigma_z_Hy[L(i,j,k)]*dt/
                (2.0*epsz));
            c2 = dt/(dz*muz*(1.0+sigma_z_Hy[L(i,j,k)]*dt/(2.0*epsz)));
            Hyz[L(i,j,k)] = c1*Hyz[L(i,j,k)]-c2*(Ex[time%2][L(i,j,k+1)]-Ex[time%2]
            [L(i,j,k)]);

            //Update Hyx

            c1 = (1-sigma_x_Hy[L(i,j,k)]*dt/(2.0*epsz))/(1+sigma_x_Hy[L(i,j,k)]*dt/
                (2.0*epsz));
            c2 = dt/(dx*muz*(1.0+sigma_x_Hy[L(i,j,k)]*dt/(2.0*epsz)));
            Hyx[L(i,j,k)] = c1*Hyx[L(i,j,k)]+c2*(Ez[time%2][L(i+1,j,k)]-Ez[time%2]
            [L(i,j,k)]);


            Hy[L(i,j,k)] = Hyz[L(i,j,k)]+Hyx[L(i,j,k)];

            break;
        }


        switch(type_Hz[L(i,j,k)])
        {
        case UPDATE_DIELECTRIC:
        case UPDATE_GAIN:
        case UPDATE_DRUDE:
    //Update Hz
            Hz[L(i,j,k)] = Hz[L(i,j,k)]+dt/(muz*dy)*(Ex[time%2][L(i,j+1,k)]-
            Ex[time%2][L(i,j,k)])-
                dt/(muz*dx)*(Ey[time%2][L(i+1,j,k)]-Ey[time%2][L(i,j,k)]);

            break;


        case UPDATE_PML:
            //Update Hzx
              c1 = (1-sigma_x_Hz[L(i,j,k)]*dt/(2.0*epsz))/(1+sigma_x_Hz[L(i,j,k)]*dt/
                (2.0*epsz));
              c2 = dt/(dx*muz*(1.0+sigma_x_Hz[L(i,j,k)]*dt/(2.0*epsz)));
              Hzx[L(i,j,k)] = c1*Hzx[L(i,j,k)]-c2*(Ey[time%2][L(i+1,j,k)]-Ey[time%2]
              [L(i,j,k)]);

    //Update Hzy
              c1 = (1-sigma_y_Hz[L(i,j,k)]*dt/(2.0*epsz))/(1+sigma_y_Hz[L(i,j,k)]*dt/
                (2.0*epsz));
              c2 = dt/(dy*muz*(1.0+sigma_y_Hz[L(i,j,k)]*dt/(2.0*epsz)));
              Hzy[L(i,j,k)] = c1*Hzy[L(i,j,k)]+c2*(Ex[time%2][L(i,j+1,k)]-Ex[time%2]
              [L(i,j,k)]);
              Hz[L(i,j,k)] = Hzx[L(i,j,k)]+Hzy[L(i,j,k)];


            break;
        }


    }
//2. Make Hx, Hz corrections over the connecting interfaces (TF/SF method)
for (k = 0;k<GLOBAL_MAX_Z;k++)
  for (i = 0;i<GLOBAL_MAX_X;i++)
  {
if(IN_LOCAL_DOMAIN(X,Y,Z,i,JLB-1,k))
  {
        double Ez_inc_L, Ex_inc_L;
```

```
            Ez_inc_L = E_inc[JLB]*cos(theta);
            Ex_inc_L = -E_inc[JLB]*sin(theta);
            Hx[P(i,JLB-1,k)] = Hx[P(i,JLB-1,k)]+dt/(dy*muz)*Ez_inc_L;//left interface
            Hz[P(i,JLB-1,k)] = Hz[P(i,JLB-1,k)]-dt/(dy*muz)*Ex_inc_L;//left interface
        }
if(IN_LOCAL_DOMAIN(X,Y,Z,i,JRB,k))
    {
            double Ez_inc_R, Ex_inc_R;
            Ez_inc_R = E_inc[JRB]*cos(theta);
            Ex_inc_R = -E_inc[JRB]*sin(theta);
            Hx[P(i,JRB,k)] = Hx[P(i,JRB,k)]-dt/(dy*muz)*Ez_inc_R;//right interface
            Hz[P(i,JRB,k)] = Hz[P(i,JRB,k)]+dt/(dy*muz)*Ex_inc_R;//right interface

    }

    }


//3. Update Hinc in the grid generating lookup table
//Update Hinc
for (j = 0;j<GLOBAL_MAX_Y-1;j++)

{
  c1 = (1.0-sigma_H_inc[j]*dt/(2.0*epsz))/(1.0+sigma_H_inc[j]*dt/(2.0*epsz));
  c2 = dt/(muz*dy)/(1.0+sigma_H_inc[j]*dt/(2.0*epsz));
  H_inc[j] = c1*H_inc[j]-c2*(E_inc[j+1]-E_inc[j]);

}

TIMER_STOP(0);
edge_update(Hx,edge_buf); /* this communicate edges with neighboring nodes */
edge_update(Hy,edge_buf);
edge_update(Hz,edge_buf);
TIMER_STOP(1);

//4.Update the polarization P and current J


TIMER_START;


for (k = 0;k<MAX_Z;k++)
  for (j = 0;j<MAX_Y;j++)
    for (i = 0;i<MAX_X;i++)
      {
          switch(type_Ex[L(i,j,k)])
          {
          case UPDATE_GAIN:
          c1 = (2.0-(omega1[L(i,j,k)]*dt)*(omega1[L(i,j,k)]*dt))/
              (1.0+0.5*gamma1[L(i,j,k)]*dt);
          c2 = (1.0-0.5*gamma1[L(i,j,k)]*dt)/(1.0+0.5*gamma1[L(i,j,k)]*dt);
          c3 = sigma1[L(i,j,k)]*dt*dt/(1+0.5*gamma1[L(i,j,k)]*dt);
          P1x[(time+1)%2][L(i,j,k)] =
c1*P1x[time%2][L(i,j,k)]-c2*P1x[(time-1+2)%2][L(i,j,k)]+c3*(N1x[time%2]
[L(i,j,k)]-N2x[time%2][L(i,j,k)])*Ex[time%2][L(i,j,k)];
          Jx[L(i,j,k)] = (P1x[(time+1)%2][L(i,j,k)]-P1x[time%2][L(i,j,k)])/dt;


            break;
          case UPDATE_DRUDE:
            //c1 = (1.0-0.5*gamma_e_x[L(i,j,k)]*dt)/(1.0+0.5*gamma_e_x[L(i,j,k)]*dt);
            //c2 = epsz*omega_pe_x[L(i,j,k)]*omega_pe_x[L(i,j,k)]*dt/(1.0+0.5*gamma
                _e_x[L(i,j,k)]*dt);
            //Jx[L(i,j,k)] = c1*Jx[L(i,j,k)]+c2*Ex[time%2][L(i,j,k)];
```

```
        c1 = 2.0/(1.0+0.5*gamma_e_x[L(i,j,k)]*dt);
        c2 = (1.0-0.5*gamma_e_x[L(i,j,k)]*dt)/(1.0+0.5*gamma_e_x[L(i,j,k)]*dt);
        c3 = epsz*(omega_pe_x[L(i,j,k)]*dt)*(omega_pe_x[L(i,j,k)]*dt)/
             (1.0+0.5*gamma_e_x[L(i,j,k)]*dt); P1x[(time+1)%2][L(i,j,k)] =
c1*P1x[time%2][L(i,j,k)]-c2*P1x[(time-1+2)%2][L(i,j,k)]+c3*Ex[time%2][L(i,j,k)];
        Jx[L(i,j,k)] = (P1x[(time+1)%2][L(i,j,k)]-P1x[time%2][L(i,j,k)])/dt;


        break;

    }


    switch(type_Ey[L(i,j,k)])

    {

        case UPDATE_GAIN:
          c1 = (2.0-(omega1[L(i,j,k)]*dt)*(omega1[L(i,j,k)]*dt))/
               (1.0+0.5*gamma1[L(i,j,k)]*dt);
          c2 = (1.0-0.5*gamma1[L(i,j,k)]*dt)/(1.0+0.5*gamma1[L(i,j,k)]*dt);
          c3 = sigma1[L(i,j,k)]*dt*dt/(1+0.5*gamma1[L(i,j,k)]*dt);
          P1y[(time+1)%2][L(i,j,k)] =
c1*P1y[time%2][L(i,j,k)]-c2*P1y[(time-1+2)%2][L(i,j,k)]+c3*(N1y[time%2]
[L(i,j,k)]-N2y[time%2][L(i,j,k)])*Ey[time%2][L(i,j,k)];
          Jy[L(i,j,k)] = (P1y[(time+1)%2][L(i,j,k)]-P1y[time%2][L(i,j,k)])/dt;


        break;
      case UPDATE_DRUDE:
        //c1 = (1.0-0.5*gamma_e_y[L(i,j,k)]*dt)/(1.0+0.5*gamma_e_y[L(i,j,k)]*dt);
        //c2 = epsz*omega_pe_y[L(i,j,k)]*omega_pe_y[L(i,j,k)]*dt/(1.0+0.5*gamma_e
             _y[L(i,j,k)]*dt);
        //Jy[L(i,j,k)] = c1*Jy[L(i,j,k)]+c2*Ey[time%2][L(i,j,k)];
        c1 = 2.0/(1.0+0.5*gamma_e_y[L(i,j,k)]*dt);
        c2 = (1.0-0.5*gamma_e_y[L(i,j,k)]*dt)/(1.0+0.5*gamma_e_y[L(i,j,k)]*dt);
        c3 = epsz*(omega_pe_y[L(i,j,k)]*dt)*(omega_pe_y[L(i,j,k)]*dt)/
             (1.0+0.5*gamma_e_y[L(i,j,k)]*dt);
        P1y[(time+1)%2][L(i,j,k)] = c1*P1y[time%2][L(i,j,k)]-c2*P1y[(time-1+2)%2]
        [L(i,j,k)]+c3*Ey[time%2][L(i,j,k)];
        Jy[L(i,j,k)] = (P1y[(time+1)%2][L(i,j,k)]-P1y[time%2][L(i,j,k)])/dt;


        break;
    }


    switch(type_Ez[L(i,j,k)])
    {
        case UPDATE_GAIN:
            c1 = (2.0-(omega1[L(i,j,k)]*dt)*(omega1[L(i,j,k)]*dt))/
                 (1.0+0.5*gamma1[L(i,j,k)]*dt);
            c2 = (1.0-0.5*gamma1[L(i,j,k)]*dt)/(1.0+0.5*gamma1[L(i,j,k)]*dt);
            c3 = sigma1[L(i,j,k)]*dt*dt/(1+0.5*gamma1[L(i,j,k)]*dt);
            P1z[(time+1)%2][L(i,j,k)] =
c1*P1z[time%2][L(i,j,k)]-c2*P1z[(time-1+2)%2][L(i,j,k)]+c3*(N1z[time%2]
[L(i,j,k)]-N2z[time%2][L(i,j,k)])*Ez[time%2][L(i,j,k)];
            Jz[L(i,j,k)] = (P1z[(time+1)%2][L(i,j,k)]-P1z[time%2][L(i,j,k)])/dt;


        break;


        case UPDATE_DRUDE:
            //c1 = (1.0-0.5*gamma_e_z[L(i,j,k)]*dt)/(1.0+0.5*gamma_e_z[L(i,j,k)]*dt);
```

```
    //c2 = epsz*omega_pe_z[L(i,j,k)]*omega_pe_z[L(i,j,k)]*dt/(1.0+0.5*gamma_e
        _z[L(i,j,k)]*dt);
    //Jz[L(i,j,k)] = c1*Jz[L(i,j,k)]+c2*Ez[time%2][L(i,j,k)];
    c1 = 2.0/(1.0+0.5*gamma_e_z[L(i,j,k)]*dt);
    c2 = (1.0-0.5*gamma_e_z[L(i,j,k)]*dt)/(1.0+0.5*gamma_e_z[L(i,j,k)]*dt);
    c3 = epsz*(omega_pe_z[L(i,j,k)]*dt)*(omega_pe_z[L(i,j,k)]*dt)/
        (1.0+0.5*gamma_e_z[L(i,j,k)]*dt);
    P1z[(time+1)%2][L(i,j,k)] = c1*P1z[time%2][L(i,j,k)]-c2*P1z[(time-1+2)%2]
    [L(i,j,k)]+c3*Ez[time%2][L(i,j,k)];
    Jz[L(i,j,k)] = (P1z[(time+1)%2][L(i,j,k)]-P1z[time%2][L(i,j,k)])/dt;
    break;


  }
}


TIMER_STOP(2);


//5. Update electric field Ex, Ey, Ez

TIMER_START;
for (k = 0; k<MAX_Z;k++)
  for (j = 0; j<MAX_Y;j++)
    for (i = 0; i<MAX_X;i++)
    {
      switch(type_Ex[L(i,j,k)])
      {
        case UPDATE_DIELECTRIC:
        case UPDATE_GAIN:
        case UPDATE_DRUDE:
        //Update Ex
          c1 = (1.0-sigmax[L(i,j,k)]*dt/(2.0*eps_Ex[L(i,j,k)]*epsz))/
              (1.0+sigmax[L(i,j,k)]*dt/(2.0*eps_Ex[L(i,j,k)]*epsz));
          c2 = 1.0/(1.0+sigmax[L(i,j,k)]*dt/(2.0*eps_Ex[L(i,j,k)]*epsz));
        Ex[(time+1)%2][L(i,j,k)] = c1*Ex[time%2][L(i,j,k)]+c2*dt/(eps_]
        Ex[L(i,j,k)*epsz*dy)*(Hz[L(i,j,k)]-Hz[L(i,j-1,k)])-
            c2*dt/(eps_Ex[L(i,j,k)]*epsz*dz)*(Hy[L(i,j,k)]-Hy[L(i,j,k-1)])-c2*dt/
            (eps_Ex[L(i,j,k)]*epsz)*Jx[L(i,j,k)];


        break;


        case UPDATE_PML:
        //Update Exy
          c1 = (1-sigma_y_Ex[L(i,j,k)]*dt/(2.0*epsz))/(1+sigma_y_Ex[L(i,j,k)]*dt/
              (2.0*epsz));
          c2 = dt/(dy*epsz*(1.0+sigma_y_Ex[L(i,j,k)]*dt/(2.0*epsz)));
          Exy[L(i,j,k)] = c1*Exy[L(i,j,k)]+c2*(Hz[L(i,j,k)]-Hz[L(i,j-1,k)]);


        //Update Exz
          c1 = (1-sigma_z_Ex[L(i,j,k)]*dt/(2.0*epsz))/(1+sigma_z_Ex[L(i,j,k)]*dt/
              (2.0*epsz));
          c2 = dt/(dz*epsz*(1.0+sigma_z_Ex[L(i,j,k)]*dt/(2.0*epsz)));
          Exz[L(i,j,k)] = c1*Exz[L(i,j,k)]-c2*(Hy[L(i,j,k)]-Hy[L(i,j,k-1)]);

          Ex[(time+1)%2][L(i,j,k)] = Exy[L(i,j,k)]+Exz[L(i,j,k)];


        break;
```

```
    }

    switch(type_Ey[L(i,j,k)])
    {
    case UPDATE_DIELECTRIC:
    case UPDATE_GAIN:
    case UPDATE_DRUDE:
    //Update Ey
    c1 = (1.0-sigmay[L(i,j,k)]*dt/(2.0*eps_Ey[L(i,j,k)]*epsz))/
         (1.0+sigmay[L(i,j,k)]*dt/(2.0*eps_Ey[L(i,j,k)]*epsz));
    c2 = 1.0/(1.0+sigmay[L(i,j,k)]*dt/(2.0*eps_Ey[L(i,j,k)]*epsz));
    Ey[(time+1)%2][L(i,j,k)] = c1*Ey[time%2][L(i,j,k)]+c2*dt/(eps_Ey[L(i,j,k)]*epsz
    *dz)*(Hx[L(i,j,k)]-Hx[L(i,j,k-1)])-
        c2*dt/(eps_Ey[L(i,j,k)]*epsz*dx)*(Hz[L(i,j,k)]-Hz[L(i-1,j,k)])-c2*dt/
        (eps_Ey[L(i,j,k)]*epsz)*Jy[L(i,j,k)];

       break;
    case UPDATE_PML:
    //Update Eyz

       c1 = (1-sigma_z_Ey[L(i,j,k)]*dt/(2.0*epsz))/(1+sigma_z_Ey[L(i,j,k)]*dt/
            (2.0*epsz));
       c2 = dt/(dz*epsz*(1.0+sigma_z_Ey[L(i,j,k)]*dt/(2.0*epsz)));
       Eyz[L(i,j,k)] = c1*Eyz[L(i,j,k)]+c2*(Hx[L(i,j,k)]-Hx[L(i,j,k-1)]);

       //Update Eyx
         c1 = (1-sigma_x_Ey[L(i,j,k)]*dt/(2.0*epsz))/(1+sigma_x_Ey[L(i,j,k)]*dt/
              (2.0*epsz));
         c2 = dt/(dx*epsz*(1.0+sigma_x_Ey[L(i,j,k)]*dt/(2.0*epsz)));
         Eyx[L(i,j,k)] = c1*Eyx[L(i,j,k)]-c2*(Hz[L(i,j,k)]-Hz[L(i-1,j,k)]);


         Ey[(time+1)%2][L(i,j,k)] = Eyz[L(i,j,k)]+Eyx[L(i,j,k)];

         break;
    }


    switch(type_Ez[L(i,j,k)])
    {
        case UPDATE_DIELECTRIC:
        case UPDATE_GAIN:
        case UPDATE_DRUDE:
        //Update Ez
          c1 = (1.0-sigmaz[L(i,j,k)]*dt/(2.0*eps_Ez[L(i,j,k)]*epsz))/
               (1.0+sigmaz[L(i,j,k)]*dt/(2.0*eps_Ez[L(i,j,k)]*epsz));
          c2 = 1.0/(1.0+sigmaz[L(i,j,k)]*dt/(2.0*eps_Ez[L(i,j,k)]*epsz));
          Ez[(time+1)%2][L(i,j,k)] = c1*Ez[time%2][L(i,j,k)]+c2*dt/(eps_Ez[L(i,j,k)]*
          epsz*dx)*(Hy[L(i,j,k)]-Hy[L(i-1,j,k)])-
              c2*dt/(eps_Ez[L(i,j,k)]*epsz*dy)*(Hx[L(i,j,k)]-Hx[L(i,j-1,k)])-c2*dt/
              (eps_Ez[L(i,j,k)]*epsz)*Jz[L(i,j,k)];

          break;

        case UPDATE_PML:
        //Update Ezx
          c1 = (1-sigma_x_Ez[L(i,j,k)]*dt/(2.0*epsz))/(1+sigma_x_Ez[L(i,j,k)]*dt/
               (2.0*epsz));
          c2 = dt/(dx*epsz*(1.0+sigma_x_Ez[L(i,j,k)]*dt/(2.0*epsz)));
          Ezx[L(i,j,k)] = c1*Ezx[L(i,j,k)]+c2*(Hy[L(i,j,k)]-Hy[L(i-1,j,k)]);

          //Update Ezy
```

```
      c1 = (1-sigma_y_Ez[L(i,j,k)]*dt/(2.0*epsz))/(1+sigma_y_Ez[L(i,j,k)]*dt/
          (2.0*epsz));
      c2 = dt/(dy*epsz*(1.0+sigma_y_Ez[L(i,j,k)]*dt/(2.0*epsz)));
      Ezy[L(i,j,k)] = c1*Ezy[L(i,j,k)]-c2*(Hx[L(i,j,k)]-Hx[L(i,j-1,k)]);

      Ez[(time+1)%2][L(i,j,k)] = Ezx[L(i,j,k)]+Ezy[L(i,j,k)];

      break;

    }
  }


//6.a. Make corrections for Ex and Ez
for (k = 0;k<GLOBAL_MAX_Z;k++)
  for (i = 0;i<GLOBAL_MAX_X;i++)
  {
    if (IN_LOCAL_DOMAIN(X,Y,Z,i,JLB,k))
    {
      double Hz_inc_L = H_inc[JLB-1]*sin(theta);
      double Hx_inc_L = H_inc[JLB-1]*cos(theta);
      Ex[(time+1)%2][P(i,JLB,k)] = Ex[(time+1)%2][P(i,JLB,k)]-dt/(dy*epsz)*Hz_
      inc_L;//left interface
      Ez[(time+1)%2][P(i,JLB,k)] = Ez[(time+1)%2][P(i,JLB,k)]+dt/(dy*epsz)*Hx_
      inc_L;//left interface
}
if (IN_LOCAL_DOMAIN(X,Y,Z,i,JRB,k))
{
      double Hz_inc_R = H_inc[JRB]*sin(theta);
      double Hx_inc_R = H_inc[JRB]*cos(theta);
      Ex[(time+1)%2][P(i,JRB,k)] = Ex[(time+1)%2][P(i,JRB,k)]+dt/(dy*epsz)*Hz_
      inc_R;//right interface
      Ez[(time+1)%2][P(i,JRB,k)] = Ez[(time+1)%2][P(i,JRB,k)]-dt/(dy*epsz)*Hx_
      inc_R;//right interface
    }
  }
//7. Update E_inc in the source and the grid generating lookup table
//in the grid generating the lookup table
for (j = 1;j<GLOBAL_MAX_Y;j++)
{
    c1 = (1.0-sigma_E_inc[j]*dt/(2.0*epsz))/(1.0+sigma_E_inc[j]*dt/(2.0*epsz));
    c2 = dt/(epsz*dy)/(1.0+sigma_E_inc[j]*dt/(2.0*epsz));
    E_inc[j] = c1*E_inc[j]-c2*(H_inc[j]-H_inc[j-1]);
}


TIMER_STOP(3);

edge_update(Ex[(time+1)%2],edge_buf);
edge_update(Ey[(time+1)%2],edge_buf);
edge_update(Ez[(time+1)%2],edge_buf);


//7.5 Update the charge distribution
if (config.cmp_charge = =1)
{
  for (k = 0;k<MAX_Z;k++)
    for (j = 0;j<MAX_Y;j++)
      for (i = 0;i<MAX_X;i++)
      {
      rou[L(i,j,k)] = epsz*(Ex[(time+1)%2][L(i,j,k)]-Ex[(time+1)%2][L(i-1,j,k)])/
      dx+ epsz*(Ey[(time+1)%2][L(i,j,k)]-Ey[(time+1)%2][L(i,j-1,k)])/dy+
          epsz*(Ez[(time+1)%2][L(i,j,k)]-Ez[(time+1)%2][L(i,j,k-1)])/dz;
      }
```

```
        }
    TIMER_STOP(4);
    //8. Update occupation numbers N0,N1,N2,N3


    TIMER_START;
    for (k = 0;k<MAX_Z;k++)
      for (j = 0;j<MAX_Y;j++)
        for (i = 0;i<MAX_X;i++)
        {
          switch(type_Ex[L(i,j,k)])
          {
            case UPDATE_GAIN:
            //Update N3
              c1 = (1.0-0.5*dt/tau32[L(i,j,k)])/(1.0+0.5*dt/tau32[L(i,j,k)]);
                c2 = dt*pump_gauss(pump_t0,pump_tau,time)*Pr_x[L(i,j,k)]/
                    (1.0+0.5*dt/tau32[L(i,j,k)]);
              N3x[(time+1)%2][L(i,j,k)] = c1*N3x[time%2][L(i,j,k)]+c2*N0x[L(i,j,k)];
            //Update N2
              c1 = (1.0-0.5*dt/tau21[L(i,j,k)])/(1.0+0.5*dt/tau21[L(i,j,k)]);
              c2 = 1.0/(2.0*HBAR*omega1[L(i,j,k)])/(1.0+0.5*dt/tau21[L(i,j,k)]);
              c3 = dt/(2.0*tau32[L(i,j,k)])/(1.0+0.5*dt/tau21[L(i,j,k)]);
              N2x[(time+1)%2][L(i,j,k)] =
    c1*N2x[time%2][L(i,j,k)]+c2*(Ex[(time+1)%2][L(i,j,k)]+Ex[time%2]
    [L(i,j,k)])*(P1x[(time+1)%2][L(i,j,k)]-P1x[time%2][L(i,j,k)])
                  +c3*(N3x[(time+1)%2][L(i,j,k)]+N3x[time%2][L(i,j,k)]);


            //Update N1
                c1 = (1.0-0.5*dt/tau10[L(i,j,k)])/(1.0+0.5*dt/tau10[L(i,j,k)]);
                c2 = 1.0/(2.0*HBAR*omega1[L(i,j,k)])/(1.0+0.5*dt/tau10[L(i,j,k)]);
                c3 = dt/(2.0*tau21[L(i,j,k)])/(1.0+0.5*dt/tau10[L(i,j,k)]);
                N1x[(time+1)%2][L(i,j,k)] =
    c1*N1x[time%2][L(i,j,k)]-c2*(Ex[(time+1)%2][L(i,j,k)]+Ex[time%2]
    [L(i,j,k)])*(P1x[(time+1)%2][L(i,j,k)]-P1x[time%2][L(i,j,k)])
                    +c3*(N2x[(time+1)%2][L(i,j,k)]+N2x[time%2][L(i,j,k)]);


            //Update N0
                N0x[L(i,j,k)] =
    (1.0-dt*pump_gauss(pump_t0,pump_tau,time)*Pr_x[L(i,j,k)])*N0x[L(i,j,k)]+0.5*dt/
    tau10[L(i,j,k)]*(N1x[(time+1)%2][L(i,j,k)]+N1x[time%2][L(i,j,k)]);
                break;
            }


            switch(type_Ey[L(i,j,k)])
            {
              case UPDATE_GAIN:
              //Update N3
                c1 = (1.0-0.5*dt/tau32[L(i,j,k)])/(1.0+0.5*dt/tau32[L(i,j,k)]);
                c2 = dt*pump_gauss(pump_t0,pump_tau,time)*Pr_y[L(i,j,k)]/
                    (1.0+0.5*dt/tau32[L(i,j,k)]);
                N3y[(time+1)%2][L(i,j,k)] = c1*N3y[time%2]
                [L(i,j,k)]+c2*N0y[L(i,j,k)];


            //Update N2
                c1 = (1.0-0.5*dt/tau21[L(i,j,k)])/(1.0+0.5*dt/tau21[L(i,j,k)]);
                c2 = 1.0/(2.0*HBAR*omega1[L(i,j,k)])/(1.0+0.5*dt/tau21[L(i,j,k)]);
                c3 = dt/(2.0*tau32[L(i,j,k)])/(1.0+0.5*dt/tau21[L(i,j,k)]);
                N2y[(time+1)%2][L(i,j,k)] =
    c1*N2y[time%2][L(i,j,k)]+c2*(Ey[(time+1)%2][L(i,j,k)]+Ey[time%2]
    [L(i,j,k)])*(P1y[(time+1)%2][L(i,j,k)]-P1y[time%2][L(i,j,k)])
                +c3*(N3y[(time+1)%2][L(i,j,k)]+N3y[time%2][L(i,j,k)]);
```

```
        //Update N1
            c1 = (1.0-0.5*dt/tau10[L(i,j,k)])/(1.0+0.5*dt/tau10[L(i,j,k)]);
            c2 = 1.0/(2.0*HBAR*omega1[L(i,j,k)])/(1.0+0.5*dt/tau10[L(i,j,k)]);
            c3 = dt/(2.0*tau21[L(i,j,k)])/(1.0+0.5*dt/tau10[L(i,j,k)]);
            N1y[(time+1)%2][L(i,j,k)] =
c1*N1y[time%2][L(i,j,k)]-c2*(Ey[(time+1)%2][L(i,j,k)]+Ey[time%2]
[L(i,j,k)])*(P1y[(time+1)%2][L(i,j,k)]-P1y[time%2][L(i,j,k)])
                +c3*(N2y[(time+1)%2][L(i,j,k)]+N2y[time%2][L(i,j,k)]);

        //Update N0
            N0y[L(i,j,k)] =
(1.0-dt*pump_gauss(pump_t0,pump_tau,time)*Pr_y[L(i,j,k)])*N0y[L(i,j,k)]+0.5*dt/
tau10[L(i,j,k)]*(N1y[(time+1)%2][L(i,j,k)]+N1y[time%2][L(i,j,k)]);


        break;
        }
        switch(type_Ez[L(i,j,k)])
        {
            case UPDATE_GAIN:

            //Update N3
            c1 = (1.0-0.5*dt/tau32[L(i,j,k)])/(1.0+0.5*dt/tau32[L(i,j,k)]);
            c2 = dt*pump_gauss(pump_t0,pump_tau,time)*Pr_z[L(i,j,k)]/
                (1.0+0.5*dt/tau32[L(i,j,k)]);
            N3z[(time+1)%2][L(i,j,k)] = c1*N3z[time%2][L(i,j,k)]+c2*N0z[L(i,j,k)];

        //Update N2
            c1 = (1.0-0.5*dt/tau21[L(i,j,k)])/(1.0+0.5*dt/tau21[L(i,j,k)]);
            c2 = 1.0/(2.0*HBAR*omega1[L(i,j,k)])/(1.0+0.5*dt/tau21[L(i,j,k)]);
            c3 = dt/(2.0*tau32[L(i,j,k)])/(1.0+0.5*dt/tau21[L(i,j,k)]);
            N2z[(time+1)%2][L(i,j,k)] =
c1*N2z[time%2][L(i,j,k)]+c2*(Ez[(time+1)%2][L(i,j,k)]+Ez[time%2]
[L(i,j,k)])*(P1z[(time+1)%2][L(i,j,k)]-P1z[time%2][L(i,j,k)])
                +c3*(N3z[(time+1)%2][L(i,j,k)]+N3z[time%2][L(i,j,k)]);

        //Update N1
            c1 = (1.0-0.5*dt/tau10[L(i,j,k)])/(1.0+0.5*dt/tau10[L(i,j,k)]);
            c2 = 1.0/(2.0*HBAR*omega1[L(i,j,k)])/(1.0+0.5*dt/tau10[L(i,j,k)]);
            c3 = dt/(2.0*tau21[L(i,j,k)])/(1.0+0.5*dt/tau10[L(i,j,k)]);
            N1z[(time+1)%2][L(i,j,k)] =
c1*N1z[time%2][L(i,j,k)]-c2*(Ez[(time+1)%2][L(i,j,k)]+Ez[time%2]
[L(i,j,k)])*(P1z[(time+1)%2][L(i,j,k)]-P1z[time%2][L(i,j,k)])
                +c3*(N2z[(time+1)%2][L(i,j,k)]+N2z[time%2][L(i,j,k)]);
        //Update N0
            N0z[L(i,j,k)] =
(1.0-dt*pump_gauss(pump_t0,pump_tau,time)*Pr_z[L(i,j,k)])*N0z[L(i,j,k)]+0.5*dt/
tau10[L(i,j,k)]*(N1z[(time+1)%2][L(i,j,k)]+N1z[time%2][L(i,j,k)]);


        break;
    }
  }


TIMER_STOP(5);
    //FDTD FIELD UPDATE DONE

    free(edge_buf);
    fdtd->time++;
}
```

# 5

# *FDTD Simulation of Trapping Microspheres and Nanowires with Optical Tweezers*

**Jing Li,[1] Chunli Zhu,[1] and Xiaoping Wu[2]**

[1]*Department of Precision Machinery and Precision Instrumentation, University of Science and Technology of China, China*

[2]*Department of Modern Mechanics, University of Science and Technology of China, China*

## CONTENTS

## 5.1 Introduction

Optical tweezers have become important and commonly used tools in the field of physics and life science for trapping and moving objects ranging in size from tens of nanometers to tens of micrometers [1–2] since their introduction in 1986 [3]. They utilize the mechanical effect of light and the interaction of strongly focused light with microparticles to form three-dimensional potential wells to trap particles.

Objects illuminated by light will be subject to a force generated by optical radiation. Since it usually appears as pressure, it is called light pressure. Scientists have known about light pressure for a long time. By the end of the 19th century, Maxwell had proved the existence of light pressure theoretically and deduced the related calculation equations. In 1901, Lebedeff succeeded in observing the existence of light pressure by experiment.

According to modern optics, light exhibits properties of both waves and particles. As light is absorbed and emitted, it exhibits particle properties and can be regarded as a particle flow composed of photons, which have both energy and momentum. As light and matter interact, energy as well as momentum carried by photons is transferred to matter. Figure 5.1 shows that the incident light is refracted and deflected by a dielectric sphere, and the momentum of the beam has changed. According to the law of momentum conservation, the changed momentum will be transferred to the sphere so as to produce force.

Momentum carried by a single photon is $p = hv/c$, where $h$ is the Planck constant, $v$ is the frequency of light, and $c$ is the speed of light. Because the momentum carried by a single photon is so small, the application and observation of the optical force are usually limited to the micro- and nanoscale.

In the 1960s, the invention of the laser greatly improved modern optical technology. The properties of laser, such as good directivity and high

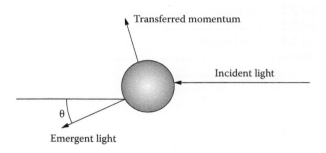

**FIGURE 5.1**
Momentum transfer between a sphere and light.

energy, provide effective tools for the study of the optical force. In 1986, Ashkin et al. invented a method of trapping a dielectric particle by a single laser beam [3]. They focused a single beam by using a high numerical aperture microscopic lens, instead of earlier solutions that used dual beams or multiple beams to simultaneously irradiate particles to ensure forces in equilibrium. Theoretical analysis and experimental observation proved that under appropriate conditions the dielectric particle is subjected to a resultant force of light that always points to the focus in the strongly focused light field, and it is stably trapped in the vicinity of the laser focus. The formed potential well is called a single-beam optical gradient force trap, which can trap dielectric particles ranging in size from tens of nanometers to tens of micrometers.

The single-beam optical gradient force trap has many advantages, such as accurate position control, high trapping ability, and a wide range of applications. On the basis of these, the function and application of optical tweezers developed rapidly. Scientists from different countries conducted research with optical tweezers technology, such as cooling and manipulating atoms, measuring DNA's tensile properties, etc. [4–8]. However, since the interaction between light and matter is affected by a lot of factors, and the optical force is associated with various parameters, such as the light source, the distribution of the optical field, the composition and shape of the particle, the environment for trapping, etc., the theory of optical forces lags behind the development of experimental technology. By now, the actual value of the optical force can only be obtained through measurement. However, it is still significant that the optical trapping process is analyzed and simulated to estimate the numeric value and character of the optical force. For the design of experiments, through theoretical calculation of the optical force, the trapping capacities of optical tweezers can be preliminarily estimated, the feasibility of experiments can be judged, and the experimental conditions and design parameters can be adjusted and optimized. In the experimental process, by comparing the difference between the experimental and theoretical results, it is helpful to judge the reliability of the experiment and make an improvement.

Currently, in the theoretical study of optical tweezers, according to the proportion of the size of an object to the wavelength of light, the optical force calculation is divided into three different models. When the diameter of the particle is much smaller than the wavelength of light, the particle is regarded as an induced dipole, and then an approximation is used for Rayleigh scattering [9]. The optical field exhibits the properties of the electromagnetic field, and the optical force is essentially the Lorentz force acting on the particle. As the diameter of the particle is much larger than the wavelength, the Mie scattering model is adopted [10, 11]. The optical force comes from the transferred momentum, which is produced by the deflection of the beam as it passes through the particle. The aforementioned models are applicable to particles with size less than or greater than the wavelength of light by one order of magnitude ($< \lambda/10$ and $> 10\lambda$, where $\lambda$ is the wavelength of light).

At present, there are relatively comprehensive models for the force acting on a particle in the optical tweezers under the aforementioned conditions. Experiments have demonstrated that when the size of the particle is larger than the wavelength of light by one order of magnitude, the results agree with the model of geometric optics. When the size of the particle is less than the wavelength of light by one order of magnitude, the results agree with the Rayleigh model well. And when the particle size is between them (i.e., $\lambda/10$–$10\lambda$), it is not accurate to use the models above to calculate the trapping force. Because the particles of this scale have a very strong diffraction effect, it is a complicated problem to calculate the optical force. It is feasible using Maxwell's equations to accurately describe the distribution and change of vector light field, analyzing the interaction of particles and light field based on electromagnetic scattering, and calculating the optical force acting on the particle through the Poynting vector or Maxwell stress tensor. In the following section, the calculation theories of optical force on the particles in three different size regimes are given.

## 5.2 Calculation of Optical Trapping Force

### 5.2.1 Rayleigh Model

The Rayleigh model is usually applied in analyzing the optical trapping force on particles much smaller than the wavelength of light. Based on this model, in 1992, Visscher et al. analyzed the transmission of the light field with a vector diffraction theory; the relationships of the axial force exerted on the particle with wavelength and numerical aperture were also analyzed [12]. In 1996, Harada used electric dipoles to describe particles approximately and analyze the force exerted on the microsphere in the non-strongly focused Gaussian laser beam [9]. In 2000, Chaumet and Nieto-Vesperinas deduced the time-averaged force expressions on Rayleigh particles based on the concept of coupling dipole. They developed the Rayleigh model and extended the calculation of the optical trapping force on the particles with larger size [13, 14]. In 2001, Rohrbach and Stelzer extended a two-component approach to absorbing or small metallic particles as well as to inhomogeneous scatterers [15, 16].

The basic principle of the Rayleigh model is shown in Figure 5.2. Although a time-varying electromagnetic field in space changes rapidly, a small enough particle can be regarded approximately as a single point to be analyzed. The electric field within the particle is regarded as uniformly distributed; the instantaneous electric field around the particle

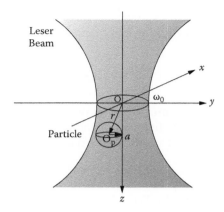

**FIGURE 5.2**
Schematic of Rayleigh model.

remains unchanged. The particle is polarized by the electric field, and it is a simple induced electric dipole. Under this assumption, the properties of the electric field are similar to those of the electrostatic field; the electrostatic formula can be used to calculate the electric dipole moment of the induced dipole. Using the international units of physics, the electric dipole moment can be expressed as

$$\mathbf{p}(r,t) = 4\pi\varepsilon_2 a^3 \left( \frac{\varepsilon_1 - \varepsilon_2}{\varepsilon_1 + 2\varepsilon_2} \right) \mathbf{E}(r,t) = 4\pi n_2^2 \varepsilon_0 a^3 \left( \frac{m^2 - 1}{m^2 + 2} \right) \mathbf{E}(r,t) \qquad (5.1)$$

Here $a$ is the radius of the particle, $m$ is the relative refractive index, and $m = n_1/n_2$, where $n_1$ is the refractive index of the particle, and $n_2$ is the refractive index of the environment medium.

In the Rayleigh model, the force of the electric dipole in an electromagnetic field can be divided into two components: scattering force and gradient force. The scattering force comes from electromagnetic scattering by the electric dipole, which leads to the change in its momentum. The gradient force is generated by Lorentz force exerted on the induced electric dipole in the electromagnetic field.

Since light is a time-harmonic oscillating electromagnetic field, the electric dipole in the field will produce synchronous oscillation with the change of the electric field and send scattered waves into its surroundings. This process will cause the change of the flow direction of the incident electromagnetic wave and its amplitude, which leads to momentum transfer between the incident wave and induced electric dipole. The effect of the momentum transfer is demonstrated as the particle is affected by the scattering force of the optical field. The scattering force is calculated as

$$\mathbf{F}_{scat}(\mathbf{r}) = \frac{C_{pr}\langle \mathbf{S}(\mathbf{r},t)\rangle_T}{c/n_2} = \vec{z}\left(\frac{n_2}{c}\right)C_{pr}\mathbf{I}(\mathbf{r}) \tag{5.2}$$

where $C_{pr}$ is the cross-section area of radiation pressure of the particle. $\vec{z}$ is the unit vector, which represents light propagation direction. $c$ is light speed in vacuum, $\mathbf{S}(\mathbf{r},t)$ is the Poynting vector, and $\mathbf{I}(\mathbf{r})$ is the time-average value of the Poynting vector.

$$\mathbf{I}(\mathbf{r}) = \langle \mathbf{S}(\mathbf{r},t)\rangle_T = \frac{1}{2}\mathrm{Re}[\mathbf{E}(\mathbf{r})\times\mathbf{H}^*(\mathbf{r})] = \vec{z}\frac{n_2\varepsilon_0 c}{2}|E(\mathbf{r})|^2 = \vec{z}I(\mathbf{r}) \tag{5.3}$$

Here $\mathbf{I}(\mathbf{r})$ represents the distribution of the light intensity in the space, which can be obtained by measurement. In the paraxial approximation, the expression of the Gaussian laser beam is

$$I(\mathbf{r}) = \left(\frac{2P}{\pi\omega_0^2}\right)\frac{1}{1+(2\tilde{z})^2}\exp\left[-\frac{2(\tilde{x}^2+\tilde{y}^2)}{1+(2\tilde{z})^2}\right] \tag{5.4}$$

Here $P$ is incident light power, $\omega_0$ is the beam waist radius, and $(\tilde{x},\tilde{y},\tilde{z}) = (x/\omega_0, y/\omega_0, z/k\omega_0^2)$ are normalized coordinates.

Considering the isotropic characteristics of light scattering by the particle, $C_{pr}$ is equal to the scattering cross section $C_{scat}$

$$C_{pr} = C_{scat} = \frac{8}{3}\pi(ka)^4 a^2\left(\frac{m^2-1}{m^2+2}\right)^2 \tag{5.5}$$

Putting formulas (5.3), (5.4), and (5.5) into formula (5.2), the scattering force exerted on the particle can be obtained. It can be seen that the direction of the scattering force is the same as that of light propagation; the force is proportional to light power and the sixth power of the radius of the particle.

The gradient force exerted on the particle comes from the Lorentz force acting on the induced electric dipole in the electromagnetic field. For a temporal electromagnetic field, static field approximation of the electromagnetic wave is used to calculate the Lorentz force. Putting the formula (5.1), which is used to calculate the induced electric dipole moment, into a formula for the calculation of the instantaneous gradient force, the gradient force is expressed as

$$\mathbf{F}_{grad}(\mathbf{r},t) = [\mathbf{p}(\mathbf{r},t)\cdot\nabla]\mathbf{E}(\mathbf{r},t) = 4\pi n_2^2\varepsilon_0 a^3\left(\frac{m^2-1}{m^2+2}\right)\frac{1}{2}\nabla\mathbf{E}^2(\mathbf{r},t) \tag{5.6}$$

The time-averaged gradient force exerted on the microsphere is

$$
\begin{aligned}
F_{grad}(r,t) &= \langle F(r,t) \rangle_T \\
&= 4\pi n_2^2 \varepsilon_0 a^3 \left( \frac{m^2 - 1}{m^2 - 2} \right) \frac{1}{2} \nabla \langle E^2(r,t) \rangle_T \\
&= \pi n_2^2 \varepsilon_0 a^3 \left( \frac{m^2 - 1}{m^2 - 2} \right) \nabla |E(r)|^2 \\
&= \frac{2\pi n_2 a^3}{c} \left( \frac{m^2 - 1}{m^2 - 2} \right) \nabla I(r)
\end{aligned}
\tag{5.7}
$$

It can be seen that the gradient force is proportional to the gradient light intensity at the position of the particle; as $m > 1$, the gradient force always points to the position with highest intensity, i.e., the beam waist. In addition, the gradient force is proportional to the third power of the radius of the particle.

In the Rayleigh model, the value and direction of the optical force exerted on the particle can be obtained through calculating the gradient force and scattering force and adding them together.

## 5.2.2 Mie Scattering Model

As the size of the particle is much larger than the wavelength of light, generally the Mie scattering model is used to calculate and analyze the optical trapping force exerted on the particle, which is based on the ray theory of geometric optics and the momentum transfer theory. After years of research, it is more comprehensively and widely applied in the biological experiments on trapping living cells and in the physical experiments on trapping larger particles. In these cases, the research objects can be regarded as transparent and uniform particles. In the Mie scattering model, due to the larger size of the particle, the diffraction phenomenon is not obvious, and the incident beam can be regarded as the composition of rays. A single ray carrying energy has its direction and polarization and propagates straightly in the homogeneous medium; the effect of diffraction can be neglected.

The Mie scattering model is based on the method proposed by Roosen and Imbert. [17] to calculate the optical trapping force exerted on the microparticle, which is generated by the surface reflection and refraction of the microparticle. In 1992, Ashkin [18] analyzed the optical trapping force of Mie's dielectric microsphere in optical tweezers and gave the force calculation formula with corresponding geometric approximation. In 1995, Gauthier put forward a calculation method with mixed geometrical and wave optics to

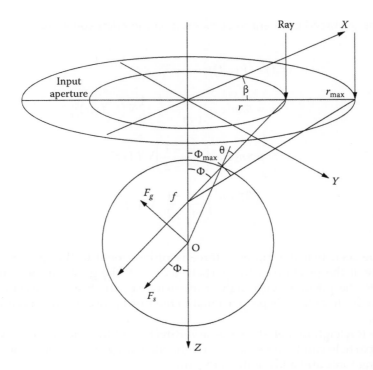

**FIGURE 5.3**
Schematic of single optical tweezers using the approximation of geometrical optics.

analyze the force exerted on a micro rotor, and verified the analytical results through the experimental measurement [19]. Since then, he improved the ray optics model and accomplished the force analysis on nonspherical particles, such as cylindrical and cubic particles [20–22].

Figure 5.3 displays the force model of single optical tweezers using the approximation of geometrical optics when the radius of the microsphere is much larger than the wavelength of the incident light. Parallel beams are converged by the lens with high numerical aperture to the focus, and the z-axis represents the optical axis. $\Phi_{max}$ is the maximum convergence angle. A single ray can be determined by using polar coordinates $r$ and $\beta$ in the entrance pupil. When the beams are directly converged to the focus, their direction and momentum don't change. If they meet the dielectric microsphere, they will be reflected and refracted at the surface of the microsphere, which leads to the change of the momentum and the generation of the optical force. The total force exerted on the microsphere is the sum of the gradient force and scattering force produced by all rays. Figure 5.3 shows the produced scattering force $F_s$ and gradient force $F_g$ by a single ray using geometric optics. The direction of the scattering force is parallel to the incident light, and the direction of the gradient force is perpendicular to the incident light.

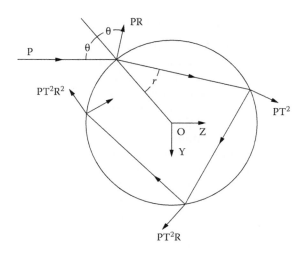

**FIGURE 5.4**
The calculation model of optical force produced by a single ray.

We assume that a single ray with power $P$ arrives at the dielectric micro-sphere with angle $\theta$, as shown in Figure 5.4. The incident momentum per second is $n_1 P/c$. $n_1$ is the refractive index of the dielectric microsphere. The ray is reflected and refracted at the surface of the microsphere. After entering the interface, it is reflected and refracted continuously within the microsphere. Assuming that the power of the ray is $PR$ after reflection from the surface of the microsphere, its power becomes $PT^2$, $PT^2R$, $PR^2T^2$, ... after each refraction. $R$ and $T$ are reflective and refractive coefficients, respectively, which change with the polarization of the incident light, as does the produced force. The net force through the origin O can be decomposed into $F_z$ and $F_y$:

$$F_z = F_s = \frac{n_1 P}{c}\left\{1 + R\cos 2\theta - \frac{T^2\left[\cos 2(\theta - r) + R\cos 2\theta\right]}{1 + R^2 + 2R\cos 2r}\right\} \tag{5.8}$$

$$F_y = F_g = \frac{n_1 P}{c}\left\{R\sin 2\theta - \frac{T^2\left[\sin 2(\theta - r) + R\sin 2\theta\right]}{1 + R^2 + 2R\cos 2r}\right\} \tag{5.9}$$

where $\theta$ is the incident angle and $r$ is the refractive angle.

The formulas (5.8) and (5.9) represent the calculation of the optical force in the model of geometrical optics. The component $F_z$ of equation (5.8), whose direction is the same as the incident light, is regarded as the scattering force $F_s$. The component $F_y$ of equation (5.9), whose direction is perpendicular to the incident light, is regarded as the gradient force $F_g$. The total gradient force and scattering force can be obtained by the vector sum of the gradient force and scattering force produced by all rays.

### 5.2.3 Region between Geometric and Rayleigh Dimensions

When the size of the particle is close to the wavelength, larger errors will be present if the above methods are used. In practice, particles of this range are usually regarded as objects to study microphenomena of interest. But in this range, there is a lack of corresponding complete and approximate theory. Therefore, at present the study of the force calculation is concentrated on the particles of this range. Since light is regarded as an electromagnetic wave, the interaction of light with matter is a process of electromagnetic scattering. Usually, the electromagnetic theory is used to calculate and analyze the trapping force exerted on the microparticles in this range.

In Maxwell's equations, the electric current $J$ and charge are the field source. In many problems, the effect of the electromagnetic field on the electric current and charge should be considered. Electromagnetic force on the electric charge can be expressed by the formula of the Lorentz force. Due to the electromagnetic force on the charge, the charged particle can be moved in the electromagnetic field and attain energy and momentum from it. Therefore, the electromagnetic field has energy and momentum. When charged particles move in the electromagnetic field, energy and momentum are exchanged between the electromagnetic field and charged particles. The expression of the energy density of the electromagnetic field is determined by the transformation between electromagnetic energy and mechanical energy of the charged particles through their interaction.

In the electromagnetic field, the Lorentz force acting on the charged particle with speed $\mathbf{v}$ is

$$\mathbf{F} = q(\mathbf{E} + \mathbf{v} \times \mathbf{B}) \tag{5.10}$$

$\mathbf{E}$ is the intensity of the electric field. $\mathbf{B}$ is the density of the magnetic flux.

Since the force is equal to the rate of change of the momentum, we have

$$\mathbf{F} = m\frac{d\mathbf{v}}{dt} = \frac{d}{dt}(m\mathbf{v}) = \frac{d}{dt}\mathbf{G}_p \tag{5.11}$$

$\mathbf{G}_p = m\mathbf{v}$ is the momentum of the charged particle. From equation (5.10) and (5.11), it is deduced that

$$\frac{d}{dt}\mathbf{G}_p = q\mathbf{E} + q\mathbf{v} \times \mathbf{B} \tag{5.12}$$

The kinetic energy of the charged particle is $W_p = \frac{1}{2}m\mathbf{v}^2$, so

$$\frac{dW_p}{dt} = m\mathbf{v} \bullet \frac{d\mathbf{v}}{dt} = \mathbf{v} \bullet \frac{d}{dt}(m\mathbf{v}) = \mathbf{v} \bullet \frac{d\mathbf{G}_p}{dt} \tag{5.13}$$

Substitute equation (5.13) with (5.12),

$$\frac{dW_p}{dt} = \mathbf{v} \bullet (q\mathbf{E} + q\mathbf{v} \times \mathbf{B}) = q\mathbf{E} \bullet \mathbf{v} \tag{5.14}$$

Equations (5.12) and (5.14) represent the time rate of momentum and energy of a single charged particle.

For a system with continuous charge distribution, its electric density is $\rho$ and electric current density is $\mathbf{J} = \rho\mathbf{v}$, and we have

$$\frac{\partial \mathbf{g}_p}{\partial t} = \rho\mathbf{E} + \mathbf{J} \times \mathbf{B} \tag{5.15}$$

$$\frac{\partial w_p}{\partial t} = \mathbf{J} \bullet \mathbf{E} \tag{5.16}$$

$\mathbf{g}_p$ and $w_p$ are the densities of the momentum and energy, respectively, namely the momentum and energy per unit volume. Equations (5.15) and (5.16) show that if $\partial \mathbf{g}_p/\partial t > 0$, $\partial w_p/\partial t > 0$, the charged system will get momentum and energy from the electromagnetic field. According to the laws of momentum and energy conservation, the electromagnetic field has momentum and energy.

The differential forms of Maxwell's equations in a homogeneous isotropic medium are

$$\nabla \times \mathbf{H} = \varepsilon \frac{\partial \mathbf{E}}{\partial t} + \sigma\mathbf{E} \tag{5.17}$$

$$\nabla \times \mathbf{E} = -\mu \frac{\partial \mathbf{H}}{\partial t} \tag{5.18}$$

$\varepsilon$ is the dielectric coefficient of the medium, $\mu$ is the magnetic permeability coefficient, $\sigma$ is the electric conductivity, and $\mathbf{H}$ is magnetic intensity.

We put the above two equations into the following vector identical equation:

$$\nabla \bullet (\mathbf{E} \times \mathbf{H}) = -\mathbf{E} \bullet \nabla \times \mathbf{H} + \mathbf{H} \bullet \nabla \times \mathbf{E} \tag{5.19}$$

It follows that

$$\nabla \bullet (\mathbf{E} \times \mathbf{H}) = -\mathbf{E} \bullet \left( \mathbf{J} + \frac{\partial \mathbf{D}}{\partial t} \right) + \mathbf{H} \bullet \left( -\frac{\partial \mathbf{B}}{\partial t} \right)$$
$$= -\mathbf{J} \bullet \mathbf{E} - \left( \mathbf{E} \bullet \frac{\partial \mathbf{D}}{\partial t} + \mathbf{H} \bullet \frac{\partial \mathbf{B}}{\partial t} \right) \tag{5.20}$$

$\mathbf{D}$ is electric flux density.

In the achromatic medium,

$$\nabla \bullet (\mathbf{E} \times \mathbf{H}) = -\mathbf{J} \bullet \mathbf{E} - \frac{\partial}{\partial t}\left(\frac{1}{2}\mathbf{E} \bullet \mathbf{D} + \frac{1}{2}\mathbf{H} \bullet \mathbf{B}\right) \tag{5.21}$$

Assume that

$$\mathbf{S} = \mathbf{E} \times \mathbf{H}$$

$$w_f = \frac{1}{2}\mathbf{E} \bullet \mathbf{D} + \frac{1}{2}\mathbf{H} \bullet \mathbf{B},$$

And combining equation (5.16), equation (5.21) is changed into

$$\frac{\partial}{\partial t}(w_p + w_f) + \nabla \bullet \mathbf{S} = 0 \tag{5.22}$$

Using the Gauss theorem to integrate the above equation, we get

$$\oint_S \mathbf{S} \bullet \mathbf{n}\, dS = -\frac{\partial}{\partial t}\int_V (w_p + w_f)dV \tag{5.23}$$

$w_p$ and $w_f$ are the energy densities of the system with continuous charge distribution and the electromagnetic field, respectively. $\mathbf{S}$ is the Poynting vector of the electromagnetic field. The above equation denotes that the power flowing out from the closed surface $S$ is equal to the total lost energy of the system of charged particles and the electromagnetic field in the volume surrounded by $S$ per unit time.

The time differential form of the Poynting vector is

$$\frac{\partial \mathbf{S}}{\partial t} = \frac{\partial}{\partial t}(\mathbf{E} \times \mathbf{H}) = \frac{\partial \mathbf{E}}{\partial t} \times \mathbf{H} + \mathbf{E} \times \frac{\partial \mathbf{H}}{\partial t} \tag{5.24}$$

Putting two differential forms of Maxwell's equations into (5.24), we get

$$\mu_0\varepsilon_0\frac{\partial \mathbf{S}}{\partial t} + \mathbf{J} \times \mathbf{B} = \mu_0(\nabla \times \mathbf{H}) \times \mathbf{H} + \varepsilon_0(\nabla \times \mathbf{E}) \times \mathbf{E} \tag{5.25}$$

According to the vector differential identical equation,

$$\nabla(\mathbf{A} \bullet \mathbf{B}) = \mathbf{A} \times (\nabla \times \mathbf{B}) + \mathbf{B} \times (\nabla \times \mathbf{A}) + (\mathbf{B} \bullet \nabla)\mathbf{A} + (\mathbf{A} \bullet \nabla)\mathbf{B} \tag{5.26}$$

The equation (5.25) can be rewritten as

$$\mu_0\varepsilon_0\frac{\partial \mathbf{S}}{\partial t} + \mathbf{J} \times \mathbf{B} = \varepsilon_0\left[-\frac{1}{2}\nabla\mathbf{E}^2 + (\mathbf{E} \bullet \nabla)\mathbf{E}\right] + \mu_0\left[-\frac{1}{2}\nabla\mathbf{H}^2 + (\mathbf{H} \bullet \nabla)\mathbf{H}\right] \tag{5.27}$$

In free space, $\nabla \bullet D = \rho$, there is

$$E\rho = \varepsilon_0 E(\nabla \bullet E) + \mu_0 H(\nabla \bullet H) \tag{5.28}$$

Add (5.27) and (5.28), and combine (5.15),

$$\frac{\partial}{\partial t}(\mu_0 \varepsilon_0 S + g_p) = -\frac{1}{2}\nabla[\varepsilon_0 E^2 + \mu_0 H^2] + \varepsilon_0[(E \bullet \nabla)E + E(\nabla \bullet E)] \tag{5.29}$$

$$+\mu_0[(H \bullet \nabla)H + H(\nabla \bullet H)]$$

Use vector-dyadic differential equations and simplify the above equation,

$$\frac{\partial}{\partial t}(\mu_0 \varepsilon_0 S + g_p) = -\nabla \bullet \left[ \left( \frac{1}{2}\varepsilon_0 E^2 + \frac{1}{2}\mu_0 H^2 \right) I - \varepsilon_0 EE - \mu_0 HH \right] \tag{5.30}$$

Assume that

$$\Phi = \left( \frac{1}{2}\varepsilon_0 E^2 + \frac{1}{2}\mu_0 H^2 \right) I - \varepsilon_0 EE - \mu_0 HH$$

$$g_f = \mu_0 \varepsilon_0 S$$

Equation (5.30) is simplified as

$$\frac{\partial}{\partial t}(g_f + g_p) = -\nabla \bullet \Phi \tag{5.31}$$

The space integral of equation (5.31) is

$$\frac{\partial}{\partial t}\int_V (g_f + g_p)dV = -\oint_S dSn \bullet \Phi \tag{5.32}$$

Here $g_p$ is the momentum density of the system of charged particles, and $g_f$ is the momentum density of the electromagnetic field in vacuum. The left side of the above equation represents the time rate of change of the total momentum of the electromagnetic field and charged system in the volume $V$, and the right side denotes the total momentum flowing into the volume $V$ through the closed surface $S$ per unit time. It can be seen that $\Phi$ represents the density of the momentum flux, named as the density tensor of the momentum flux of the electromagnetic field.

If we let $T = -\Phi$, the equations (5.31) and (5.32) can be changed into

$$\frac{\partial}{\partial t}(g_f + g_p) = \nabla \bullet T \tag{5.33}$$

$$\frac{\partial}{\partial t}\int_V (g_f + g_p)dV = \int_V \nabla \bullet T dV \tag{5.34}$$

The left part of equation (5.34) represents the time rate of change of the total momentum in the volume $V$. Compared with the law of conservation of momentum in classical mechanics, it can be seen that the right side of equation (5.34) should be equal to the total force from the outside of the volume $V$. $\nabla \bullet T$ is the volume force density and only contains components of the electromagnetic field; therefore, it is the force produced by the electromagnetic field. Equation (5.34) is the force-momentum conservation equation of the electromagnetic field.

The total force exerted by the electromagnetic field on the inner portion of volume $V$ is

$$\mathbf{F} = \int_V \nabla \bullet \mathbf{T} = \oint_S \mathbf{T} \bullet \mathbf{n} \, d\mathbf{S} \tag{5.35}$$

The volume force acting on the inner portion of volume $V$ is equivalent to the tension on the closed surface $S$ surrounding volume $V$. $\mathbf{T}$ is the stress tensor of the electromagnetic field acting on a unit area. The force acting on the surface element is $d\mathbf{F} = \mathbf{T} \bullet d\mathbf{S}$.

The dyadic expression of the stress tensor of the electromagnetic field in vacuum is

$$\mathbf{T} = \varepsilon_0 \mathbf{EE} + \mu_0 \mathbf{HH} - \left( \frac{1}{2} \varepsilon_0 E^2 + \frac{1}{2} \mu_0 H^2 \right) \mathbf{I} \tag{5.36}$$

It can be deduced that $\mathbf{T}$ is a symmetrical tensor with nine components, and it can be expressed in the following matrix:

$$\mathbf{T} = \begin{bmatrix} \varepsilon_0 E_1^2 + \mu_0 H_1^2 - \frac{1}{2}(\varepsilon_0 E^2 + \mu_0 H^2) & \varepsilon_0 E_1 E_2 + \mu_0 H_1 H_2 & \varepsilon_0 E_1 E_3 + \mu_0 H_1 H_3 \\ \varepsilon_0 E_2 E_1 + \mu_0 H_2 H_1 & \varepsilon_0 E_2^2 + \mu_0 H_2^2 - \frac{1}{2}(\varepsilon_0 E^2 + \mu_0 H^2) & \varepsilon_0 E_2 E_3 + \mu_0 H_2 H_3 \\ \varepsilon_0 E_3 E_1 + \mu_0 H_3 H_1 & \varepsilon_0 E_3 E_2 + \mu_0 H_3 H_2 & \varepsilon_0 E_3^2 + \mu_0 H_3^2 - \frac{1}{2}(\varepsilon_0 E^2 + \mu_0 H^2) \end{bmatrix} \tag{5.37}$$

where 1, 2, 3 in the Cartesian coordinate system represent $x, y, z$, respectively. The component can be expressed as

$$\mathbf{T}_{ij} = \varepsilon_0 E_i E_j + \mu_0 H_i H_j - \frac{1}{2} \delta_{ij} (\varepsilon_0 E^2 + \mu_0 H^2) \tag{5.38}$$

where

$$\delta_{ij} = \begin{cases} 0 & i \neq j \\ 1 & i = j \end{cases}$$

$\mathbf{T}_{ij}$ denotes $i$ axial component of the electromagnetic force, which acts on the unit area perpendicular to the $j$-axis.

For the time-harmonic field, the time-averaged tensor of the electromagnetic field can be expressed by complex amplitude as

$$\langle \mathbf{T} \rangle = \frac{1}{2} \mathrm{Re} \left[ \varepsilon_0 \mathbf{E}\mathbf{E}^* + \mu_0 \mathbf{H}\mathbf{H}^* - \frac{1}{2}(\varepsilon_0 \mathbf{E}^* + \mu_0 \mathbf{H}^*)\mathbf{I} \right] \tag{5.39}$$

Therefore, according to the spatial distribution of the electromagnetic field, using equations (5.35) and (5.38), the time-average value of the electromagnetic force acting on any area in the time-harmonic field can be calculated.

## 5.3 FDTD Methods

In recent years, this has become the main way that the comprehensive electrodynamics theory and Maxwell stress tensor method are employed to calculate the strongly focused laser beam and the force exerted on the particle with size between $\lambda/10$ and $10\lambda$. It requires that six components of the electromagnetic field on the surface of the particle be calculated. Usually, the numerical methods for solving electromagnetic field distribution include the finite element method (FEM) [23], the T-matrix method [24, 25], and the finite-difference time-domain (FDTD) method [26–30]. Among them, the use of the finite-element method is limited by the minimum of the space grid and time interval, and the T-matrix method is appropriate for highly symmetric particles. Nevertheless, the FDTD method can solve problems, such as the electromagnetic scattering, radiation, etc., for objects with complicated shapes. Meanwhile, it can give the time evolution process of the electromagnetic field. Thus, this method is adopted to get the optical trapping force exerted on microparticles of this dimension.

In regard to the FDTD method, in 2003, Collett et al. calculated the electromagnetic field of the scattered light on the surface of the micro-rotor using FDTD, obtained the corresponding force exerted on the rotor using the Maxwell stress tensor method, and then got the rotor torque [26]. In 2004, Zhang et al. investigated the transfer procedure of orbital angular momentum from a focused optical vortex to the particle by FDTD simulation [27]. In 2005, Gauthier calculated the optical trapping force exerted on dielectric objects by using 2D FDTD and the "two-step" method [28]. Zakharian et al. calculated the electromagnetic field distribution in and around dielectric media of various shapes and optical properties and then employed the Lorentz law of force to determine the distribution of force density within the regions of interest [29]. In 2008, Sung simulated the focused laser beam by using the non-paraxial Gaussian formula, calculated the electromagnetic field distribution on the surface of a micro-bubble by the FDTD method, and then calculated the optical trapping force by using the Maxwell stress tensor [30].

The FDTD method is a numerical method to calculate the electromagnetic field and was proposed by Yee in 1966. The basic conception of FDTD is that the three-dimensional space containing the studied object is discretized into small hexahedral grids along the coordinate axis. The electrical parameters of each grid are described with $\varepsilon$ (electric permittivity), $\mu$ (magnetic permeability), $\sigma$ (electric conductivity), and $\sigma_m$ (magnetic conductivity). The studied object is simulated by the combination of these small grids with specific medium parameters; the calculating time is discretized into the order of time step. The algebraic equations of the field are obtained by using the time differential equations based on three spatial axes and one time axis to approximate the Maxwell differential equations; then adding the appropriate boundary conditions and exciting conditions according to specific problems; and at last getting the time-domain electromagnetic field distribution by using iterative calculations to solve the algebraic equations.

The usage of the FDTD method can be divided into the following steps:

1. Analyze the requirement.
2. Build the FDTD model, which includes meshing, the selection of the algorithm and boundary conditions, and the setup of the exciting source.
3. Use time-iterative calculation to get useful information from the obtained time-domain waveform according to the requirement.

## 5.4 Application

### 5.4.1 FDTD Numerical Simulation of the Trapping Force of Microsphere in Single Optical Tweezers [31]

In this section, the model of trapping force on microsphere near focus in single optical tweezers is built by the three-dimensional FDTD and Maxwell stress tensor methods. A fifth-order Gaussian beam based on spherical vector wave function (VSWF) is adopted as a simulation light source; the correct light field transmission is obtained. The influences of the wavelength, waist, and polarization of light sources and the radius and refractive index of the microsphere on the optical trapping force are discussed. The influence of nearby micro-sphere and beam polarization on the trapping force of the trapped microsphere in single optical tweezers is analyzed. The effect of beam polarization working on the trapping force of the trapped microsphere is especially analyzed.

### 5.4.1.1 Calculation Method

#### 5.4.1.1.1 The Setting of the Gaussian Light Source

To ensure the accuracy of the simulation, the size of the grid is not greater than $\lambda/20$. The PML absorbing boundary [32] is used.

According to the principle of equivalence, the light source is added by updating the electric and magnetic field values on a plane. In many texts, usually, the formula of the zero-order Gaussian beam is used as an exciting source [33]:

$$A = A_0 e^{2\left(\frac{r}{W}\right)} \tag{5.40}$$

where $A$ represents the amplitude of a point on the plane, $A_0$ is the ampli-tude of the center point on the beam's cross section, $r$ is the distance from a point to the center, and $W$ is the waist of the beam. Formula (5.40) is not the strict solution of Maxwell's equations. It is derived from the paraxial wave equation. However, the optical tweezers are strongly focused and the amplitude will change drastically. Therefore, it is not accurate enough to use formula (5.40) to simulate the propagation of electromagnetic fields in FDTD. As an alternative, the following spherical vector wave functions (VSWF) [34, 35] are used to express the fifth-order Gaussian light source as the incident field.

$$\mathbf{E}_{\text{inc}} = \sum_{n=1}^{\infty} \sum_{m=-n}^{n} [a_{mn} Rg\mathbf{M}_{mn}(kr,\theta,\phi) + b_{mn} Rg\mathbf{N}_{mn}(k\mathbf{r},\theta,\phi)] \tag{5.41}$$

$$\mathbf{H}_{\text{inc}} = -j\sqrt{\frac{\varepsilon}{\mu}} \sum_{n=1}^{\infty} \sum_{m=-n}^{n} [b_{mn} Rg\mathbf{M}_{mn}(kr,\theta,\phi) + a_{mn} Rg\mathbf{N}_{mn}(k\mathbf{r},\theta,\phi)] \tag{5.42}$$

where $\mathbf{E}_{\text{inc}}$ and $\mathbf{H}_{\text{inc}}$ are the electric and magnetic field components of the inci-dent field, respectively, $k$ is the wave number, $a_{mn}$ and $b_{mn}$ are the expansion coefficients, and $Rg\mathbf{M}_{mn}$ and $Rg\mathbf{N}_{mn}$ are vector wave functions in spherical coordinates. When using $x$ linearly polarized beams, $m$ is 1 and –1, and the expansion coefficients are in the following forms [36],

$$a_{1n} = a_{-1n} = -b_{1n} = b_{-1n} = -(-i)^{n+1}[4\pi(2n+1)]^{1/2} g_{5,n} \tag{5.43}$$

where

$$g_{5,n} = \exp[-s^2(n-1)(n+2)]\left\{1+(n-1)(n+2)s^4\left[3-(n-1)(n+2)s^2\right]\right.$$

$$\left.+(n-1)^2(n+2)^2s^8\left[10-5(n-1)(n+2)s^2+0.5(n-1)^2(n+2)^2s^4\right]\right\}$$

(5.44)

$$s = 1/k\omega_0 \tag{5.45}$$

$\omega_0$ is the waist radius of the laser; and $n$ is the number of the spherical wave. As $n \geq 30$, the change of the field value can be ignored [37]. So, $n = 30$ is taken. The obtained converging field intensity distribution with the VSWFs is shown in Figure 5.5(a). The wavelength of incident light is 600 nm, the waist radius is 300 nm, the light source plane is located at $z = 0$, and the focus point is located at $z = 0.525$ μm. It can be seen that light is focused on the focal position accurately in the FDTD simulation.

In the FDTD simulation, the size of the light source plane affects the calculation of the transmitted light field. The larger the area of light source is, the more accurate the distribution of the optical field is. Figure 5.5(b) shows the amplitude of the electric field distribution on the optical axis, and $z = 0$ represents the focal position. The solid line indicates the source area of 2.25 μm² and the dotted line denotes the source area of 9 um². It is apparent that the converging effect of the solid line is not as accurate as that of the dashed line. Thus, the area of the light source should be as large as possible in the FDTD simulation.

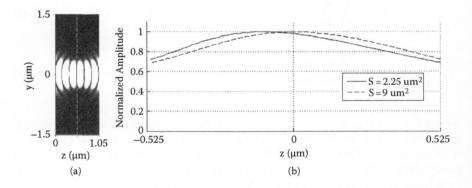

**FIGURE 5.5**
(a) The obtained converging field intensity distribution with VSWFs. (b) The influence of the size of the source plane on the simulation accuracy.

*5.4.1.1.2 Calculation of the Optical Trapping Force*

The electromagnetic field force of a dielectric microsphere immersed in liquid can be obtained from the following Maxwell stress tensor integral formula:

$$\mathbf{F} = \int_S \langle \mathbf{T} \rangle \bullet d\mathbf{S} \tag{5.46}$$

where $\mathbf{T}$ is the Maxwell stress tensor.

$$\mathbf{T} = \varepsilon(\mathbf{E} \bullet \mathbf{n})\mathbf{E} + \mu(\mathbf{H} \bullet \mathbf{n})\mathbf{H} - \frac{1}{2}(\varepsilon E^2 + \mu H^2)\mathbf{n} \tag{5.47}$$

where $\varepsilon$ and $\mu$ are the dielectric permittivity and magnetic permeability of the background medium, respectively; $\mathbf{n}$ is the normal vector of the outer surface of the microsphere; $d\mathbf{S}$ is the micro-area of the microsphere on the outward surface along the normal direction; $\mathbf{E}$ and $\mathbf{H}$ represent the total field values of the electric and magnetic fields, respectively; and the angle bracket < > indicates the time-average value.

The computer is configured as following: Intel dual-core E5200 CPU, main frequency 2.52 GHz, 3.50 GB memory. MATLAB language is used to program. For the limitation of the computer memory, the computing space in FDTD is divided into $250 \times 250 \times 120$ grids.

In the simulation, the wavelength of light source is 1200 nm, its waist radius is 600 nm, and power is 10 mw, which is $x$ linearly polarized. The radius of the silicon microsphere is 600 nm, whose refractive index is $n_s = 1.46$. The microsphere is immersed in water (the refractive index of water is $n_b = 1.33$). The light propagates along the z-axis, in z positive direction.

### 5.4.1.2 Numerical Simulation Results

*5.4.1.2.1 The Influence of the Parameters of Light Source and*
*the Microsphere on the Optical Trapping Force*

The simulation calculation region is shown in Figure 5.6. Table 5.1 lists the obtained trapping stiffness coefficients when the microsphere is moved along the three coordinate directions near the focus respectively. As shown in Table 5.1, when the light source is x linearly polarized, the trapping stiffness in $x$- and $y$-directions are almost the same, except that in the $y$-direction it is slightly larger. However, the trapping stiffness of the z-direction is much smaller than that along the $x$- and $y$-directions. It shows that the axial trapping capacity of the optical tweezers is weaker than the radial. Therefore, in the following, the influence on axial trapping force $F_z$ is discussed. In the following figures, the abscissa represents the distance from the center of the microsphere to the laser focal point, which is at $z = 0$. Accordingly, the ordinate represents the axial force $F_z$ exerted on the microsphere. The positive value indicates that the optical trapping force and the light propagation

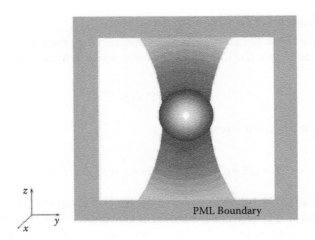

**FIGURE 5.6**
Schematic of simulation domain.

are in the same direction, and the microsphere is subject to a pushing force. Accordingly, the negative value indicates that the optical trapping force and the light propagation are in the opposite direction, and the microsphere is subject to a pulling force.

The influence of the size of the waist radius on the axial force $F_z$ is shown in Figure 5.7(a). As shown, before the focus, the silicon microsphere is subject to the force that pushes it to the focus; behind the focus, the silicon microsphere is subject to the force that pulls it to the focus. These results are consistent with other reports [3]. When the waist radius is 600 nm, the balance position of the axial force on the microsphere is near the focus. While the waist radius is 700 nm, the balance position moves significantly backward, and the maximum axial pushing and pulling forces decrease. That is, the smaller the waist is, the stronger the light field convergence will be, which causes the light intensity gradient and the corresponding gradient force to increase. That easily balances with the scattering force worked as the pushing force, and a stable trapping is achieved in the light propagation direction. On the contrary, while the waist increases, the light field converges weakly, and the light intensity gradient decreases. If the light

**TABLE 5.1**

The Trapping Stiffness on Three Coordinates Directions

| Direction | Trapping Stiffness Coefficient (N/m) |
|---|---|
| $x$ | 1.0814e-005 |
| $y$ | 1.1544e-005 |
| $z$ | 4.5145e-006 |

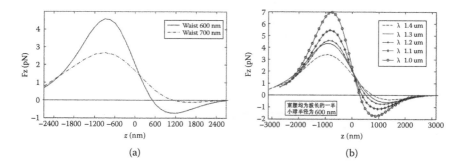

**FIGURE 5.7**

(a) The axial forces change with Gaussian beam's waist; (b) the axial forces change with Gaussian beam's wavelength.

field convergence is very weak, the microsphere will be mainly subjected to the scattering force in the direction of light propagation, which leads to an unstable trapping in the axis direction. This is also the result of Ashkin's light suspension experiment [38], which was completed in 1970.

Figure 5.7(b) shows that the axial forces exerted on the microsphere change with the wavelength of laser. The waist radius is half as big at the corresponding wavelength. It can be seen that, as the wavelength is in the range of 1000 to 1400 nm, with the increase of the wavelength, the axial force exerted on the microsphere decreases, and the corresponding balance position is located farther behind the focus.

The effect of the microsphere's refractive index on the optical trapping force is shown in Figure 5.8(a). It can be seen that, as the refractive index increases, the pushing force exerted on the microsphere increases, but the pulling force decreases. As the refractive index increases to a certain value (such as the refractive index of 2 shown in Figure 5.8), the microsphere is only subject to the pushing force, and the optical tweezers cannot capture

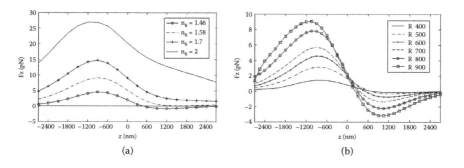

**FIGURE 5.8**

(a) Dependence of the axial forces on the refractive index. (b) Dependence of the axial force on the radius.

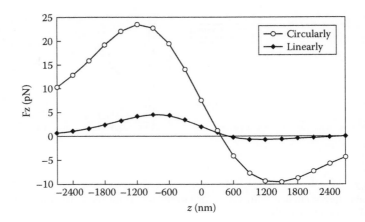

**FIGURE 5.9**
The axial force with different polarization of light.

the microsphere. In addition, the axial trapping balance position moves far-ther away from the focus with the increase of the refractive index.

The axial force also changes with the radius of the microsphere. The dif-ferent line types shown in Figure 5.8(b) represent the axial forces of the microspheres with the radius of 400 nm, 500 nm, 600 nm, 700 nm, 800 nm, and 900 nm, respectively. It can be seen that, as the radius increases, the maximum pushing and pulling force also increase, and the balance posi-tion moves closer to the focus.

In recent years, more and more attention has been paid to the effect of the polarization state of light source on the optical trapping force. The optical trapping force is often regarded as the sum of the gradient force, the scatter-ing force, and the force generated by polarization [39]. Figure 5.9 shows the obtained simulation results when the light source is left-handed circularly polarized and linearly polarized, respectively. It can be seen from the figure that when the light source is left-handed circularly polarized, the maximum axial force is an order of magnitude higher than that of the linearly polarized.

### 5.4.1.2.2 *The Effect of an Adjacent Microsphere on the Optical Trapping Force Exerted on the Captured Microsphere*

Usually, the sample pool contains many samples. When a particular sam-ple is trapped for manipulation, other samples in the vicinity will affect the force acted on the captured sample. This situation is simulated, and the effect of the polarization on the optical trapping force is also analyzed.

The simulation area is shown in Figure 5.10. $z$ is the propagation direction of light. The origin point is located at the center of microsphere 1 in the force equilibrium position alone without being disturbed. Microsphere 2, with the same geometrical and optical parameters as microsphere 1, moves along the positive direction of the $x$-, $y$-, and $z$-axis. $d$ represents the distance between

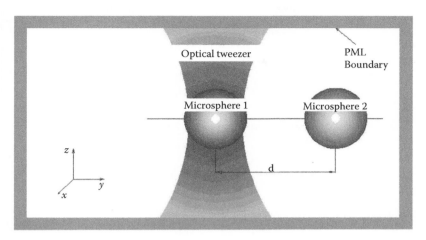

**FIGURE 5.10**
The schematic of the simulation domain with dual microspheres and single optical tweezers.

the two microspheres. The obtained force exerted on microsphere 1 is shown in Figure 5.11. Figure 5.11(a–c) show the corresponding forces $F_x$, $F_y$, and $F_z$ exerted on microsphere 1 by $x$ linearly polarized beams when microsphere 2 moves away from microsphere 1 along the positive direction of the $x$-, $y$-, and $z$-axis, respectively. Figure 5.11(d–f) show the corresponding force exerted on microsphere 1 by left-handed circularly polarized beams. The abscissa represents the distance between two microspheres (i.e., $d$ as microsphere 2 moves in each direction). The ordinate represents the force corresponding to the moving direction; a positive value indicates that microsphere 1 is subject to an attractive force from microsphere 2, and a negative value indicates a repulsive force. From Figure 5.11(a) and (b), it can be seen that when microsphere 2 moves along the $x$- and $y$-axis directions, microsphere 1 is subjected to the repulsive force from microsphere 2, which oscillates to zero with the increase of the distance between two microspheres. The maximum force in the $x$-direction is slightly greater than that of the $y$-direction. Figure 5.11(c) shows the axial force $F_z$ of microsphere 1 when microsphere 2 moves along the $z$-axis direction. With the increasing of the distance between the two microspheres, $F_z$ fluctuates and the maximum axial force of microsphere 1 is significantly less than that in Figure 5.11(a) and (b). From Figure 5.11(a–c), it can be seen that when the optical tweezers capture a microsphere, other microspheres around will affect the force exerted on the captured microsphere, and the effects are different in different directions, which results in losing microsphere 1's balance. Figure 5.11(d–f) are the corresponding forces when the light source is left-handed circularly polarized. Due to the symmetry of the circularly polarized light, Figure 5.11(d) and (e) have the same force, which changes from attractive force to repulsive force and then approaches zero. Figure 5.11(f) shows that as microsphere 2 moves away, the repulsive

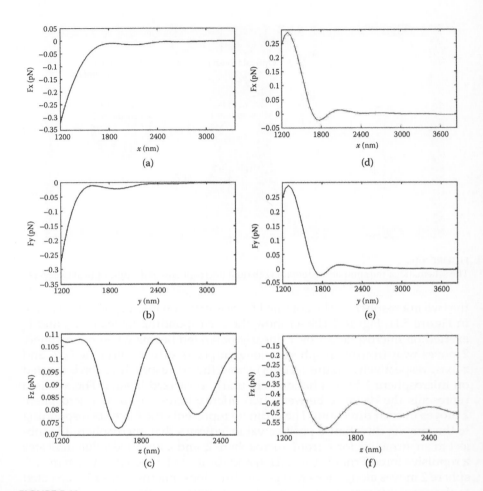

**FIGURE 5.11**
The obtained optical trapping force exerted on the captured microsphere 1 as microsphere 2 moves. The wavelength is 1200 nm, the beam waist is 600 nm, the radius of the microspheres is 600 nm, and $ns = 1.46$. With $x$ linearly polarized light beams: (a) microsphere 2 moves along $x$-direction, (b) microsphere 2 moves along $y$-direction, (c) microsphere 2 moves along $z$-direction; with left-handed circularly polarized beams: (d) microsphere 2 moves along $x$-direction (e) microsphere 2 moves along $y$-direction (f) microsphere 2 moves along $z$-direction.

force exerted on microsphere 1 becomes larger at the beginning and then fluctuates. Compared with Figure 5.11(c), it is obvious that the effect of the polarization on the optical trapping force is very significant.

### 5.4.1.3 Conclusions

An accurate transmission light field is obtained by using the three-dimensional FDTD method and fifth-order Gaussian beams based on spherical vector wave functions. And then, using the Maxwell stress tensor formula, the optical trapping force is obtained. The simulation results show that the settings of the light source, the waist radius, wavelength, and polarization of beams, and the radius and refractive index of the microsphere all have effects on the optical trapping force. When the microsphere is stably trapped, the force is affected by adjacent microspheres and the polarization of beams. The effect of the polarization of beams in single optical tweezers on the optical trapping force is especially discussed. The capture ability of the circularly polarized beam is higher than that of the linearly polarized beam. When there are microspheres around the captured microsphere, the effect of the circularly polarized beam on the value and direction of the optical trapping force is significantly different from the linearly polarized beam.

### 5.4.2 FDTD Simulation of Trapping Nanowires with Linearly Polarized and Radially Polarized Optical Tweezers [40]

In this section, the trapping capacities on nanoscale-diameter nanowires are discussed in terms of a strongly focused linearly polarized beam and radially polarized beam.

#### 5.4.2.1 Introduction

Recently, optical tweezers have been used to trap and process nanowires and other nanomaterials to build new types of micro-nano devices [41–43]. However, most materials used for nanowires have high refractive indices, which cause difficulty in trapping by optical tweezers.

The radially polarized beam is a kind of center hollow beam, and the energy of the beam is distributed at the fringe of the beam, which is an advantage for the reduction of the scattering force on the nanowire and the improvement of the gradient force. In addition, the radial polarization also contributes to the improvement of gradient force. Here the trapping capacity on nanowires with the radially polarized beam is simulated and compared with that of a conventionally linearly polarized beam.

The diameter and length of nanowires in the simulation model both are between $0.1\lambda$ and $10\lambda$ ($\lambda$ is beam wavelength). Direct solutions of Maxwell's equations are suitable to get the electromagnetic field, and then the force on a nanowire can be calculated from the Maxwell stress tensor or Poynting's vector method. In this regard the T-matrix method has been used to simulate laser trapping of nanorods with a circularly polarized beam [44] and linear nanostructures composed of identical nanospheres with a linearly polarized beam [45]. The three-dimensional FDTD method is also a robust approach to solve Maxwell's equations. Although this method is time-consuming, it is very powerful with the potential to model arbitrary shaped objects and different traps. Therefore, in our work a three-dimensional FDTD is used to set up a numerical simulation model.

### 5.4.2.2 Calculation

#### 5.4.2.2.1 Setting of Beam

The vector spherical wave functions (VSWFs) are a complete and orthogonal set of solutions to the vector Helmholtz equation [44, 39]. For tightly focused beams, the fifth-order Gaussian beam description provides a significantly improved solution to Maxwell's equations in comparison with commonly used paraxial Gaussian beam descriptions [46]. To accurately simulate the transmission form of the electromagnetic field, the VSWFs are adopted to express fifth-order Gaussian beams as incident light [37, 46–48].

$$\mathbf{E}_{\text{inc}} = \sum_{n=1}^{\infty} \sum_{m=-n}^{n} \left[ a_{mn} Rg\mathbf{M}_{mn}(k\mathbf{r}, \theta, \varphi) + b_{mn} Rg\mathbf{N}_{mn}(k\mathbf{r}, \theta, \varphi) \right] \tag{5.48}$$

$$\mathbf{H}_{\text{inc}} = -j\sqrt{\frac{\varepsilon}{\mu}} \sum_{n=1}^{\infty} \sum_{m=-n}^{n} \left[ b_{mn} Rg\mathbf{M}_{mn}(kr, \theta, \varphi) + a_{mn} Rg\mathbf{N}_{mn}(k\mathbf{r}, \theta, \varphi) \right] \tag{5.49}$$

In the above formulas, $k$ is the wave number, $a_{mn}$ and $b_{mn}$ are shape coefficients, and $Rg\mathbf{M}_{mn}$ and $Rg\mathbf{N}_{mn}$ are vector wave functions based on spherical coordinates. In the case of $x$ linearly polarized beam, $m = 1$ and $-1$,

$$a_{1n} = a_{-1n} = -b_{1n} = b_{-1n} = -(-i)^{n+1}[4\pi(2n+1)]^{1/2} g_{5,n} \tag{5.50}$$

$$g_{5,n} = \exp\left[-s^2(n-1)(n+2)\left\{1+(n-1)(n+2)s^4\left[3-(n-1)(n+2)s^2\right]\right.\right.$$
$$\left.\left.+(n-1)^2(n+2)^2 s^8\left[10-5(n-1)(n+2)s^2 +0.5(n-1)^2(n+2)^2 s^4\right]\right\}\right] \tag{5.51}$$

$$s = 1/k\omega_0 \tag{5.52}$$

Here $g_{5,n}$ are coefficients for fifth-order Davis beam [39], $s$ is the dimensionless beam shape parameter, $\omega_0$ is the beam waist radius, and $n$ is the number of the spherical wave. As $n \geq 30$, the numerical error is negligible [37], so $n = 30$.

In the case of radially polarized beam,

$$\mathbf{E}_{\text{inc}}^r = \mathbf{E}_{\text{inc}} \cdot \mathbf{r}_{\text{inc}} \tag{5.53}$$

$$\mathbf{H}_{\text{inc}}^r = \mathbf{H}_{\text{inc}} \cdot \mathbf{r}_{\text{inc}} \tag{5.54}$$

Here $\mathbf{r}_{\text{inc}}$ is a radial position vector matrix in the incident plane of the beam. The TM components of formulas (5.53) and (5.54) represent a radially polarized beam [48, 49]. Figure 5.12 shows the normalized intensity distribution

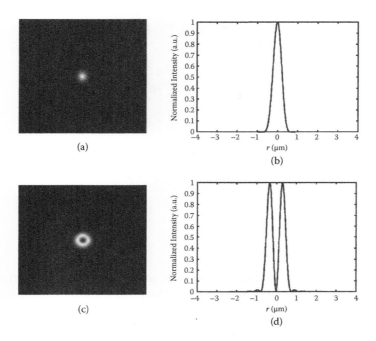

**FIGURE 5.12**
The normalized intensity distribution in the focal plane: (a) linearly polarized beam; (b) the radial cross section of linearly polarized beam; (c) radially polarized beam; (d) the radial cross section of radially polarized beam.

in the focal plane of linearly polarized beam and radially polarized beam, respectively.

### 5.4.2.2.2 Calculation of Optical Trapping Force

The electromagnetic force on a nanowire in a medium can be obtained by the Maxwell stress tensor:

$$\mathbf{F} = \int_S \langle \mathbf{T} \rangle \cdot d\mathbf{S} \qquad (5.55)$$

$\mathbf{T}$ is the Maxwell stress tensor,

$$\langle \mathbf{T} \rangle = \frac{1}{2} \mathrm{Re} \left[ \varepsilon \mathbf{E}\mathbf{E}^* + \mu \mathbf{H}\mathbf{H}^* - \frac{1}{2}(\varepsilon \mathbf{E}^2 + \mu \mathbf{H}^2)\mathbf{n} \right] \qquad (5.56)$$

The bracket $\langle \ \rangle$ denotes time-averaged value, $d\mathbf{S}$ is the unit normal to the nanowire's surface, $\varepsilon$ is the permittivity of the background medium, $\mu$ is the permeability of the background medium, $\mathbf{n}$ is the normal vector of the nanowire's outer surface, and $\mathbf{E}^*$ and $\mathbf{H}^*$ represent the complex conjugates of the electric and magnetic field.

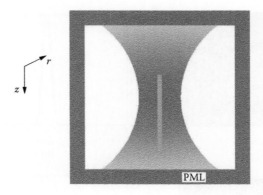

**FIGURE 5.13**
Computational domain.

### 5.4.2.3 Results

Simulations were performed on the computer with dual-core CPU E5200 and 4.0 GB memory. The MATLAB language is used for programming. In the model the size of the lattice can be $250 \times 250 \times 220$, the cell size is not greater than $\lambda/20$, and PML (perfectly matched layer) absorbing boundary is adopted. Figure 5.13 demonstrates a calculation region for single optical tweezers to trap a nanowire in water with the beam propagating along the $z$-axis, in the $z$-direction. The radial direction $r$ is perpendicular to the $z$-axis. The nanowire has a cylindrical shape with a radius of 60 nm and a height of 2100 nm; its refractive index is 1.6. The trapping beam with a wavelength of 600 nm, a waist radius of 300 nm, and a power of 10 mW is linearly polarized and radially polarized, respectively.

As single optical tweezers trap a nanowire, the nanowire orients itself to the optical axis of the beam (Figure 5.14). The axial force efficiency $Q_z$ ($Q = Fc/(np)$,

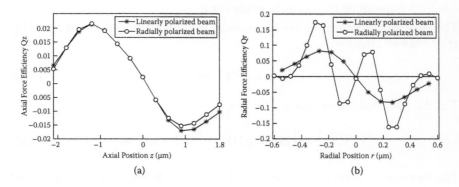

**FIGURE 5.14**
Axial (a) and radial (b) forces on nanowire with refractive index of 1.6 as a function of nanowire's displacement from the focus. The asterisk represents linearly polarized beam; the circle denotes radially polarized beam.

$c$ is the speed of light, $F$ is the trapping force, $n$ is the refractive index of the background medium, and $p$ is the beam power) and radial force efficiency $Q_r$ are shown in Figure 5.14. In the following figures the abscissa denotes the displacement of the geometrical center of nanowire from the focus and the ordinate represents force efficiency. Figure 5.14(a) shows that there is little difference between the maximal axial forces of the linearly polarized beam and those of the radially polarized beam; the axial trapping positions in both beams are located behind the focus. The maximal radial forces of the radially polarized beam are obviously greater than those of the linearly polarized beam (Figure 5.14[b]); the radial forces of the radially polarized beam near the focus appear to fluctuate and there exists more than one radial equilibrium position.

As the nanowire's refractive index is 2.5, the obtained forces are shown in Figure 5.15. It can be seen that the axial forces and radial forces of radially polarized beams increase greatly compared with those of linearly polarized beams. This is due to the hollow beam and polarization singularity of radially polarized beams, which decrease radiation pressure forces on nanowire and improve gradient forces.

As the nanowire's refractive index is 3.0, it is infeasible for a linearly polarized beam to trap nanowires along the axial direction in the vicinity of the focus (shown in Figure 5.16). However, trapping is achievable for radially polarized beams and the axial trapping position is located 1.5 μm behind the focus. The axial trapping potential corresponding to the area under the curve $Q_z$ (z) is defined as $U(z) = -\int F_z dz = -(np/c)\int Q_z dz$ [44], which represents the energy required for nanowires to escape from the trap. Although the axial trapping potential of the radially polarized beam in Figure 5.16 is of pronounced asymmetry with respect to the equilibrium position $Q_z = 0$, the energy barrier in the beam propagating direction is significantly larger than the thermal energy $k_B T$ ($k_B$ is the Boltzmann constant, $T$ is the temperature of

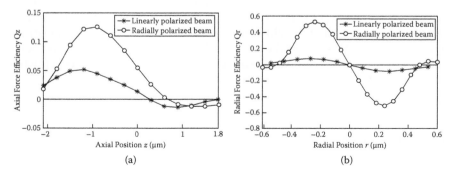

(a)

(b)

**FIGURE 5.15**
Forces on nanowire with refractive index of 2.5.

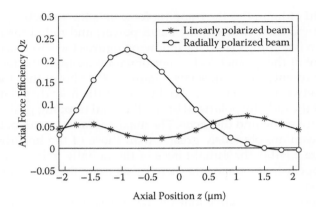

**FIGURE 5.16**
Axial forces on nanowire with refractive index of 3.0.

the background medium, $T = 300$ K). Therefore, it is difficult for nanowires to escape from the trap by the Brownian motion.

### 5.4.2.4 Conclusions

The three-dimensional FDTD method and Maxwell stress tensor are used to set up the simulation model and calculate the trapping force. By using VSWFs the transmission fields are obtained with linearly polarized and radially polarized beams in the vicinity of the focus point. The trapping capacities on nanoscale-diameter nanowires based on both the linearly polarized beam and the radially polarized beam are simulated and compared with each other. When the radially polarized beam is adopted for lower refractive index nanowires, the multiple trapping equilibrium positions beyond the focal plane exist. With the increase of the refractive indices of nanowires, the axial and radial forces of the radially polarized beam both increase greatly. The radially polarized beam, compared with the linearly polarized beam, demonstrates higher trapping efficiency on the higher refractive index nanowire. It is shown that the radially polarized beam is suitable for trapping those higher refractive index nanowires.

### Acknowledgments

We are grateful to the National Natural Science Foundation of China (Grant No. 50975271 and 91023049) and the Major Project of Chinese

National Programs for Fundamental Research and Development (Grant No. 2012CB937500) for financial support of this work. We thank Fred Firstbrook for critical reading.

## References

1. D. G. Grier, "A revolution in optical manipulation." *Nature* 424, 810–816 (2003).
2. A. Ashkin, "History of optical trapping and manipulation of small-neutral particle, atoms, and molecules," *IEEE J.Sel.Top.Quantum Elec.* 6, 841–856 (2000).
3. A. Ashkin, J. M. Dziedzic, J. E. Bjorkholm, and S. Chu, "Observation of a single-beam gradient force optical trap for dielectric particles," *Opt.Lett.* 11, 288–290 (1986).
4. S. Chu, J. E. Bjorkholm, A. Ashkin, and A. Cable, "Experimental observation of optically trapped atoms," *Phys. Rev. Lett.* 57, 314–317 (1986).
5. S. Chu, "Laser manipulation of atoms and particles," *Science* 253, 861–866 (1991).
6. A. Horst, A. Campbell, L. K. Vugt, D. A. M. Vanmaekelbergh, D. Dogterom, and A. Blaaderen, "Manipulating metal-oxide nanowires using counter-propagating optical line tweezers," *Opt. Exp.* 15, 11629–11639 (2007).
7. P. J. Pauzauskie, A. Radenovic, E. Trepagnier, H. Shroff, P. Yang, and J. Liphardt, "Optical trapping and integration of semiconductor nanowire assemblies in water," *Nature* 5, 97–101 (2006).
8. Y. L. Zhang, Y. Q. Zhao, Q. W. Zhan, and Y. P. Li, "Study of 3D optical chain with highly focused vector beam," *Acta Phys. Sin.* 55, 1253–1258 (2004).
9. Y. Harada and T. Asakura, "Radiation forces on a dielectric sphere in the Rayleigh scattering regime," *Opt. Commun.* 124, 529–541 (1996).
10. A. Ashkin, "Forces of a single-beam gradient laser trap on a dielectric sphere in the ray optics regime," *J. Biophys.* 61, 569–582 (1992).
11. R. Gussgard and T. Lindmo, "Calculation of the trapping force in a strongly focused laser beam," *J. Opt. Soc. Am. B* 9, 1922–1930 (1992).
12. K. Visscher and G. J. Brakenhoff, "Theoretical study of optically induced forces on spherical particles in a single beam trap I: Rayleigh scatters," *Optik.* 89, 174–180 (1992).
13. P. C. Chaumet and M. Nieto-Vesperinas, "Time-averaged total force on a dipolar sphere in an electromagnetic field," *Opt. Lett.* 25(15), 1065–1067 (2000).
14. P. C. Chaumet and M. Nieto-Vesperinas, "Coupled dipole method determination of the electromagnetic force on a particle over a flat dielectric substrate," *Phys. Rev. B* 61(20), 14119–14127 (2000).
15. A. Rohrbach and E. H. Stelzer, "Trapping forces, force constants, and potential depths for dielectric spheres in the presence of spherical aberrations," *Appl. Opt.* 41(13), 2494–2507 (2002).
16. A. Rohrbach and E. H. Stelzer, "Optical trapping of dielectric particles in arbitrary fields," *J. Opt. Soc. Am. A* 18, 839–853 (2001).
17. G. Roosen and C. Imbert, "Optical levitation by means of two horizontal laser beams: A theoretical and experimental study," *Physics Letters A* 59(1), 6–8 (1976).
18. A. Ashkin, "Forces of a single-beam gradient laser trap on a dielectric sphere in the ray optics regime," *Biophys.* 61(2), 569–582 (1992).

19. R. C. Gauthier, "Ray optics model and numerical computations for the radiation pressure micromotor," *Applied Physics Letters* 67, 2269–2271 (1995).
20. R. C. Gauthier, "Optical trapping: A tool to assist optical machining," *Optics & Laser Technology* 29, 389–399 (1997).
21. R. C. Gauthier, "Theoretical investigation of the optical trapping force and torque on cylindrical micro-objects," *Opt. Soc. Am. B* 14, 3323–3333 (1997).
22. R. C. Gauthier, "Theoretical investigation of the optical trapping properties of a micro-optic cubic glass structure," *Applied Optics* 39, 3060–3070 (2000).
23. D. A. White, "Vector finite element modeling of optical tweezers," *Computer Physics Communications* 128, 558–564 (2000).
24. T. A. Nieminen, H. Rubinsztein-Dunlop, N. R. Heckenberg, and A. I. Bishop, "Numerical modeling of optical trapping," *Computer Physics Communications* 142, 468–471 (2001).
25. S. H. Simpson and S. Hanna, "Numerical calculation of interparticle forces arising in association with holographic assembly," *Optical Society of America* 23(6), 1419–1430 (2006).
26. W. L. Collett, C. A. Ventrice, and S. M. Mahajan, "Electromagnetic wave technique to determine radiation torque on micromachines driven by light," *Applied Physics Letters* 82(16), 2730–2732 (2003).
27. D. Zhang, X. C. Yuan, S. C. Tjin, and S. Krishnan, "Rigorous time domain simulation of momentum transfer between light and microscopic particles in optical trapping," *Optics Express* 12(10), 2220–2230 (2004).
28. R. C. Gauthier, "Computation of the optical trapping force using an FDTD based technique," *Optics Express* 13(10), 3707–3718 (2005).
29. A. R. Zakharian, M. Mansuripur, and J. V. Moloney, "Radiation pressure and the distribution of electromagnetic force in dielectric media," *Optics Express* 13(7), 2321–2336 (2005).
30. S. Y. Sung and Y. G. Lee, "Trapping of a micro-bubble by non-paraxial Gaussian beam: Computation using the FDTD method," *Optics Express* 16(5), 3463–3473 (2005).
31. G. J. Hu, J. Li, Q. Long, T. Tao, G. X. Zhang, and X. P. Wu, "FDTD numerical simulation of the trapping force of microsphere in single optical tweezers," *Acta Phys. Sin.* 60, 30301 (2011).
32. A. Taflove and S. C. Hagness, *Computational Electrodynamics: The Finite-Difference Time-Domain Method*, 3rd ed. Norwood, MA: Artech House, 2005.
33. R. C. Gauthier, "Computation of the optical trapping force using an FDTD based technique," *Opt. Exp.* 13, 3707–3718 (2005).
34. R. G. Yang, D. Z. Cheng, and P. C. Liu, *Electromagnetic Theory*. Xi'an: Xi'an Jiaotong University Press, p. 53, 1991.
35. Y. P. Han, Y. G. Du, and H. Y. Zhang, "Radiation trapping forces acting on a two-layered spherical particle in a Gaussian beam," *Acta Phys. Sin.* 55, 4557–4562 (2006).
36. S. H. Simpson and S. Hanna, "Optical trapping of spheroidal particles in Gaussian beams," *J. Opt. Soc. Am. A* 24, 430–443 (2007).
37. S. H. Simpson and S. Hanna, "Numerical calculation of inter-particle forces arising in association with holographic assembly," *J. Opt. Soc. Am. A* 23, 1419–1431 (2006).
38. A. Ashkin, "Acceleration and trapping of particles by radiation pressure," *Phys. Rev. Lett.* 24, 156–159 (1970).
39. D. C. Benito, S. H. Simpson, and S. Hanna, "FDTD simulations of forces on particles during holographic assembly," *Opt. Exp.* 16, 2942–2957 (2008).

40. L. Jing and W. Xiaoping, "FDTD simulation of trapping nanowires with linearly polarized and radially polarized optical tweezers," *Optics Express* 19, 20736–20742 (2011).

41. P. J. Pauzauskie, A. Radenovic, E. Trepagnier, H. Shroff, P. D. Yang, and J. Liphardt, "Optical trapping and integration of semiconductor nanowire assemblies in water," *Nature Materials* 5, 97–101 (2006).

42. T. Yu, F. C. Cheong, and C. H. Sow, "The manipulation and assembly of CuO nanorods with line optical tweezers," *Nanotechnology* 15, 1732–1736 (2004).

43. R. Agarwal, K. Ladavac, Y. Roichman, G. H. Yu, C. M. Lieber, and D. G. Grier, "Manipulation and assembly of nanowires with holographic optical traps," *Opt. Express* 13, 8906–8912 (2005).

44. P. B. Bareil and Y. L. Sheng, "Angular and position stability of a nanorod trapped in an optical tweezers," *Opt. Express* 18, 26388–26398 (2010).

45. F. Borghese, P. Denti, R. Saija, M. A. Iatì, and O. M. Maragò, "Radiation torque and force on optically trapped linear nanostructures," *Phys. Rev. Lett.* 100, 163903 (2008).

46. J. P. Barton and D. R. Alexander, "Fifth-order corrected electromagnetic field components for a fundamental Gaussian beam," *J. Appl. Phys.* 66, 2800–2802 (1989).

47. G. Gouesbet, J. A. Lock, and G. Grehan, "Partial-wave representations of laser beams for use in light-scattering calculations," *Appl. Opt.* 34, 2133–2143 (1995).

48. N. Passilly, R. de S. Denis, K. Aït-Ameur, F. Treussart, R. Hierle, and J. F. Roch, "Simple interferometric technique for generation of a radially polarized light beam," *J. Opt. Soc. Am. A* 22, 984–991 (2005).

49. T. A. Nieminen, N. R. Heckenburg, and H. Rubinsztein-Dunlop, "Forces in optical tweezers with radially and azimuthally polarized trapping beams," *Opt. Lett.* 33, 122–124 (2008).

# 6

# *Nanostructured Photodetector for Enhanced Light Absorption*

**Narottam Das**

*Department of Electrical and Computer Engineering, Curtin University, Australia;*
*School of Engineering, Edith Cowan University, Australia*

## CONTENTS

## 6.1 Introduction

The photoconductor had been an attractive device because of its easy fabrication, but it has high dark current and associated Johnson noise, so it was not suitable for high performance or speed communication systems. Therefore, with proper design, an APD PIN diode can provide high responsivity and good quantum efficiency, and thus it can satisfy most of the detection requirements for optical communication systems. However, certain drawbacks—such as requirements for high bias values, together with the requirement of bias stability, and considerable noise associated with avalanche process—have renewed interest in other devices, such as phototransistors, modulated barrier photodiodes, Schottky diodes, and metal-semiconductor-metal (MSM) diodes. For simplicity of operation and ease of fabrication, Schottky diodes and MSM diodes are mostly preferred nowadays in communication applications. Since the MSM diodes have simple technology, any improvement in responsivity, the current limiting factor to their widespread application, will allow this type of photodetector to supplant existing photodetectors in the marketplace. The MSM photodetectors (MSM-PDs) offer an attractive benefit over alternative photodetectors, such as conventional PIN diodes. An MSM-PD is inherently planar and requires only a single lithography step, which is compatible with the existing field effect transistor technology; MSM-PDs have very low dark current.

Metal-to-semiconductor contacts have great importance since they are present in every semiconductor device. They can behave like a Schottky barrier or an ohmic contact dependent on the characteristics of the interface. A planar MSM-PD consists of interdigitated metal fingers forming Schottky diodes on a semiconductor surface as a substrate. These detectors are very attractive for many future optoelectronic applications, particularly because of their high-frequency capability combined with simple, integrated circuit (IC) compatible processing technology, which enables multi-gigabit optical communication systems and easy system integrations. The planar structure of MSM-PDs also results in the inherently low capacitance, which is beneficial for the detector sensitivity. The operation of MSM-PDs can be classified into two categories according to whether its intrinsic speed is limited by the transit time or recombination time. In the first category, an attempt is made to increase the detector speed by minimizing the finger spacing, which decreases the transit time of the photo-generated charge carriers. In the latter or second category, the carrier recombination time is shortened by introducing recombination centers into the semiconductor. However, this technique to increase the device speed has some drawbacks: it is not an IC compatible technique and it decreases the sensitivity of the detector. The fastest transit-time-limited GaAs-MSM-PDs have been fabricated and reported in [1]. By using e-beam lithography they could make the detectors having the smallest finger spacing and finger width of 25 nm resulting in an estimated impulse response as short as 250 fs. Compound semiconductors have been the most studied materials for MSM-PDs, but recently silicon has also attracted significant interest [2] due to the low-cost substrate material and the IC compatibility.

In recent years, subwavelength plasmonic nanostructure gratings have been identified as the promising candidates for realizing high-speed improved-responsivity MSM-PDs [3]. A subwavelength plasmonic nanograting is able to interact strongly with the incident light and potentially traps it inside the sub-surface region of semiconductor substrates. The MSM-PDs are very promising candidates for a wide range of applications, such as optical fiber communication systems, high-speed chip-to-chip interconnects, and high-speed sampling. An MSM-PD simply consists of two back-to-back Schottky diodes and has interdigitated metal fingers deposited onto a semiconductor surface; it detects photons by collecting the electric signals generated by photo-excited charge carriers in the semiconductor region, which drift under the electrical field applied between the fingers. The MSM-PDs can be classified according to whether their speed is intrinsically limited by the carrier transit time between the electrodes or by the carrier recombination effects. The implementation of interdigitated electrodes in MSM-PDs has led to a huge increase in bandwidth and reduction in dark current, in comparison to the ordinary PIN photodiodes with active areas of similar size [4–5]. The MSM-PDs have a much smaller capacitance per unit area due to their lateral geometry. They have response times in the range of a few tens of picoseconds (limited by the transit time of photo-generated carriers to the metallic contact pads), due to the nanoscale spacing between the electrode fingers. However, the downsizing of the electrode spacing leads to a decreased active area, resulting in a degraded responsivity. Recently, the use of surface plasmon-assisted effects for the design of such photodetectors has led to the development of nanostructured MSM-PDs having a high responsivity-bandwidth product, well beyond that of the ordinary PIN photodetectors, thus attracting a great deal of interest in future research [6].

The application of surface plasmon polaritons (SPPs) [7] for the light absorption enhancement using subwavelength metal nanogratings has promised a large improvement in light collection efficiency. In the past decade, there have been several theoretical and experimental research and development activities reported on the extraordinary optical transmission through subwavelength metallic apertures [8–12] as well as through periodic metal nanograting structures [13–15]. The simulation results based on the finite-difference time-domain (FDTD) method have shown a significant boost or improvement of light absorption through interaction with the SPPs, and this has straight application to the design of MSM-PDs [3,6,8,16,17]. Hence, an accurate modeling of the MSM-PDs will result in the optimization of the metal finger patterns, thus opening the way for the development of ultra-high responsivity-bandwidth-product photodetectors that are attractive for many future practical applications. It is important to note that this approach to light absorption enhancement is based on the engineering of a plasmonic light-capture system that enables a greater fraction of the incident optical power to reach the absorbing active regions of the device, rather than on altering the absorption coefficient of the semiconductor or changing any absorption path lengths. To quantify this absorption enhancement, calculate the normalized power spectra of the optical radiation integrated over the active regions of the device cross section located directly under the semiconductor surface and under the subwavelength apertures, which collect the incident light. Then, compare the spectra of optical power captured by the active regions for the cases of a nanograting-assisted device and a conventional or standard type device. Therefore, any light capture or absorption improvement occurring outside the active and photo-response device regions does not change the validity of these calculations.

The FDTD analysis is used to predict the performance of an MSM-PD structure employing a single metal (gold: Au) layer deposited onto a GaAs substrate and a nano-engineered metal grating etched lithographically into the metal (Au) layer above an unperturbed part of the same metal layer based on the extraordinary optical absorption (EOA) model [18–23]. Researchers have experimentally investigated the cross-sectional profiles of nanograting-assisted MSM-PD samples patterned by the focused ion beam (FIB) lithography and an atomic force microscopy (AFM). From the nanograting imaging experiments, it was observed that the real nanograting profile shapes were not a rectangular (which would represent an ideal profile in terms of the achievable absorption enhancement) groove type, even when researchers aimed to achieve a rectangular-shaped profile, but rather were more trapezoidal shaped. Therefore, the practical manufacturing constraints have a strong impact on the resulting nanograting grooves or profile shapes, which in turn affect the maximum achievable light absorption enhancement in MSM-PDs and in all other devices relying on subwavelength-feature-size grating-type structures for enhancing the light capture performance. For this reason, it was particularly focused on the nanograting groove-shape-dependent properties of these MSM-PDs, in order to evaluate the effects of device manufacturing technology-related limitations on the performance of these nanograting patterns. The simulation results show that the enhancement in light trapping near 850 nm of up to about 50 times is possible for an idealized rectangular-shaped metal nanograting cross-sectional geometry in comparison to the standard-designed MSM-PDs, mainly due to the extraordinary optical wave propagation through the metal nanogratings [24–38].

## 6.2 FDTD Software for Simulation

### 6.2.1 The Basics of the FDTD Simulation Method

Optiwave FDTD (OptiFDTD) is a powerful simulation tool, highly integrated software that allows computer-aided design and simulation of advanced passive photonic components. The OptiFDTD software package is based on the FDTD method. The FDTD method has been established as a powerful engineering tool for integrated and diffractive optics device simulations. This is due to its unique combination of features, such as the ability to model light propagation, scattering and diffraction, and reflection and polarization effects. The method allows for the effective and powerful simulation and analysis of submicron (nanoscale) devices with very fine structural details.

A submicron or nanoscale device implies a high degree of light confinement and, correspondingly, the large refractive index difference of the materials (mostly semiconductors) to be used in a typical device design [39–41].

The OptiFDTD software package was developed by Optiwave Inc. This software is established to aid users in the design and simulation of advanced passive photonic components based on the FDTD method. In addition to the FDTD method's aforementioned features, it can also model material anisotropy and dispersion without any assumption of field behavior, such as the slowly varying amplitude approximation.

The FDTD method is a simple and intuitive way to solve numerically the partial differential equations. The FDTD method utilizes the central difference approximation to discretize the two Maxwell's curl equations—namely, Faraday's and Ampere's laws—both in time and spatial domains, and then it solves the resulting equations numerically to derive the electric and magnetic field distributions at each time step using an explicit leapfrog scheme. The FDTD solution, thus derived, is second-order accurate and is stable if the time step satisfies the Courant condition. One of the most important attributes of the FDTD algorithm is that it is absurdly parallel in nature because it only requires exchange of information at the interfaces of the subdomains in the parallel processing scheme.

## 6.2.2 The Equations of 2D FDTD Method

The FDTD approach is based on a direct numerical solution of the time-dependent Maxwell's curl equations. As such, the photonic device is laid out in the $x$-$z$ plane. The propagation is along the $z$-direction. The $y$-direction is assumed to be infinite in 2D simulation. This assumption removes all the derivatives from Maxwell's equations and splits them into two (transverse electric [TE] and transverse magnetic [TM]) independent sets of equations. The 2D computational domain is shown in Figure 6.1. The space

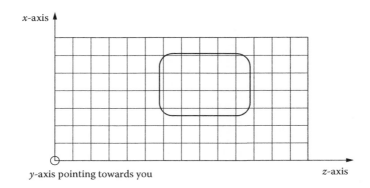

**FIGURE 6.1**
Numerical representation of a 2D computational domain for simulation [41].

steps are $\Delta x$ in the $x$-direction and $\Delta z$ in the $z$-direction of the $x$-$z$ plane. Each mesh point is associated with a specific type of material and contains information about its properties, such as refractive index and dispersion parameters [41].

### 6.2.2.1 TE Waves for FDTD Simulation

In the 2D TE wave case ($H_x$, $E_y$, $H_z$ nonzero components, propagation along the $z$-axis, transverse field variations along the $x$-direction) in lossless media, Maxwell's equations can be written in the following form [41]:

$$\frac{\partial E_y}{\partial t} = \frac{1}{\varepsilon}\left(\frac{\partial H_x}{\partial z} - \frac{\partial H_z}{\partial x}\right), \qquad \frac{\partial H_x}{\partial t} = \frac{1}{\mu_0}\frac{\partial E_y}{\partial z}, \qquad \frac{\partial H_z}{\partial t} = -\frac{1}{\mu_0}\frac{\partial E_y}{\partial x} \qquad (6.1)$$

where $\varepsilon = \varepsilon_0\varepsilon_r$ is the dielectric permittivity and $\mu_0$ is the magnetic permeability of the vacuum. The refractive index is represented by $n = \sqrt{(\varepsilon_r)}$.

Each field is represented by a 2D array—$E_y$ $(i,k)$, $H_x$ $(i,k)$, and $H_z$ $(i,k)$—corresponding to the 2D mesh grid as given in Figure 6.1. The indices $i$ and $k$ account for the number of space steps in the $x$- and $z$- direction, respectively. For the case of TE waves, the location of the fields in the mesh (computational domain) is shown in Figure 6.2.

The TE fields stencil can be explained as follows. The $E_y$ field locations coincide with the mesh nodes given in Figure 6.1. In Figure 6.2, the solid lines represent the mesh given in Figure 6.1. The $E_y$ field is considered to be the center of the FDTD space cell. The dashed lines form the FDTD cells. The

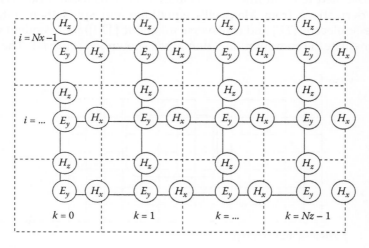

**FIGURE 6.2**
Location of the TE fields in the computational domain [41].

magnetic fields $H_x$ and $H_z$ are associated with cell edges. The locations of the electric fields are associated with integer values of the indices $i$ and $k$. The $H_x$ field is associated with integer $i$ and $(k + 0.5)$ indices. The $H_z$ field is associated with $(i + 0.5)$ and integer $k$ indices. The numerical analogue in Eq. 6.1 can be derived from the following relationship [41]:

$$\frac{\partial E_y}{\partial t} = \frac{1}{\varepsilon}\left(\frac{\partial H_x}{\partial z} - \frac{\partial H_z}{\partial x}\right) \tag{6.2}$$

### 6.2.2.2 TM Waves for FDTD Simulation

In the 2D TM wave case ($E_x$, $H_y$, $E_z$ nonzero components, propagation along the $z$-axis, transverse field variations along the $x$-axis) in lossless media, Maxwell's equations take the following form [41]:

$$\frac{\partial H_y}{\partial t} = -\frac{1}{\mu_0}\left(\frac{\partial E_x}{\partial z} - \frac{\partial E_z}{\partial x}\right), \qquad \frac{\partial E_x}{\partial t} = \frac{1}{\mu_0}\frac{\partial H_y}{\partial z}, \qquad \frac{\partial E_z}{\partial t} = \frac{1}{\varepsilon}\frac{\partial H_y}{\partial x} \tag{6.3}$$

The location of the TM fields in the computational domain (mesh) follows the same philosophy as shown in Figure 6.3.

Now, the electric field components $E_x$ and $E_z$ are associated with the cell edges, while the magnetic field $H_y$ is located at the cell center.

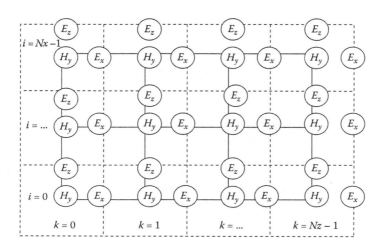

**FIGURE 6.3**
Location of the TM fields in the computational domain [41].

### 6.2.3 Lorentz-Drude Model

#### 6.2.3.1 Lorentz-Drude Model in Frequency Domain

It has been shown in Ref. [19] that a complex dielectric function for some metals and surface plasmas can be expressed in the following form [41]:

$$\varepsilon_r(\omega) = \varepsilon_r^f(\omega) + \varepsilon_r^b(\omega) \tag{6.4}$$

This form separates explicitly the intraband effects (usually referred to as free electron effects) from interband effects (usually referred to as bound-electron effects). The intraband part of the dielectric function is described by the well-known free-electron or Drude model in Ref. [20,21]:

$$\varepsilon_r^f(\omega) = 1 + \frac{\Omega_p^2}{j\omega\Gamma_0 - \omega^2} \tag{6.5}$$

The interband part of the dielectric function is described by the simple semiquantum model resembling the Lorentz results for insulators:

$$\varepsilon_r^b(\omega) = \left( \sum \frac{\Omega_p^2}{\omega_m^2 - \omega^2 + j\omega\Gamma_m} \right) \tag{6.6}$$

where $\omega_p$ is the plasma frequency, $m$ is the number of oscillators with frequency $\omega_m$ and lifetime $1/\Gamma_m$, $\Omega_p = \sqrt{(G_m)}\omega_m$ is the plasma frequency associated with intraband transitions with oscillator strength $G_0$, $G_m$ is related to the oscillator strengths, and the damping constant is $\Gamma_0$.

The above Lorentz-Drude model can be expressed as the more general equation

$$\varepsilon_r(\omega) = \varepsilon_{r,\infty} + \sum_{m=0}^{M} \frac{G_m\Omega_p^2}{\omega_m^2 - \omega^2 + j\omega\Gamma_m} \tag{6.7}$$

where $\varepsilon_{r,\infty}$ is the relative permittivity in the infinity frequency, $\Omega_p$ is the plasma frequency, $\omega_m$ is the resonant frequency, and $\Gamma_m$ is the damping factor or collision frequency.

In this general equation, if only the term $m = 0$ exists, and $\omega_0 = 0$, then the general equation describes the Drude model as in (6.5). If only the $m = 1,\ldots, M$ term exists, and $\Omega_1 = \Omega_2 = \Omega_3 = \cdots = \Omega_M$, then the general model becomes the Lorentz model as in (6.6). This model can also work as the separate Drude and Lorentz models.

Ref. [19] also provided Lorentz-Drude (LD) parameters for 11 noble metals; their unit is in electron volts. Lorentz-Drude parameters for selected materials contains parameters compiled by Optiwave that describe noble metals.

### 6.2.3.2 Lorentz-Drude Model in Time Domain

The Lorentz-Drude model shown in (6.7) is in the frequency domain form. However, the FDTD is a time-domain method, and therefore it would be suitable for broadband simulations. Therefore, we need to transform (6.7) to the time-domain form, so that the FDTD can handle the full-wave analysis for the Lorentz-Drude material. This transformation to time domain is accomplished by using the polarization philosophy within Maxwell's equation as given below. The Lorentz-Drude model in the time domain can be expressed as follows [41]:

$$\mu_0 \frac{\partial \vec{H}}{\partial t} = \nabla \times \vec{E} \tag{6.8}$$

$$\varepsilon_{r,\infty}\varepsilon_0 \frac{\partial \vec{E}}{\partial t} + \sum_{m=0}^{M} \frac{\partial \vec{P}_m}{\partial t} = -\nabla \times \vec{H} \tag{6.9}$$

$$\frac{\partial^2 \vec{P}_m}{\partial t^2} + \Gamma_m \frac{\partial \vec{P}_m}{\partial t} + \omega_m^2 \vec{P}_m = \varepsilon_0 G_m \Omega_m^2 \vec{E} \tag{6.10}$$

The FDTD algorithm is derived based on the above equation.

However, the FDTD simulation results have demonstrated significant enhancement of light absorption for the design of ultrafast MSM-PDs [8,14,16,27,28]. The FDTD algorithm was originally proposed in 1966 by K. S. Yee [39], who introduced a modeling technique to solve the Maxwell equations applying the finite difference approach (or mathematics). The FDTD algorithm can directly calculate the value of $E$ (electric field intensity) and $H$ (magnetic field intensity) at different points of the computational domain $(i,j,k)$ [29]. Since then, it has been used for several applications and many extensions of the basic algorithm have been developed [40,41]. Now, the FDTD method is very efficient, accurate, and widely used for electromagnetism computations, such as fields and resonant modes.

## 6.3 Metal-Semiconductor-Metal Photodetectors

Recently, the use of surface plasmon-assisted effects has been reported for the development of MSM-PDs with high responsivity [16]. An MSM-PD comprises two Schottky contacts (i.e., interdigitated metallic electrodes on a semiconductor material), in contrast to a p-n junction as in a photodiode. When an electric field (or a voltage) is applied between the electrodes and the device active region is illuminated, the electric carriers (electrons and holes) are generated; they then drift toward opposite electrodes due to the electric field and form a photocurrent. They can enhance the light absorption in the subwavelength apertures due to the electrons and holes recombination [8,14,25–28]. These MSM-PDs are promising candidates for future-generation optoelectronic systems, such as optical fiber communications, high-speed chip-to-chip interconnects, and high-speed sampling. For sufficiently small grating periods, higher-order diffractions are suppressed, leaving only the zero-order diffraction [3,7]. Thus, the subwavelength nanogratings can be represented as a homogeneous medium with optical properties determined by the nanograting geometry [27, 28]. When the nanograting period is on the order of the light wavelength, the light wave may be resonant and reflects in the structure, so that a resonant reflection occurs. The subwavelength structure does not generate real diffracted light waves other than the zero-order diffraction waves [24]. In recent years, due to the advances in nanotechnology, there has been a wide interest in exploiting the dielectric response of metals to make photonic materials. For instance, the photonic insulating properties of metals can be used to trap incident light concentrating in very small volumes (or areas) [12]. Over the past decade, researchers have discovered an interesting effect of light interacting with metal structures [28]; the light through the subwavelength slits can be transmitted at a higher order of magnitude than the standard aperture theory (or conventional type). Several theoretical and experimental results have been reported about the extraordinary optical transmission through the subwavelength metallic apertures/slits, as well as the periodic metal nanograting structures [8,12,27,28]. An MSM-PD with the conventional design is shown in Figure 6.4. Impinging the light from the top on the device surface, there will be huge light reflections, which can reduce the light absorption. This phenomenon can be avoided by the modeling and fabrication of nanogratings on the metal fingers. Hence, the individual subwavelength apertures can exhibit notable light transmission when surrounded by a periodic structure that harvests the incident light (the mechanism is shown in Figure 6.5). The experiments demonstrate that

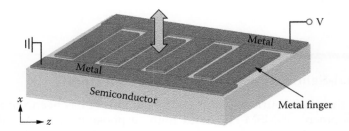

**FIGURE 6.4**
A simple symmetric diagram of a conventional MSM-PD structure with interdigitated electrodes and semiconductor substrate (GaAs).

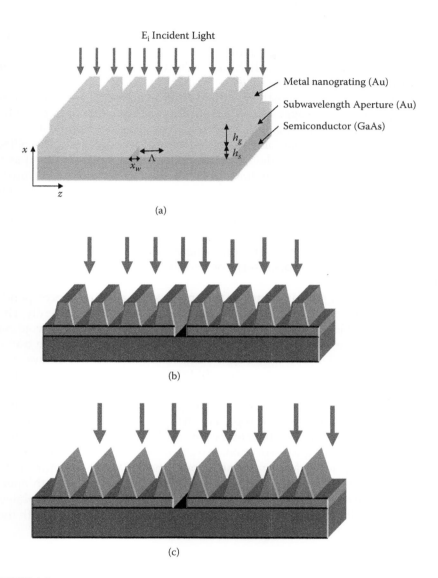

**FIGURE 6.5**
Schematic diagram of different types of nanostructured MSM-PDs. Here the nanogratings are etched inside the top part of the gold (Au) layer. (a) The metal (Au) nanogratings are shown as rectangular-shaped or having a rectangular profile in their cross section (where the aspect ratio is 1), (b) trapezoidal-shaped nanograting profile, and (c) the metal nanograting having a triangular cross-sectional profile (where the aspect ratio is 1).

the incident light on a subwavelength aperture cannot be diffracted in all angles [31], and if such nanostructure is embedded on top of the electrodes, then, remarkably, the emerging light can be beamed into the structure rather than diffracted out (or reflected back).

In addition, the light is mostly reemitted from a very small area surrounding the aperture. This suggests that a well-directed source of light could be generated using a subwavelength aperture, an exciting development that is being pursued as a source for a variety of optical technologies. The SPPs, which are supported by the device active region, show great potential as a new type of subwavelength photonic phenomenon. Excitation of SPP waves causes resonance absorption, which can be observed as partial or total absorption of incident light. Particularly, SPPs attract their near-field characters and associated field enhancements [12]. The absorption enhancement caused by the excitation of SPPs is associated with the incident photons and their interaction with the nanogratings, whereas Fabry-Perot-like resonances are included in transmission absorption process through the subwavelength slits.

There has been a growing interest to explain all phenomena corresponding to the improvement in absorption via nanoscale structuring recently. The main motivation is the need to come up with a practical method for avoiding the undesirable reflection from the electrodes surface to enhance the transmission efficiency. Therefore, the FDTD simulation tool is used to analyze the light absorption enhancement in an MSM-PD structure with rectangular-shaped nanogratings profile. We investigated the effects of nanograting structures on the MSM-PD device. The nanogratings are deposited on a thin metal layer of the same kind or type. We changed the geometric properties of the device and analyzed the simulation results to obtain the best condition of the device in light absorption. We calculated the normalized power spectra of the optical wave propagation over the MSM-PD structure with several rectangular nanograting grooves around the subwavelength aperture, which are able to capture the light in the semiconductor region located just under the slit. We also compared the normalized power spectra with the one obtained for a conventional type device.

## 6.4  Nanostructured MSM-PD Design

The proposed design of nanostructured MSM-PD consists of three separate parts. These are, namely, (1) top layer (metal nanograting), (2) unperturbed metal layer (underlayer) containing the standard or conventional subwavelength apertures, and (3) semiconductor (GaAs) substrate, as shown in

Figure 6.5. The metal nanogratings are formed by etching or FIB milling the lines inside a metal, such as gold or silver (Au or Ag) layer with the grooves being perpendicular to the $x$-direction, and its dimensions and geometry are optimized to in-couple light near the design wavelength and trigger the SPPs propagating along the $z$-direction. For a metal nanograting period of $\Lambda$, the wave vector of the excited SPPs is given by the following equation [3,7,8, 27,28,32].

$$K_{sp} = \frac{\omega}{c} \sin\theta \pm \frac{2\pi l}{\Lambda} = \frac{\omega}{c} \sqrt{\frac{\varepsilon'_m \varepsilon_d}{\varepsilon'_m + \varepsilon_d}} \qquad (6.11)$$

where $\omega$ is the angular frequency of the incident light wave, $c$ is the speed of light in vacuum, $\theta$ is its angle of incidence with respect to the device normal, and $l$ is an integer (i.e., $l = 1, 2, 3,..., N$). In the analysis, the complex dielectric permittivity of the metal is denoted as $\varepsilon_m = \varepsilon'_m + i\varepsilon''_m$, where $\varepsilon'_m$ and $\varepsilon''_m$ are the real and imaginary part of the dielectric permittivity, respectively, and $i$ is imaginary unity, i.e., $i = \sqrt{-1}$; the dielectric permittivity of air (or the incidence medium) is denoted as $\varepsilon_d$. Equation (6.1) expresses mathematically the wave-vector matching condition (momentum conservation condition) for a SPP excited by the light wave of angular frequency $\omega$ incident at an angle $\theta$ onto a grating-corrugated (period $\Lambda$) metal-dielectric interface. Here, the finite nanograting height effects are neglected and also any plasmon scattering effects caused by the surface roughness features, since the nanograting heights used in the models are much smaller than the expected SPP propagation lengths of several micrometers. Using Eq. (6.11) to design the absorption enhancement peak spectrally located near 850 nm for the Au grating/ Au/GaAs material system and air as incidence medium, it was found that a nanograting periodicity of $\Lambda = 810$ nm was optimum. Following the approach reported in [3], the duty cycle of nanograting corrugations was selected to be 50%. Here, the dependency of the plasmon-assisted light absorption enhancement on the geometric shape (or groove) of the nanograting corrugations cross sections was investigated using FDTD analysis. Figure 6.5 illustrates schematically several types of nanograting profile grooves that were investigated.

Figure 6.5(a) shows the schematic diagram of an MSM-PD with plasmonic rectangular-profile nanogratings etched inside the top part of the gold (Au) layer. The thickness of the unperturbed metal layer containing subwavelength apertures is shown as $h_s$ (typically about 10–50 nm) and the height of the metallic nanogratings, $h_g$ (varied between 50 and 100 nm in this modeling). A single nanograting period ($\Lambda$) of 810 nm (fixed in this design and modeling) contains a metal line and a cutout (free space) with a duty cycle of 50%. The subwavelength aperture width is $x_w$ and its value was varied between 50 and 300 nm in this modeling.

Figure 6.5(b) illustrates a hatchtop-cone (trapezoidal-shaped) nanograting profile that was used for the simulation. It is the closest cross-sectional profile (groove) to that generated in practice by the FIB lithography in the case of small-feature-size (sub-500 nm) nano-patterned grooves. The trapezoidal shapes modeled had several different top-to-base aspect ratios of 0.8, 0.5, and 0.2, being the ratios between the trapezoid-top and base lengths.

Figure 6.5(c) illustrates an alternative geometry, the triangular-shaped nanograting profile. Figures 6.5(a–c) all show that a metal (Au) layer is deposited on top of the GaAs substrate; the thickness of this layer was kept constant at 100 nm throughout the analysis. For this analysis, a range of nanograting profile-shape variations was considered, keeping the unperturbed part of metal layer (underlayer) thickness at 10 nm and the metal nanograting height $h_g$ at 90 nm. The light is normally incident on top of the plasmonic nanogratings, which, as shown later, assist absorption within the subsurface GaAs regions.

Figure 6.6 shows the measured results of nanograting imaging obtained experimentally. Figure 6.6(a) shows a scanning electron microscope (SEM) image of a typical plasmonic nanograting etched inside the top part of Au layer using FIB milling technology. A 10 nm Cr adhesion layer is deposited followed by a 190 nm thick Au layer on top of the GaAs substrate using an e-beam evaporator to form the Schottky pads (cathode and anode). It was observed that the FIB lithography had superior performance for cutting the sharp-edge profiles compared to the conventional photolithography systems. The modern FIB systems enable the nano-patterning of materials with a minimum feature size of about 10 nm, which is required for quantum devices operating near room temperature and is of increasing importance for nanofabrication tools. In further investigation, it has been observed through the SEM imaging that the nanogratings shape is about trapezoidal (i.e., like a V-grooved shape), rather than the rectangular one. This trapezoidal shape results from the redeposition of gold atoms and the truncation of nanograting edges by the etching ion beam. Further to confirm the shape of the nanogratings, an atomic force microscopy (AFM) system (XE-100 from Park Systems) was used to analyze the surface topography of the nanogratings. Figure 6.6(b) shows a 3D view of the nanograting profile of the MSM-PD structure. It was clearly observed that the top-side length of the nanograting groove is smaller than its bottom-side length (of the nanograting). Figure 6.6(c) shows a 2D nanograting profile, which was trapezoidal-shaped with an aspect ratio of 0.5–0.6. Based on these basic experimental results, the dependency of the nanograting shape on the light absorption enhancement in MSM-PDs was subsequently investigated.

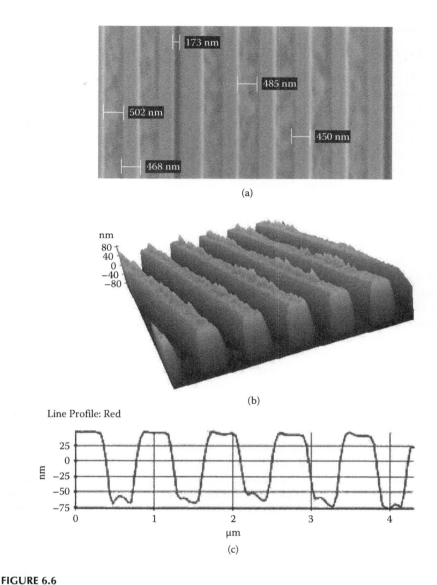

**FIGURE 6.6**
Experimentally measured results: (a) SEM image of a nanograting profile, which is made by
the FIB lithography process; (b) a 3D view of nanostructured MSM-PD observed by an AFM
system for typical nanograting profiles; and (c) an AFM image of typical nanograting profiles
of FIB-made nanostructured MSM-PD.

## 6.5 FDTD Simulation of Nanostructured MSM-PDs

In FDTD simulation models, the normally incident light passes through the subwavelength apertures and reaches to the GaAs substrate, thus generating electron-hole pairs. Furthermore, the light absorption within the GaAs is assisted by the SPPs excited near the metal-semiconductor interface in the nanograting regions. The energy of light incident on the metal nanogratings is coupled partially into propagating SPPs that can improve the light absorption efficiency in thin subsurface regions of semiconductor under the subwavelength apertures. The absorption-improving effects of nanogratings are due to the SPP-generated localized regions of high-intensity electric field (as shown in Figure 6.7). The nanogratings can therefore act as light concentrators (such as plasmonic lenses), which is essential for triggering the EOA of light inside the active regions of the MSM-PDs.

Two-dimensional FDTD models of different designs of MSM-PDs (with and without the metal nanogratings) were generated using the Opti-FDTD software package developed by Optiwave Inc. For simulations, a mesh step size of $\delta x = 5$ nm (even though when testing the computation accuracy with a refined model of 1 nm mesh size, it obtained the same computational results as in these models) and a time step of $\delta t < 0.1 \delta x/c$. The excitation field was modelled as a Gaussian-modulated continuous wave. The incident light wave was TM-polarized (its electric field oscillation direction was along the $z$-axis, perpendicular to the nanograting profiles or grooves). The anisotropic perfectly matched layer (APML) boundary conditions were applied in both the $x$- and $z$-directions. In all FDTD simulations, the gold (Au) dielectric permittivity was defined by the popular Lorentz-Drude model [19] and the GaAs dielectric permittivity data was taken from the Ref. [22,23]. Poynting vector integration was performed numerically for the cases of the nanograting-assisted as well as conventional designs over the rectangular-shaped planar areas defined within the substrate subsurface regions of the device. These integration areas were defined using the so-called observation lines in the 2D FDTD model. The integration areas were geometrically located directly underneath the subwavelength apertures and had the same widths as these apertures.

For the calculation of total transmitted electric field strength within the substrate cross section, a custom-designed MATLAB algorithm is developed. Figure 6.7 shows the density plot of the total transmitted electric field strength within the substrate cross section, as a result of vector summation of the modeled complex electric field component distributions along the $x$-direction ($E_x$) and $z$-direction ($E_z$). It shows that the significant amount of light passed through the region of subwavelength aperture width, which is mainly due to the plasmon-assisted effects, results in propagating SPPs excited by the incident light waves and the energy concentration within small material volumes near the photodetector's active region. The light transmission inside the substrate depends on the nanograting profile or groove shapes. Figure 6.7(a) shows the

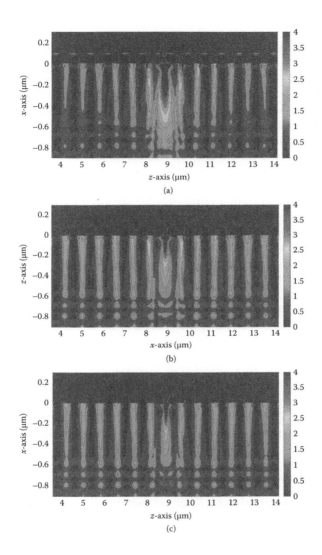

**FIGURE 6.7**

Total electric field intensity distribution in the cross section of computational volume behind the nanograting with different profiles or grooves cross section and the following nanograting parameters: nanograting height of 90 nm, unperturbed gold layer thickness of 10 nm, period 810 nm, slit (subwavelength aperture) width of 50 nm. The design parameters used for this computation are the same as those resulting in the maximum predicted (almost 50 times when the nanogratings shape is rectangular) light absorption enhancement due to plasmon-assisted light intensity concentration under the subwavelength apertures. The color scale has been optimized for representing the small weak-field intensity variations. (a) For a rectangular-shaped nanograting profile with the aspect ratio of 1; (b) for a trapezoidal-shaped nanograting profile (for this case, the aspect ratio is 0.5), the absorption enhancement is about 11 times; and (c) for a triangular-shaped nanograting profile with the aspect ratio of 0 (zero), the absorption enhancement is about 5 times. (See color figure.)

light transmission map (generated using the computed electric field strength distribution) inside the substrate cross section for a rectangular-shaped nanograting. Figure 6.7(b) shows the light transmission map inside the substrate for a trapezoidal-shaped (for this case, the aspect ratio is 0.5) nanograting profile. It can be noted that the light transmission inside the substrate is reduced compared with the case of a rectangular-shaped nanograting profile. The light transmission through the nanogratings and subwavelength apertures is also dependent on the effective refractive index of the periodic structure and it can be analyzed using the effective medium theory as reported in Ref. [24]. The reduced light transmission through the metallic structures having features (such as profiles or grooves) of non-rectangular cross-sectional profile compared to that of the rectangular-profiled structures is due to the increased reflectivity of the nano-patterned metal layers as well as changes in the effective refractive index of such nanograting layers. A trapezoidal cross-sectional profile or groove shape (with the aspect ratio of 0.5) agrees closely with the AFM measured imaging results as shown in Figure 6.6(b). For further analysis and investigation, the performance of triangular-shaped [25,26] nanogratings was also modeled. Figure 6.7(c) shows the light transmission into the substrate for a triangular-shaped nanograting profile. In this case, the light transmission has reduced further as compared with the rectangular- or trapezoidal-shaped nanogratings of the same height (as shown in Figures 6.7[a] and [b]). Therefore, these modeling results obtained emphasize the importance of establishing the lithography processes that would lead to achieving near-vertical etched-wall profiles. These results confirm that the nanograting manufacturing processes based on the FIB milling technology are superior to all conventional photolithography techniques in this respect; however, the goal is to achieve the near-rectangular nanograting grooves or profiles in subwavelength nanogratings.

## 6.6 Simulation Results of Nanostructured MSM-PDs

This section describes the simulation results of different nanostructured MSM-PDs having several types of groove-geometry designs as well as the light absorption enhancement dependency on the subwavelength aperture width. The groove shape dependency of light absorption enhancement was comprehensively modeled, as well as the aperture width dependency of the light absorption enhancement factor, defined as the ratio of the normalized power transmittance of MSM-PD with metal nanogratings to the normalized power transmittance of the MSM-PD without the metal nanogratings [8,27, 28]. The propagation of light through the nanostructured MSM-PD of the described type was modeled for a range of subwavelength aperture widths while keeping the values of $h_s = 10$ nm and $h_g = 90$ nm (optimized separately for devices operating near the wavelength range 830–850 nm as shown in Figure 6.8) constant.

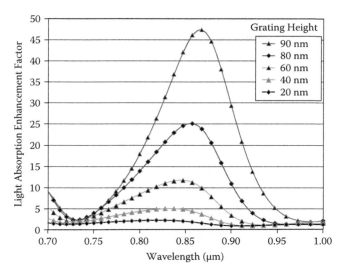

**FIGURE 6.8**
Light absorption enhancement factor spectra for nanostructured MSM-PDs with several metal nanograting heights. For this simulation, the metal nanograting heights were varied from 20 nm to 90 nm, while keeping the subwavelength aperture width constant at 50 nm; the shapes of all grating corrugations were rectangular type. (See color figure.)

Figure 6.8 shows the light absorption enhancement factor spectra for several optimized nanograting heights of nanostructured MSM-PDs with a rectangular-shaped cross section. The nanograting heights were varied ($h_g$) from 20 nm to 90 nm and the subwavelength aperture height ($h_s$, which is also the unperturbed metal layer thickness) was kept constant at 10 nm. The subwavelength aperture width was kept constant at 50 nm and the nanograting period kept constant at 810 nm. The obtained results show that the light absorption enhancement factor increases rapidly with increasing the nanograting heights. From the simulation, the light absorption enhancement factor was about 50 times when the nanograting height was 90 nm but about 5 times when the nanograting height was 40 nm. It is clear that the light absorption enhancement factor significantly depends on the nanograting heights of the nanostructured MSM-PDs. Therefore, for the case of thick metal layers without any nanogratings inscribed, there would be almost no light transmission into the semiconductor regions in between the subwavelength aperture locations. Inscribing the nanogratings inside the metal layers creates the surface plasmon-assisted light energy transfer mechanism into the semiconducting regions, thus removing a large part of absorption inside the metal layer and engineering the concentration of the incident light wave energy inside the photo-generated charge-carrier separation regions.

Figure 6.9 shows the calculated light absorption enhancement factor spectra of several optimized nanostructured MSM-PDs. In the spectra, all the

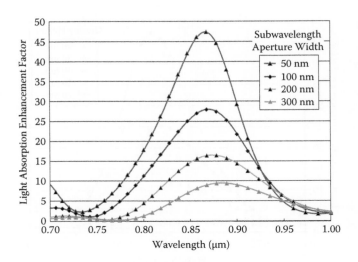

**FIGURE 6.9**
The light absorption enhancement factor spectra for nanostructured MSM-PDs with different subwavelength aperture widths. The subwavelength aperture widths were varied between 50 to 300 nm, and the groove shapes were all rectangular-type for this result. (See color figure.)

nanogratings are rectangular-shaped in their cross section. For this simulation, the subwavelength aperture widths were varied from 50 nm to 300 nm and the nanograting period was kept constant at 810 nm. The obtained results show that the light absorption enhancement factor decreases rapidly with increasing the aperture widths. In this case, the obtained light enhancement factor is about 50 times for a 50 nm wide aperture and about 10 times for a 300 nm aperture. The simulation results shown in Figures 6.7 and 6.8 are in excellent agreement with the results reported in references [14,18,27, 32], thus verifying the FDTD simulation model used in this analysis.

Figure 6.10 shows the light absorption enhancement factor versus the wavelength characteristics for different groove shapes (or profiles) when the subwavelength aperture width was kept constant at 100 nm. From this plot, it was found that in the case of using the (ideal) rectangular-shaped profile, the maximum light absorption enhancement of about 28 times could be obtained. For the same nanostructured MSM-PD, if using a trapezoidal-shaped profile with the aspect ratio of 0.8 (here, the aspect ratio is defined as the top-side to the bottom-side length of the metal nanograting corrugations), it was possible to obtain the light absorption enhancement of about 16 times. Similarly, for a triangular-shaped corrugation profile (where the aspect ratio is zero) nanograting, a light absorption enhancement factor of about 4.4 times was achieved.

Figure 6.11 shows the spectral distribution of the light absorption enhancement factor for nanostructured MSM-PDs with different profiles (or groove shapes). The subwavelength aperture width was kept constant at 50 nm (as

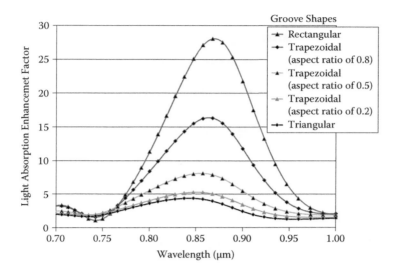

**FIGURE 6.10**
Light absorption enhancement factor versus the wavelength characteristics for different groove shapes or profiles when the subwavelength aperture width was fixed at 100 nm and the nanograting heights were kept constant at 90 nm. (See color figure.)

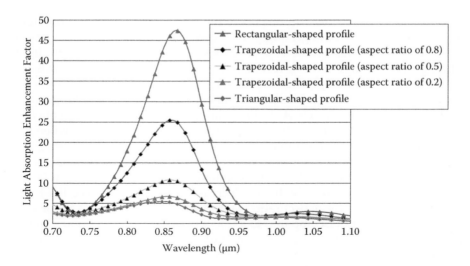

**FIGURE 6.11**
Light absorption enhancement factor spectra for different groove shape (or profile) geometries of nano-patterned gratings. For the simulation, the subwavelength aperture width was kept constant at 50 nm. (See color figure.)

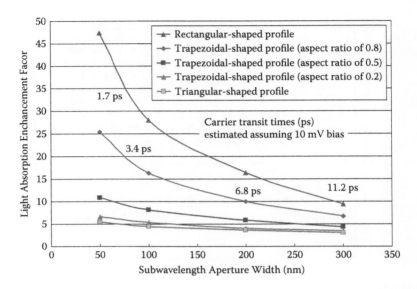

**FIGURE 6.12**
Maximum light absorption enhancement factor versus the subwavelength aperture width characteristics for different groove geometries (or profiles). The lower limits (drift time-limited) of the carrier transit time at 10 mV bias conditions are shown as a rough-guide order-of-magnitude estimate only. (See color figure.)

already optimized value with respect to the device bandwidth and the light absorption), and the nanograting period was fixed at 810 nm. Five types of selected nanograting geometries (or grooves) were modeled on the light propagation and the following results were obtained. The selected models are (1) rectangular-shaped profile (or groove), (2) trapezoidal-shaped profile (or groove) with the aspect ratio of 0.8, (3) trapezoidal-shaped profile (or groove) with the aspect ratio of 0.5, (4) trapezoidal-shaped profile (or groove) with the aspect ratio of 0.2, and (5) triangular-shaped profile (or groove), when the aspect ratio is 0 (zero). For the rectangular-shaped nanograting profile, the predicted absorption enhancement factor was at its maximum at about 50 times, and for the trapezoidal-shaped nanogratings, the enhancement factor was reduced progressively with the decreasing aspect ratios. For the trapezoidal profile with the aspect ratio of 0.8, the light enhancement factor was reduced by a factor of 2 compared to its "ideal-case" (i.e., rectangular-shaped profile) maximum limit. In practical manufacturing situations, this groove-shape dependency is therefore of primary importance. The triangular-shaped groove geometry, though impractical, results in the minimum light absorption enhancement factor of about 5 but it is still significant. Therefore, while the rectangular-shaped profile results in the highest light absorption enhancement factor, realistically such a profile would invariably be closer to some trapezoidal-shaped one in practical device manufacturing situations and the light absorption enhancement factor varies from 10 to 25.

Figure 6.11 shows the peak light absorption enhancement factor versus the aperture width for different groove shapes. For all analyzed profiles or groove shapes, the absorption enhancement factor was decreasing exponentially with increasing the aperture width (however, decreasing at different rates). The estimated lower limits for the carrier transit time for different subwavelength aperture widths were calculated using a simplified kinematics-based carrier drift model. From the transit-time estimates obtained (as shown in Figure 6.10), it is clear that by using very short subwavelength aperture widths, carrier drift-limited device operation in the sub-terahertz frequency range is possible.

Figure 6.13 shows the maximum light absorption enhancement factor versus the aspect ratio characteristics for nanostructured MSM-PDs. For this simulation, the nanograting period was considered to be 810 nm. In this analysis, the subwavelength aperture widths were varied from 50 nm to 300 nm for different nanograting groove-shaped profiles and the aspect ratios were varied from 0 (triangular-shaped profile) to 1 (rectangular-shaped profile). It is evident that when the nanograting's aspect ratio approaches 0 (zero), resulting in a triangular-shaped profile, the light absorption enhancement factor drops to about 5 times, which is only 10% of its potentially achievable ideal value (within the model limitations used).

Figure 6.14 shows the peak wavelength at maximum light absorption versus the subwavelength aperture width characteristics for different profiles

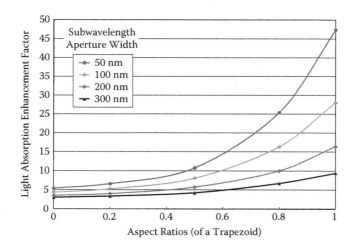

**FIGURE 6.13**
Maximum light absorption enhancement factor versus the aspect ratios characteristics for different subwavelength aperture widths and groove-shapes (or profiles). Here, the subwavelength aperture widths are varied from 50 nm to 300 nm and the aspect ratios are from 0 (triangular-shaped) to 1 (rectangular-shaped). (See color figure.)

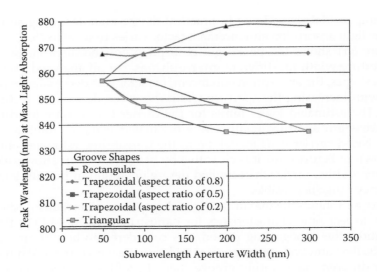

**FIGURE 6.14**

Light absorption-enhancement peak wavelength versus the subwavelength aperture width characteristics for different groove shapes (or profile) of metal nanogratings of height 90 nm. (See color figure.)

(or groove shapes) of metal nanogratings. For this analysis, the constant nanograting period of 810 nm is considered and the subwavelength aperture widths were varied from 50 nm to 300 nm for a constant nanograting height of 90 nm. For rectangular-shaped (when the aspect ratio is 1) and near-rectangular (i.e., trapezoidal-shaped with a high aspect ratio of 0.8) profiles, the light absorption enhancement peak drifts between ~860 and ~880 nm. For the trapezoidal-shaped nanogratings (with low aspect ratios of 0.5 and 0.2) and triangular-shaped gratings (i.e., an aspect ratio of 0), the peak locations were varied between ~840 and ~860 nm. These variations, as well as the absorption-enhancing performance differences between different groove shapes, are likely due to the shape-dependent differences in the effective refractive indices of the nano-patterned metal layers.

## 6.7 MSM-PD Structure Design with Nanograting Phase Shift

In this section, the FDTD method is used to simulate the light absorption enhancement of nanostructured MSM-PD, where the metal fingers are structured to form a metal nanograting above an unperturbed part of the metal fingers. The cross-sectional profiles of nanograting-assisted MSM-PD samples patterned by FIB milling and the surface topography of the device are experimentally investigated using an AFM. The experimental results show that the real nanograting profile shapes are not rectangular (which would represent an ideal profile in terms of the achievable absorption enhancement) for grooves formed aiming at achieving a rectangular-shaped profile, but rather close to trapezoidal. In Ref. [27], the effects of metal nanograting groove shapes or profiles were investigated and their impact (or influence) on the light absorption enhancement in MSM-PDs was analyzed. Here, the FDTD simulation is carried out on metal nanogratings with phase shifts [33,34] and confirmed that, due to the combined effects of the metal nanograting groove shapes and the nanograting phase shift, the light absorption enhancement decreases and the wavelength, at which the light absorption enhancement is maximum, is red shifted.

### 6.7.1 Nanograting Phase-Shift Observation in MSM-PD

Figure 6.15 shows the SEM images of two different nanostructured MSM-PDs etched inside the top part of the metal (gold: Au) layer using the FIB milling system. In this case, a 10 nm Cr adhesion layer and a 190 nm thick Au layer were deposited on top of the GaAs substrate using an e-beam evaporator to form the Schottky-diode pads (cathode and anode). The FIB lithography (which has superior performance for cutting sharp-edge profiles compared to conventional photolithography systems) was used to partially etch the Au layer and fully etch a 100 nm aperture, as shown in Figure 6.15. The FIB-etched aperture widths were 145 nm and 168 nm, respectively, causing nanograting phase-shifts of (145 – 100) nm = 45 nm (~20°) and (168 – 100) nm = 68 nm (~31°) as shown in Figure 6.15(a) and (b), respectively. It was also observed, through the SEM imaging, that the nanograting profile is trapezoidal (or closely like a V-grooved shape), rather than a rectangular shape. This trapezoidal shape is due to the re-deposition of gold atoms and truncation of the nanograting edges by the etching ion beam.

In addition to the SEM images (shown in Figure 6.15), and to further confirm the nanograting groove shapes or profiles, an AFM system (XE-100 from Park system) was used to analyze the surface topography of the nanogratings. The AFM images shown in Figure 6.16(a) and (b) confirmed that the shape of the nanograting is nearly trapezoidal and the aspect ratios are between 0.5 to 0.8 (it is not to scale, as measured approximately). The aspect ratio is defined as the trapezoid top-to-bottom base ratio. Based on these experimental results (the SEM and AFM images), the groove shape of the nanograting was taken into account in conjunction with the phase shift of the nanogratings to simulate the light absorption enhancement in plasmonics-based nanostructured MSM-PDs.

### 6.7.2 Modelling of MSM-PDs with Nanograting Phase-Shifts

This section discusses the design of plasmonics-based high-speed MSM-PD structures having different types of nanograting groove profiles and

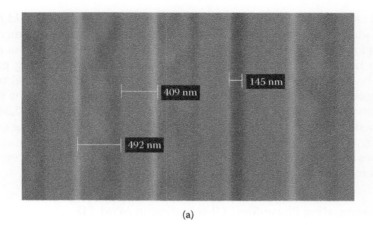

(a)

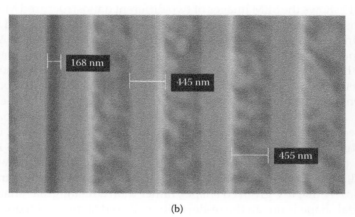

(b)

**FIGURE 6.15**
Typical SEM images of the metal nanogratings fabricated with FIB milling process for an estimated (predesigned/selected) subwavelength width of 100 nm. (a) Nanograting phase shift is
(145 – 100) nm = 45 nm (~20°) and (b) nanograting phase shift is (168 – 100) nm = 68 nm (~31°).

nanograting phase shifts. Figure 6.17 illustrates a plasmonic-based MSM-PD
structure, which consists of three separate layers, namely (1) a top layer (metal
nanograting), (2) an unperturbed metal layer (underlayer) containing the conventional subwavelength apertures, and (3) a semiconductor (GaAs) substrate.

Using Eq. (6.11), a nanograting period of $\Lambda = 810$ nm for the Au/air interface
corresponds to a GaAs MSM-PD with an edge of absorption of around 830 nm
for $\theta = 0$. The duty cycle of nanograting corrugations was selected to be ~50%
[3]. A typical nanostructured MSM-PD with a rectangular Au nanograting is
shown in Figure 6.17. The dependency of the plasmon-assisted light absorption enhancement on the geometric shape of the nanograting corrugations
cross sections and nanograting phase shift was investigated using the FDTD

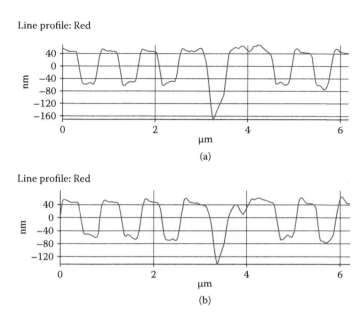

**FIGURE 6.16**
AFM images of two different nanostructured MSM-PDs having different nanograting profiles by the FIB milling process.

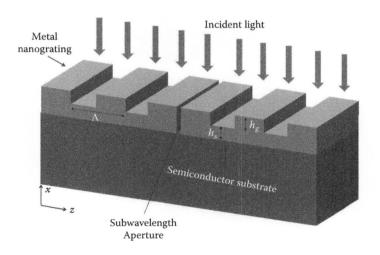

**FIGURE 6.17**
Schematic diagram of a typical nanostructured MSM-PD with plasmonic nanogratings etched inside the top part of the Au layer. Here, the metal nanograting profile is rectangular-shaped.

simulation method. In this simulation, the thickness of the unperturbed metal layer containing subwavelength apertures shown as $h_s$ was 10 nm and the height of the metallic nanogratings, $h_g$, was 90 nm. The subwavelength aperture width $x_w$ was varied from 50 to 200 nm.

Figure 6.18 shows the schematic diagrams of the nanograting without (a) and with (b) a phase shift defined as $\delta \equiv 2\pi x_{gp}/\Lambda$, respectively, where $x_{gp}$ is half the difference between the widths of the first metal ridges of the nanograting, and $\Lambda$ is the metal nanograting period.

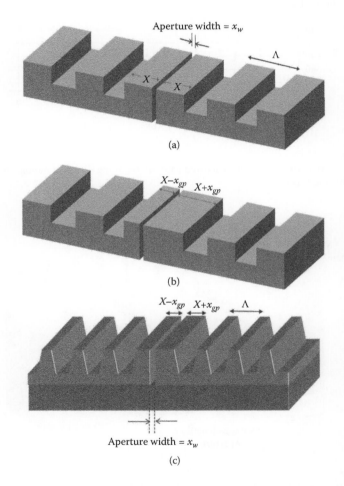

**FIGURE 6.18**
Schematic diagrams of different Au nanogratings (a) without the phase shift and (b) with the phase shift, $\delta \equiv 2\pi x_{gp}/\Lambda$. (c) General trapezoidal-shaped nanograting profile, with the aspect ratio defined as the trapezoid top-to-bottom base ratio.

The general trapezoidal-shaped nanograting profile was considered more practical than the ideal rectangular-shaped profile, because it is the closest cross-sectional profile to what was developed using FIB milling in the case of small-feature-size (sub-100 nm) nano-patterned grooves. Here, two trapezoidal shaped nanogratings were simulated, which had the aspect ratios (trapezoid-top to base length) of 0.8 and 0.5, respectively. Figure 6.18(c) illustrates a trapezoidal-shaped nanograting profile with the nanograting phase shift. In the simulation, the nanograting phase was varied from 0° to 90°.

The incident light normally passes through the subwavelength apertures and reaches the GaAs substrate, thus generating electron-hole pairs. In addition, the light absorption within GaAs is assisted by the SPPs excited near the metal-semiconductor interface near the nanograting region. The energy of light incident on the metal nanograting is partially coupled into propagating SPPs that increase the light coupled into the subwavelength aperture, thus improving the light absorption efficiency in the semiconductor substrate. The improvement in light absorption is due to SPP-generated localized regions of high-intensity electric field distribution (as illustrated in Figure 6.19). Therefore, the nanograting acts as a light concentrator (plasmonic lens) or collector triggering extraordinary optical transmission (and therefore absorption) within the active regions of the photodetector.

A density plot of the total transmitted electric field strength within the substrate cross section is shown in Figure 6.19. For this calculation, a custom-designed MATLAB algorithm is used to calculate the total electric field intensity distribution, which results from the vector summation of the modeled complex electric field component distributions along the x-direction ($E_x$) and the z-direction ($E_z$). Figure 6.19 clearly shows that the significant amount of light passing through the region of subwavelength aperture width is mainly due to the plasmon-assisted effects, which result in propagating SPPs excited by the incident light waves and concentrated (by plasmonic lenses) within the photodetector's active region. The light transmission inside the substrate depends on the aperture size and the nanograting profile. Figure 6.19(a) shows the total electric field intensity distribution (generated using the computed complex electric field distribution) inside the substrate cross section for an ideal rectangular-shaped nanograting with the grating phase shift 0°. Figures 6.19(b) and (c) show the total electric field intensity distribution inside the substrate for more practical trapezoidal-shaped (aspect ratio of 0.8 and 0.5) nanograting profiles with the nanograting phase shift of 45° and 90°, respectively. It can be noted that the light transmission inside the substrate of the trapezoidal-shaped nanograting MSM-PD is lower than that of a rectangular-shaped nanograting MSM-PD. This reduction is due to the combined effects of the nanograting groove shapes and the nanograting phase shift, as discussed in more detail in a later section.

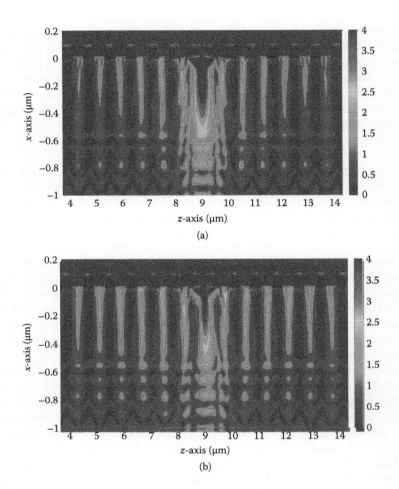

**FIGURE 6.19**
Total electric field intensity distribution in the cross section of computational volume behind the nanograting with different grooves cross section and the following grating parameters: the nanograting height is 90 nm, unperturbed gold (Au) layer thickness is 10 nm, grating period is 810 nm, and the subwavelength aperture width is 100 nm. The color scale has been optimized for representing small weak-field intensity variations. The nanograting profiles are (a) rectangular-shaped (aspect ratio 1) with the nanograting phase shift of 0°, (b) trapezoidal-shaped (with an aspect ratio of 0.8) with a nanograting phase shift of ~45°, and (c) trapezoidal-shaped (with an aspect ratio of 0.5) with a nanograting phase shift of ~90°. (See color figure.) (*Continued*)

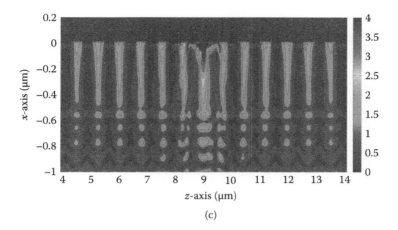

(c)

**FIGURE 6.19** (*Continued*)
Total electric field intensity distribution in the cross section of computational volume behind the nanograting with different grooves cross section and the following grating parameters: the nanograting height is 90 nm, unperturbed gold (Au) layer thickness is 10 nm, grating period is 810 nm, and the subwavelength aperture width is 100 nm. The color scale has been optimized for representing small weak-field intensity variations. The nanograting profiles are (a) rectangular-shaped (aspect ratio 1) with the nanograting phase shift of 0°, (b) trapezoidal-shaped (with an aspect ratio of 0.8) with a nanograting phase shift of ~45°, and (c) trapezoidal-shaped (with an aspect ratio of 0.5) with a nanograting phase shift of ~90°. (See color figure.)

## 6.8 Results of MSM-PDs with Nanograting Phase Shifts

This section discusses the simulation results of different nanostructured MSM-PDs having different types of nanograting groove-geometry designs as well as the light absorption enhancement dependency on the subwavelength aperture widths and the nanograting phase shift. Here, the impact of the nanograting groove shape, the nanograting phase shift, and the aperture width on the light absorption enhancement are specifically modeled. The light absorption enhancement factor is defined as the ratio of the normalized power transmittance of MSM-PD with the metal nanograting to the normalized power transmittance of the same device without a metal nanograting [8,27,28].

Figure 6.20 shows the simulated light absorption enhancement factor spectra of nanostructured MSM-PD with different aperture widths, namely, 50 nm, 100 nm, and 200 nm. The grooves of the nanograting were rectangular-shaped in their cross sections. The simulation results show that the light absorption enhancement factor decreases rapidly with the increasing of subwavelength aperture width. The results show a light absorption enhancement factor of about 50 times for a 50 nm wide subwavelength aperture

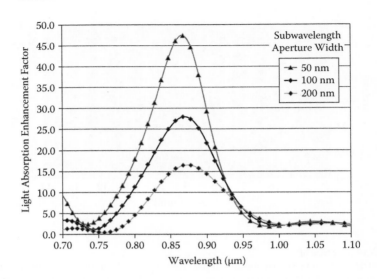

**FIGURE 6.20**
Light absorption enhancement factor spectra for nanostructured MSM-PDs with plasmon-assisted operation for different subwavelength aperture widths, such as 50 nm, 100 nm, and 200 nm. For this case, the nanograting grooves shape was rectangular type. (See color figure.)

width, about 28 times for a 100 nm wide subwavelength aperture, and about 16 times for a 200 nm wide subwavelength aperture width.

From the above results (shown in Figure 6.20), the subwavelength aperture width of 100 nm can be realized with an FIB milling system. Figure 6.21 shows the simulated light absorption enhancement factor spectra of several optimized nanostructured MSM-PDs with different nanograting phase shifts (ranges are varied from 0° to 90°). These results show that the light absorption enhancement factor decreases rapidly with increasing the nanograting phase shift. The light absorption enhancement factor is about 28 times for a nanograting phase shift of 0°. However, when the grating phase shift is increased to 90° (for the same aperture width of 100 nm), the light absorption enhancement factor drops to around 10 times (as shown in Figure 6.21). Also note from Figure 6.21 that the peak wavelength position is shifted from 850 nm to 837 nm when the nanograting phase shift is increased from 0° to 90°.

Figures 6.22(a) and (b) show the spectral distribution of the light absorption enhancement factor for trapezoidal nanograting groove shapes with aspect ratios of 0.8 and 0.5, respectively, for different nanograting phase shifts. Figure 6.22 reveals that the light absorption enhancement factor decreases slowly with increasing the nanograting phase shift. However, for a smaller aspect ratio the light absorption enhancement factor is smaller. Also, when the grating phase shift is 90° the degradation in light absorption enhancement drops by 50% in comparison to the case of a 0° grating phase shift. Note that, from Figures 6.22(a) and (b), the combined impact (or influence) of

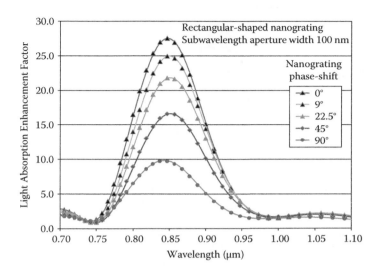

**FIGURE 6.21**
Light absorption enhancement factor spectra for nanostructured MSM-PDs with plasmon-assisted operation for various nanograting phase shifts. Here, the nanograting phase shift was varied from 0° to 90°. The subwavelength aperture width was kept constant at 100 nm and the nanograting groove shape was rectangular type. (See color figure.)

the groove shape and the nanograting phase shift is a reduction in the light absorption enhancement factor from 15 to around 5.

The maximum light absorption enhancement factor versus the nanograting phase shift is shown in Figure 6.23, for rectangular and trapezoidal grating profiles. It is clear that the light absorption enhancement factor decreases rapidly with increasing the nanograting phase shift for rectangular-shaped nanogratings. However, for trapezoidal-shaped nanogratings, the rate at which the light absorption enhancement factor decreases is small in comparison with that of the rectangular-shaped nanograting profile.

Figure 6.24 shows the peak wavelength at maximum light absorption versus the nanograting phase-shift characteristics for different nanograting profiles. In this simulation, the constant grating period of 810 nm was considered and the nanograting phase shifts were varied from 0° to 90° for a constant grating height of 90 nm. For the rectangular-shaped (when the aspect ratio is 1) nanograting profiles the maximum light absorption enhancement peak was about 847 nm and it was not affected by the nanograting phase shifts—i.e., the peak wavelength was independent on the nanograting phase shift. However, for a trapezoidal-shaped nanograting profile with an aspect ratio of 0.8, the peak wavelength shift drifted from 847 nm to 837 nm when the nanograting phase shift was increased from 0° to 90°. For the same conditions, the peak wavelength shift drifted from 837 nm to 827 nm, when the aspect ratio was 0.5 for a trapezoidal-shaped nanograting profile.

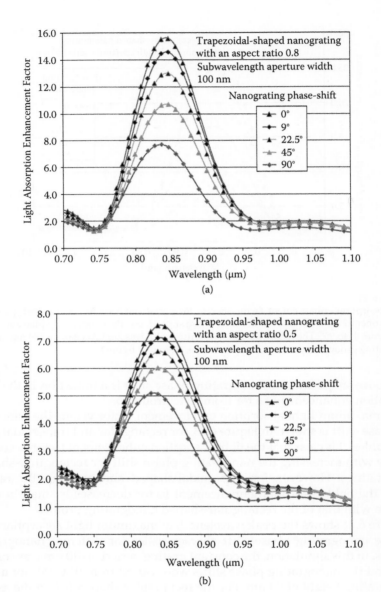

**FIGURE 6.22**
Light absorption enhancement factor spectra for trapezoidal-shaped nanograting with the aspect ratios of 0.8 (a) and 0.5 (b) for different nanograting phase shifts. In this case, the sub-wavelength aperture width was kept constant at 100 nm. (See color figure.)

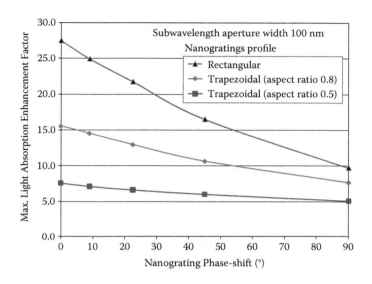

**FIGURE 6.23**
Maximum light absorption enhancement factor versus the grating phase shift for rectangular- and trapezoidal-shaped (aspect ratios are 0.8 and 0.5) nanograting geometries. In this case, the subwavelength aperture width was kept constant at 100 nm.

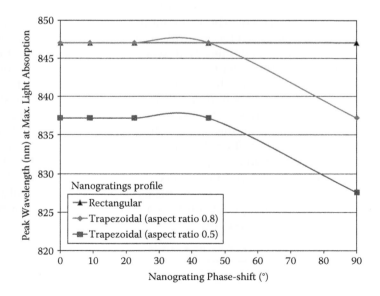

**FIGURE 6.24**
Peak wavelength at maximum light absorption enhancement versus the nanograting phase-shift characteristics for different nanogratings profile, where the nanograting height is 90 nm. Here, the nanograting profiles are rectangular- and trapezoidal-shaped (aspect ratios are 0.8 and 0.5).

These variations, as well as the light absorption-enhancing performance differences between the different nanograting profiles, are likely due to the shape-dependent (profile dependent) and phase-shift differences in the effective refractive indices of the nano-patterned metal layers.

## 6.9 Conclusions

In summary, this chapter presented several examples of nanostructured MSM-PDs manufactured using the FIB lithography and characterized typical nanograting groove profiles obtained using the AFM imaging. It also modeled the light-capture performance of these nanostructured MSM-PD designs employing the metal nanogratings of different cross-sectional profiles or groove shapes in terms of the light absorption enhancement and the responsivity-bandwidth products. In addition, it has discussed the application of an extended EOA model for accurately predicting the light absorption as a result of light interaction with both the subwavelength apertures and the metal nanogratings. The FDTD technique has been used to analyze the performance of nanostructured MSM-PDs with different geometric parameters, such as subwavelength aperture width, nanograting heights, grating phase shift, and other structure parameters (corrugation shapes) for maximized light absorption enhancement. The simulation results have shown that the nanostructured MSM-PDs can theoretically attain a maximum light absorption near 850 nm of almost 50 times (with rectangular-shaped nanogratings and 50 nm aperture width) better than that achieved with a standard or conventionally designed MSM-PDs; however, 15–25 times enhancement is more realistic for practical devices. Another set of simulation results have shown that an optimized MSM-PD structure can attain a maximum light absorption enhancement of about 28 times (with 100 nm wide subwavelength aperture width) better than conventional MSM-PDs. Furthermore, the simulation results of practical trapezoidal-shaped nanogratings have shown that for a 90° phase shift, the light absorption enhancement drops by around 50%. The results are expected to be useful for the design and development of nanostructured MSM-PDs with high responsivity-bandwidth products for future high-speed communication systems.

## Acknowledgment

This research is supported by the Centre for Smart Grid and Sustainable Power Systems, the Faculty of Science and Engineering, Curtin University, Australia.

# References

1. S. Y. Chou, Y. Liu, and P. B. Fischer, "Ultra-fast nanoscale metal-semiconductor-metal photodetectors on bulk and low-temperature grown GaAs," *Appl. Phys. Lett.* 61(477), 819–821 (1992).
2. B. F. Levine, J. D. Wynn, F. P. Klemens, and G. Sarusi, "1 Gb/s Si high quantum efficiency monolithically integrable $\lambda$ = 0.88 μm detector," *Appl. Phys. Lett.* 66, 2984–2986 (1995).
3. J. A. Shackleford, R. Grote, M. Currie, J. E. Spanier, and B. Nabet, "Integrated plasmonic lens photodetector," *Appl. Phys. Lett.* 94, 083501 (2009).
4. J. B. D. Soole and H. Schumacher, "InGaAs metal-semiconductor-metal photodetectors for long wavelength optical communication," *IEEE J. Quantum Electron.* 27(3), 737–752 (Mar. 1991).
5. M. Ito and O. Wada, "Low dark current GaAs metal-semiconductor-metal (MSM) photodiodes using WSi contacts," *IEEE J. Quantum Electron.* 22(7), 1073–1077 (Jul. 1986).
6. J. Hetterich, G. Bastian, N. A. Gippius, S. G. Tikhodeev, G. von Plessen, and U. Lemmer, "Optimized design of plasmonic MSM photodetector," *IEEE J. Quantum Electron.* 43(10), 855–859 (2007).
7. R. Raether, *Surface Plasmons on Smooth, Rough Surfaces, and Gratings.* Berlin: Springer-Verlag, 1988.
8. C. L. Tan, V. V. Lysak, K. Alameh, and Y. T. Lee, "Absorption enhancement of 980 nm MSM photodetector with a plasmonic grating structure," *Opt. Commun.* 283, 1763–1767 (2010).
9. F. J. García-Vidal and L. Martín-Moreno, "Transmission and focusing of light in one-dimensional periodically nanostructured metals," *Phys. Rev. B* 66, 155412 (2002).
10. G. Lévêque, O. J. F. Martin, and J. Weiner, "Transient behavior of surface plasmon polaritons scattered at a subwavelength groove," *Phys. Rev. B* 76, 155418 (2007).
11. B. Sturman, E. Podivilov, and M. Gorkunov, "Theory of extraordinary light transmission through arrays of subwavelength slits," *Phys. Rev. B* 77, 075106 (2008).
12. L. Martín-Moreno, F. J. García-Vidal, H. J. Lezec, K. M. Pellerin, T. Thio, J. B. Pendry, and T. W. Ebbesen, "Theory of extraordinary optical transmission through subwavelength hole arrays," *Phys. Rev. Lett.* 86, 1114 (2001).
13. F. J. García-Vidal, H. J. Lezec, T. W. Ebbesen, and L. Martín-Moreno, "Multiple paths to enhance optical transmission through a single subwavelength slit," *Phys. Rev. Lett.* 90, 213901 (2003).
14. L. Martín-Moreno, F. J. García-Vidal, H. J. Lezec, A. Degiron, and T. W. Ebbesen, "Theory of highly directional emission from a single subwavelength aperture surrounded by surface corrugations," *Phys. Rev. Lett.* 90, 167401 (2003).
15. H. J. Lezec and T. Thio, "Diffracted evanescent wave model for enhanced and suppressed optical transmission through subwavelength hole arrays," *Opt. Express* 12, 3629–3651 (2004).
16. Z. Yu, G. Veronis, S. Fan, and M. L. Brongersma, *Appl. Phys. Lett.* 89, 151116 (2006).
17. R. D. R. Bhat, N. C. Panoiu, S. R. J. Brueck, and R. M. Osgood, *Opt. Express* 16, 4588 (2008).

18. J. S. White, G. Veronis, Z. Yu, E. S. Barnard, A. Chandran, S. Fan, and M. L. Brongersma, "Extraordinary optical absorption through subwavelength slits," *Optics Letters*, 34(5), 686–688 (2009).

19. A. D. Rakić, A. B. Djurišić, J. M. Elazar, and M. L. Majewski, "Optical properties of metallic films for vertical-cavity optoelectronic devices," *Appl. Opt.* 37, 5271–5283 (1998).

20. M. I. Markovic and A. D. Rakic, "Determination of reflection coefficients of laser light of wavelength from the surface of aluminum using the Lorentz-Drude model," *Appl. Opt.* 29, 3479–3483 (1990).

21. M. I. Markovic and A. D. Rakic, "Determination of optical properties of aluminum including electron reradiation in the Lorentz-Drude Model," *Opt. Laser Technol.* 22, 394–398 (1990).

22. http://www.luxpop.com (viewed in May 2012).

23. E. D. Palik, "Galllium arsenide (GaAs)," in *Handbook of Optical Constants of Solids*, E. D. Palik, ed. San Diego: Academic, 1985.

24. H. Kikuta, H. Toyota, and W. J. Yu, "Optical elements with subwavelength structured surfaces," *Opt. Review* 10, 63–73 (2003).

25. T. Søndergaard and S. I. Bozhevolnyi, "Surface-plasmon polariton resonances in triangular-groove metal gratings," *Phys. Rev. B* 80, 195407 (2009).

26. A. K. Sharma, S. H. Zaidi, P. C. Logofătu, and S. R. J. Brueck, "Optical and electrical properties of nanostructured metal-silicon-metal photodetectors," *IEEE J. Quantum Electron.* 28(12), 1651–1660 (2002).

27. N. Das, A. Karar, M. Vasiliev, C. L. Tan, K. Alameh, and Y. T. Lee, "Analysis of nanograting assisted light absorption enhancement in metal-semiconductor-metal photodetectors patterned using focused ion beam lithography," *Opt. Commun.* 284, 1694–1700 (March 2011).

28. N. K. Das, A. Karar, M. Vasiliev, C. L. Tan, K. Alameh, and Y. T. Lee, "Groove shape dependent absorption enhancement of 850-nm MSM photodetectors with nano-gratings," *Proc. of the 10th Int'l. Conf. on Nanotechnology (IEEE NANO 2010)*, Seoul, South Korea, Aug.17–20, 2010.

29. D. Dhakal, "Analysis of the sub-wavelength grating in OptiFDTD simulator," M.Sc. thesis, School of Eng. and Science, Jacobs Univ., Bremen, Germany, 2009.

30. A. Battula, Y. Lu, R. J. Knize, K. Reinhardt, and S. Chen, "Extraordinary transmission and enhanced emission with metallic gratings having converging-diverging channels," *Act. Passive Electron. Compon.* (Nov. 2007).

31. D. Z. Lin et al., "Beaming light from a subwavelength metal slit surrounded by dielectric surface gratings," *Opt. Express* 14(8), 3503–3511 (2006).

32. A. Karar, N. K. Das, C. L. Tan, K. Alameh, and Y. T. Lee, "High-responsivity plasmonics-based GaAs metal-semiconductor-metal photodetectors," *Appl. Phys. Lett.* 99, 133112 (Sept. 2011).

33. R. D. R. Bhat, N. C. Panoiu, R. M. Osgood, Jr., and S. R. J. Brueck, "Enhancing infrared photodetection with a circular metal grating," Conference on Lasers and Electro-Optics (CLEO), Baltimore, Maryland, May 6, 2007, Photodetectors (CMY6).

34. N. K. Das, A. Karar, C. L. Tan, K. Alameh, and Y. T. Lee, "Impact of nano-grating phase-shift on light absorption enhancement in plasmonics-based metal-semiconductor-metal photodetectors," *Journal of Advances in Optical Technologies* 2011, Article ID 504530, 8 pages (May 2011). DOI: 10.1155/2011/504530.

35. L.-H. Laih, T.-C. Chang, Y.-A. Chen, W.-C. Tsay, and J.-W. Hong, "Characteristics of MSM photodetectors with trench electrodes on P-type Si wafer," *IEEE Trans. Electron Devices* 45(9), 2018–2023 (Sep. 1998).

36. S. Collin, F. Pardo, R. Teissier, and J. Pelouard, "Efficient light absorption in metal–semiconductor–metal nanostructures," *Appl. Phys. Lett.*, vol. 85(2), July 2004.

37. S. A. Maier, *Plasmonics: Fundamentals and Applications.* New York: Springer Science, 2007.

38. T. W. Ebbesen, H. J. Lezec, H.F. Ghaemi, T. Thio, and P. A. Wolff, "Extraordinary optical transmission through sub-wavelength hole arrays," *Nature* 391, 667 (1998).

39. H. D. Ko, "Surface plasmon coupled sensor and nanolens," Ph.D. dissertation, Dept. Elect. Eng., Texas A&M Univ., College Station, TX, 2009.

40. A. Bondeson, T. Rylander, and P. Ingelstrom, *Computational Electromagnetics,* 1st ed. New York: Springer, 2005.

41. http://www.optiwave.com/products/fdtd_overview.html. *Finite Difference Time Domain Photonics Simulation Software*, OptiFDTD Technical Background and Tutorials. Version 8.

# 7

# Finite-Difference Time-Domain Method Application in Nanomedicine

**Viroj Wiwanitkit (Retired)**

*Hainan Medical University, China; Visiting Professor, University of Niš, Serbia; Adjunct Professor, Joseph Ayo Babalola University, Nigeria*

## CONTENTS

## 7.1 Introduction

*Charge* is an important physical property. Talking about charge, one can imagine its relationship to electricity. Indeed, electricity is a kind of energy. It is not possible to touch charge, but charge really exists. Charge is the basic property that can be seen in everything in our world. It is the result of the compositions within every molecule. Hence, the charge is the basic fundamental thing around us. As already mentioned, every object has charge. It can be simply said that electricity is the property of anything. This physical property can be measured and can be classified as a specific property of each individual. Any objects, small or large, have this property.

At present, talking about small things is a new concept. Small is an actual new focus in science. The scientific society has turned its attention to small things. Smaller and smaller objects are being studied. Very tiny objects have become challenges in science. This has led to the new wave in the scientific world. In the past, humans were limited to studying objects in the visible scale. However, when the development of the microscope was complete, the scientific society turned its attention to a smaller scale called the microscale. With use of the microscope, small objects at $10^{-9}$ m level can be visualized. This led to a jump in scientific understanding. Many new things were identified such as cells and bacteria. However, the science never stops. Scientists continued to seek and challenge the human ability to a smaller scale.

Scientists have successfully reached a smaller level, an extremely small scale, at $10^{-9}$ m level. At this level, the mentioned scale is called nanoscale. At present, new science focuses and acts on this extremely small level. This means that scientists presently deal with the nanoscale size and this new area of science is called *nanoscience*. This is a real challenge of reaching smaller and smaller things. At present, nanoscience is widely studied around the world. In fact, nanoscience can be useful. The application of nanoscience can be seen in many aspects. Through nanoscience, many new facts are discovered. Many new things can be seen and observed. From nanoscience applications, the manipulation and understanding of the extremely small object is not a dream but a reality. As already mentioned, nanoscience involves the study of nanoscale objects. The nanoscale object is as real as objects in other scales. However, there are some interesting properties of nanoscale objects that should be mentioned.

When an object decreases in size, many aspects of that object change. At the very least, the size is changed. This means it poses and presents many new properties. The differences of properties are the important concept, and this is the point to be hit when one refers to nanoscience. The differences can be seen in many ways. Compared to a large-scale object, a nanoscale object has many interesting physical and chemical properties. The new properties of the nano-object may be useful. The application based on the changed physical and chemical property brings many interesting new nanoscience-based tools. When an object becomes a nano-object, new electrical and biochemical properties can be expected. The first change in surface area is the fundamental change that leads to many other changes, including electricity.

In fact, nano-objects can be either biological objects or not. Hence, nanoscience is not a pure physical or biological science; nanoscience is a hybrid that merges both physical and biological sciences. The answer to physical, biological, or hybrid questions can be gotten based on the new concept of nanoscience. At present, there is no doubt that nanoscience is widely studied and practiced. The questions on extremely small objects that were difficult to answer in the past become simple and successfully manipulated based on nanoscience. Hence, nanoscience acts as a new alternative approach for solving many hard-to-solve scientific queries. It is a useful novel science.

As already mentioned, nanoscience is convergent and it bridges difference aspects of sciences. This is the solution to the present requirement for consolidation of scientific knowledge. It matches with the present trend on integration. Many high-level education institutions and universities around the world include nanoscience in their curriculums. Due to the nature of multidisciplinary science, nanoscience can be applied for usage in many situations. As already noted, due to integration and combination of both physical and biological concerns, the wide scope and holistic approach can be expected. In fact, the modern hybrid nanosciences were introduced a few years ago, but rapid progress can be seen. Modern hybridizations in many forms of many modern sciences can be seen at present. Good examples of new modern hybridization include nanomaterial, nanobiology, nanophysics, nanochemistry, nanoengineering, nanomedicine, and so on. Several advantages of those modern sciences can be seen and no one can deny the usefulness of continuous development of hybrid nanoscience.

An important new hybrid, a new branch of applied biological science, nanomedicine, has been studied for less than 10 years. Nanomedicine is a good example of a hybrid nanoscience that can offer several advantages to our world. In brief, nanomedicine is the newest branch of medicine. It is the medicine that focuses on nanoscale objects. In fact, there are many nanoscale objects in the medical view. Germs, biochemistry, and drugs are examples of the nanoscale things in medicine. The introduction of nanomedicine challenges classical medicine. All activities in medicine, including diagnosis, treatment, and prevention, can be modified into the more modern faces with implementation of nanomedicine. Any aspects in medicine can be more effective and easier based on nanomedicine. This means newly effective, accurate, and reliable medical service can be expected. At present, it is not just a pipe dream; nanomedicine is already in use in many medical settings. Nanomedicine approach is the new trend in medical society. Several new nanomedicine experiments and developments have been introduced and many ongoing pieces of work are in progress. Applied diagnostic and therapeutic means based on nanomedicine that are presently in use are high achievements and are the pride of many modern medical centers.

Nanomedicine applications can be used in many ways including *in vivo*, *in vitro*, or *in silico* (computational) applications [1]. Classically, the long-used technique in medicine is the study in living things (*in vivo*) or outside living things in the laboratory (*in vitro*). For sure, those two classical approaches are based on several requirements and have a high unit cost. The present concept turns to the use of a computational modeling approach. This is the fundamental concept of the *in silico* approach. For nanomedicine, the two classical approaches, the *in vivo* and *in vitro* methods, are available. However, the basic limitations (e.g., they are expensive and time-consuming) can still be seen. Hence, the newest method, the *in silico* approach, is implemented in nanomedicine. The new *in silico* approach helps clarify and manipulate many questions in nanomedicine, which usually deals with very small objects and phenomena that are hard to study by either *in vivo* or *in vitro* means. Computational technology is the fundamental concept for the *in silico* approach in nanomedicine. The so-called computational nanomedicine is the good example. There is no doubt that the *in silico* approach is not a real approach but its reliability has been proven. The new *in silico* study can replace the classical *in vivo* and *in vitro* approaches with significant reduction of time and cost. Solving of difficult nanomedicine questions can be performed within a short time at a low unit cost. Also, since it is a simulating condition, the problem of confounding interferences can be solved. It can be said that the *in silico* approach is a practical approach without any interference effects. Focusing on the *in silico* approach, the so-called *omics* sciences are good applications at present [2]. Several new omics sciences can be seen in the present day. The omics is an interesting scientific technique. It has been in use for only a few years, but it is rapidly growing. Several new omics sciences have led to actual progress in science. The present era in science is an actual era of omics [2]. Similar to any fields of scientific study, applied computational technology in nanomedicine with help of omics technique is available. This can be the solutions to many advanced problems in nanomedicine. The *in silico* technique can be useful in solving the nanoscale problems; both clarification of the scenario and prediction of imaginary phenomenon in nanomedicine [3,4]. The imaginary approach with the computational approach in nanomedicine is a specific focus in a subbranch of nanomedicine that is called nanoinformatics. Much data has been gleaned from nanoinformatics and this subbranch of nanomedicine is an actual modern medicine that offers new hope to medical society.

At present, there are many new approaches for use in nanomedicine research. Of several basic techniques, the author hereby discusses the nanoscale finite-difference time-domain (FDTD) method, a new modality of computational nanoscience. Briefly, the FDTD method deals with *finite* and *time*. Hence, the dimension of size and period has to be considered. By definition, the FDTD method is a kind of mathematical application. This technique is based on numerical analysis technique and it can be further used for modeling computational electrodynamics. The fundamental mathematics as differential equation is the basis that is used in the numerical analysis. As already noted, the FDTD technique consists of a part of time domain. The solution in FDTD has to cover a frequency range in each simulation. This is simply used in treating nonlinear material properties in a natural way. Applied in medicine, the study of electrodynamics can be useful in many areas, such as radiology.

FDTD is a new scientific approach. Briefly, it is a specific assessment with special interest on electrodynamics issues. Classically, FDTD is an important topic in electricity. The old work is usually on the study of electrical systems. Focusing on the modern era of the FDTD technique, the implementation of several new computational approaches turns it into an advanced technique. Basically, computational modeling can help solve the difficult scientific and engineering problems. Also, it can extend to the medical field. Basic work is any piece dealing with electromagnetic wave interactions with material structures. For sure, if the materials being studied are nanomaterials, the topic becomes a consideration in nanoscience. Furthermore, if it is relating to medicine, it is also the nanomedical science. In general, there are many kinds of electromagnetic waves such as microwave and visible light. The application in medicine of those waves can be expected, such as in medical imaging, diagnostics, and therapy. In nanomedical science, the examples are nanobiomedical imaging, nanoelectromagnetic therapy, medical nanophotonic crystals, nanoplasmonics, and medical nanobiophotonics.

Indeed, the talk on the electromagnetic wave is not new in medicine. Visible light is a basic wave that has been generally focused in medicine for a long time. There is a well-known medical science on visible light: medical optics. With improvement of medical and scientific knowledge, the focus shifts from simple visible light to other nonvisible electromagnetic waves. Several rays have been introduced for medical usage over the centuries. The interaction is an important focus in medical study. Spectrometry is used for getting spectroscopic data, which can represent the interaction. Indeed, the spectrometry can be useful in several ways. It is the basic technique for analysis in science. Specifically, several kinds of spectrometry can be seen in scientific usage. The examples are ion-mobility spectrometry [5–7], mass spectrometry [8–10], neutron triple-axis spectrometry [11–13], and optical spectrometry [14–16]. Ion-mobility spectrometry [5–7] is the specific spectrometry technique that focuses on the movement of ions. Briefly, ion-mobility spectrometry is based on separation and identification of small molecules, which can include nanomolecules, based on movement of their ions. The use of buffer gas phase carrier is required for analytical process in ion-mobility spectrometry. Mass spectrometry [8–10] is another widely used spectrometry technique. As its name suggests, mass spectrometry is a specific spectrometry technique based on the mass analysis. Indeed, mass is the basic property of and method for identifying any substance. The mass is an actual specific property. Accompanied with mass, several other physical properties of substances can be further described and specifically matched. Mass spectrometry is a specific basic spectrometry technique that determines mass and relationship to charge. The charge of the substance is another important specific property of substance. In nanoscience, mass is very low and the alteration of charge at this stage can be expected. So the application of mass spectrometry, measuring mass to charge ratio based technique, can be very useful. It can be applied in many kinds of scientific branches. In medicine, mass spectrometry is widely used in biochemio-analytical processes. The good examples are in medical toxicological and pharmacological studies. Neutron triple-axis spectrometry [11–13] is another important kind of spectrometry. As its name implies, neutron triple-axis spectrometry focuses on the neutron. Briefly, the neutron is a neutral submolecular charge. The measurement of the neutron is the basic principle of neutron triple-axis spectrometry. This kind of spectrometry starts from activation, and the neutron scattering will be further determined in analytical process. The inelastic neutral scattering is the first detected result, which will be further interpreted into spectrometry data. A similar spectrometry technique to neutron triple-axis spectrometry is Rutherford backscattering spectrometry [17–20]. This is another spectrometry technique that involves measurement of scattering particle. Rutherford backscattering spectrometry specifically used the measurement of backscattering of a high-energy ion beam for analytical process. This is the basic point for determination of the composition of the studies sample. Similar to neutron triple-axis spectrometry, Rutherford backscattering spectrometry can be useful in several branches of science. Optical spectrometry [14–16] is another widely used spectrometry technique. What this kind of spectrometry measures is distributing light across optical spectrum. Hence, optical spectrometry includes the region ranging from infrared through visible light to ultraviolet. The measurement of this wide spectrum is the basic process during analysis by the optical spectrometry technique.

As already mentioned, the following characteristics are focused on in the study of electromagnetic wave in medicine: (1) interaction, (2) interrelationship of objects and rays, and (3) tools for assessment. Electromagnetic waves can be measured in the nanometer scale; hence, there is no doubt that dealing with electromagnetic waves in medicine can be classified as an actual medical nanoscience study [21–25]. Also, focusing on the interrelationship of objects and rays, the object in focus can be very small, within nanoscale; hence, the role of nanoscience cannot be avoided. To simply manipulate with electromagnetic waves in nanomedicine, a specific nanoscience, nanophotonics, can be useful. Further computational manipulation by FDTD can extend the knowledge in nanomedicine. The FDTD technique can be useful in many activities in medicine especially for the study of electromagnetic waves, nanoscale cells or particles, and the linking interrelationship. In this chapter, the author will focus on the application of nanoscale FDTD in medicine with special focus in nanomedicine. Examples of reports based on this technique will be discussed.

## 7.2 Overview of FDTD and Its Application in Medicine

The FDTD method can be useful in several branches of science. Indeed, electromagnetic waves are all around us, and they are usually studied in several ways. In general, as already mentioned, visible light is a kind of well-known electromagnetic wave. The electromagnetic wave is an example of a nano-level phenomenon. The wavelength is commonly presented in nanometer level. So the application of nanoscience to deal with electromagnetic waves is feasible. The nanoscale electromagnetic wave is a simple thing in the scientific world. Dealing with electromagnetic waves with a nanotechnology approach is a specific new nanoscience focus. As noted, nanophotonics, the new branch of nanoscience, is leading the way in this endeavor. In fact, nanophotonics is the study of photons of light; hence, it is the study of an actual nanoscale thing. In the past, nanophotonics was an advanced approach relating to optical science and engineering. But it can be extended to the medical field as already mentioned. This means nanophotonics poses both basic and applied usefulness. The fundamental nature and principles of the electromagnetic photon and its interaction with objects at the nanoscale is the main issue. Considering the applied issues, the development of new nanodevices is the main concern. Studying and developing new tools by nanoengineering is the first step, and their use in medical science can be an on-top process. Indeed, many new medical nanodevices are good examples of applied medical nanophotonics tools, such as medical nanospectrometry.

With regard to specific manipulation techniques on nanoscale electromagnetic wave, the details of FDTD should be mentioned. Application is an actual advanced science bridging between physical and biological sciences, resulting in a very useful new complex science. As a basic explanation, the FDTD method is a kind of general class of grid-based differential time domain numerical modeling methods. Gridding is the first step to be mentioned. Generally, visualization or "ability to see" is mainly gotten by visible electromagnetic wave or visible light. Light is basic need for visualization. An object can be visible only if it interrelates with light. For sure, interrelation must have specific site of interaction and this can be determined as a specific space. The space can be gridded when one focuses and applies FDTD. Also, such interaction must have period. This is the basic required thing, time. Hence, in one interrelationship, both space and time have direct involvement. In principle, the time-dependent Maxwell's equations are used in the FDTD method. The equations can be discretized based on space-based central difference estimations and time-based partial derivatives. Because it deals with time as well as difference, the finite-difference equation is the result. With use of advanced computational technology, the finite difference equations can be solved with help of computational software and hardware.

Step-by-step instructions for solving the finite-difference equation should be mentioned. After successful gridding of the interaction between space and time, it is the time for solving of the equation. To solve, the electromagnetic wave has to be dissected into two views of interest: electrical and magnetic issues. Two separated issues will be further separately manipulated. As a rule, first, the electric field vector components will be focused and manipulated. All components displaying in a specific interested volume of space will be solved based on an assigned instant in time. Next, the corresponding magnetic field vector components displaying in the same interested focused spatial volume will be solved based on the next instant in time. These processes will be repeatedly done. It will be run over and over again to get the final desired steady-state electromagnetic field behavior. Once the computational approach on the gridded materials is established, the described step-by-step processes can be done. One can imagine the already described technique. The simulating of electromagnetic wave interaction on objects including nano-objects can be performed.

Naturally, the electromagnetic wave possesses wave properties. It can transmit as well as reflect. Scattering of wave after impact to object can also be expected. To collect the basic data on photonic interaction on the object is therefore an important step. Simply, the study on the wave can be done via nanoscale spectrometry as already mentioned. The nanoscale spectrometry is a specific tool for measuring nano-spectrum. With use of nanoscale spectrometry, the electromagnetic wave can be determined for its function of wavelength. This can be basic information for further gathering the data on electrical and magnetic issues. In medicine, the good examples of tools for spectrometry are spectrometers, spectrophotometers, spectrographs, and spectralanalyzers. In medicine, spectrometry is very simple and widely used. Further manipulating of the derived spectrum data by the FDTD method should be further mentioned. Indeed, such applications of FDTD can be seen in several medical areas, including clinical microscopy, ophthalmology, radiology, oncology, and laboratory medicine.

Focusing on clinical microscopy, the medical subject on the use of the microscope, the usefulness of the method can be expected. Indeed, clinical microscopy has been used in medical society for many years and its importance in medical practice cannot be overstated. Indeed, clinical microscopy is a medical science dealing with electromagnetic waves. There are many well-known tools in clinical microscopy and several kinds of microscopes (both visible and nonvisible wave-based types). The actual aim of clinical microscopy is the viewing or visualizing of small objects (this includes the nanoscale objects). Hence, there is no doubt that clinical microscopy has directly dealt with novel nanophotonics techniques [26]. Spectroscopy and spectrum analysis become the simple thing in clinical microscopy. Several new spectrometry tools have been developed for use in nanophotonics. This can yield much new knowledge in clinical microscopy. The specific new microscope spectrometers are available. Those new microscope spectrometers usually aim at measuring UV-visible-NIR spectrum of microscopic focused spaces (sole small objects or areas of larger objects) [27]. At present, as already mentioned, the new microscope spectrometer is widely applied in nanophotonics study. There are two main kinds of microscope spectrometer. In general, a microspectrometer is used for measuring the spectrum of microscopic samples based on transmission, absorbance, reflectance, fluorescence, polarization, and emission spectrometry. With microspectrometer, high solution digital imaging can be expected. For sure, all activities that are mentioned are the assessment on the interaction between electromagnetic wave and small object; hence, the role of the FDTD method can be expected. The applied computational technology adding to the basic simple microspectrometer can result in better manipulation of the microscope data. To help the reader better understand the present application, some interesting reports on the FDTD method applications in clinical microscopy will be further shown.

- Chung et al. studied a compact light concentrator by the use of plasmonic faced folded nanorods [28]. In this study, Chung et al. developed a compact nanometallic structure for enhancing and concentrating far-field transmission and used the FDTD method for confirming the holographic microscopy result [28]. Chung et al. found that their new tools could provide a good three-dimensional map of the complex wave fronts [28].

- Kawakami et al. studied micromirror arrays to assess luminescent nano-objects [29]. In this report, Kawakami et al. constructed a new array of submicrometer mirrors that could successfully collect luminescence [29]. The FDTD method also confirmed the experimental finding in this report [29].

- Foteinopoulou et al. investigated optical excitations on single silver nanospheres and nanosphere composites with the FDTD method [30]. Foteinopoulou et al. proposed using the new nanomaterials to construct a nanolens that could further improve scanning near-field optical microscopy detection and the excitation of surface plasmon-based guiding devices [30].

- Liu et al. used the FDTD method to study the near-field effects on coherent anti-Stokes Raman scattering (CARS) microscopy on nanoparticles [31]. Liu et al. concluded that "employing the perpendicular polarization component of CARS signals can significantly improve the contrast of CARS images, and be particularly useful for revealing the fine structures of bio-materials with nano-scale resolutions" [31].

- Lu et al. studied polarization-dependent Talbot effect by scanning near-field optical microscopy with additional help of the FDTD method [32]. Lu et al. concluded that "the polarization-dependent Talbot effect should help us to understand more clearly the diffraction behavior of a high-density grating in nano-optics and contribute to wide application of the Talbot effect" [32].

- H'dhili et al. studied nano-patterning photosensitive polymers using local field enhancement at the end of apertureless SNOM tips [33]. In this report, H'dhili et al. used the FDTD method to confirm the experiment performed by atomic force microscopy [33]. H'dhili et al. concluded that the SNOM tips could be applied for further usage in near-field optical lithography [33].

Based on the given examples, there is no doubt that the nanoscale FDTD method can be useful for clinical microscopy. It can be used for confirmation of microscopic experiment as well as improvement of the microscopic spectrum result. At present, the application of the nanoscale FDTD method in clinical microscopy is well accepted.

Radiology is another medical science that deals with the electromagnetic waves. The x-ray is also considered a useful wave in biomedical science. In nanomedicine, nanoradiology is quite a new face of biomedical radiology. Based on the nanophotonics tool, x-ray spectrometry is available. The new nano-x-ray tool is accepted as an accurate and economical analytical method for the determination of many chemicals. Several applications of the new tool can be seen in present advanced medicine. The medical crystallography study and medical radiation research are the best examples. The nano-technique can be applied for study of new developed therapeutic agents and biomolecules. For sure, those medical studies are based on the assessment on the interaction between electromagnetic wave and nano-object. The role of the FDTD method is apparent. To help the reader better understand the present application, some interesting reports on FDTD method applications in radiology will be further shown.

- Chung et al. reported on a compact light concentrator by the use of plasmonic faced folded nanorods [34]. Chung et al. used three-dimensional FDTD calculation in their study and concluded that the new compact nanometallic structure was successful in enhancing and concentrating far-field transmission [34].

- Su et al. investigated transmission properties of two crossing dielectric slot waveguides (Si-Air-Si) [35]. It was found that the most of low transmission was generated by the reflection and radiation loss rather than the crosstalk [35].

- Zhao et al. investigated the light focusing by using dielectric nano-waveguides array with its length in micron via the FDTD method [36]. Zhao et al. concluded that "the unique focusing behavior is contributed to the radiation mode with longer decay length and the large evanescent field which appears in the nano-waveguide array" [36].

- Ahmadi et al. discussed the concept and design of a reflectarray nanoantenna at optical frequencies [37]. In their report [37], Ahmadi et al. used FDTD numerical method and dipole-modes scattering theory for characterizing and tuning the reflectarray design.

Based on the given examples, there is no doubt that the nanoscale FDTD method can be widely applied in nanoscaled biomedical radiology. It is usually used for the study of the structure of nano-object. Confirmation of experiment is the general application. Also, the technique can be used as a basic step for developing new nanoradiology apparatus. The application of the nanoscale FDTD method in nanoradiology is presently accepted.

As already noted, it should be said that the electromagnetic wave in medicine is not only the nonvisible ray. The visible ray, which is the basis for normal visualization, should also be mentioned. Indeed, in nanoscience, there is a specific branch called nanophotonics. Nanophotonics directly deals with photons including the visible light. In biomedicine, the specific medical science focusing on normal visualization is ophthalmology. Nanophotonics is an actual new concept in ophthalmology. There is no doubt that nanophotonics can be applied in ophthalmology because ophthalmology deals with eyes, light, and visualization, which are all related to visible light [21–25]. Many new techniques based on nanophotonics can also help improve nano-ophthalmology. The specific concern is usually on the nano-optics. Optics is a specific consideration on scattering, absorption, and radiative transfer. At the nano-level, it is simply called nano-optics. Many new research projects dealing with nano-electro-optics and chips are in progress. There is no doubt that optics is an actual subject dealing with electromagnetic waves. Hence, nano-optics deals with nano-objects and their interrelationship to the electromagnetic wave. Therefore, the role of the FDTD method can be imagined. To help the reader better understand the present application, some interesting reports on FDTD method applications in biomedical optics will be further shown.

- Lu et al. used scanning near-field optical microscopy accompanied with the FDTD method to study polarization-dependent Talbot effect [32].
- Chung et al. used three-dimensional FDTD calculation to evaluate a new compact light concentrator constructed based on the use of plasmonic faced folded nanorods [34].
- Kawakami et al. studied the use of micromirror array for assessment of luminescent nano-objects [38]. Kawakami et al. proved that reflection of luminescence from nano-objects within MMAs toward the Si (001) surface normal could increase the probability of optics collecting luminescence [38]. Kawakami et al. used the FDTD method to confirm the experiments [38].

- Li et al. reported on a new superlens nano-patterning technology based on the distributed polystyrene spheres [39]. Li et al. used the FDTD method for numerical simulation on a typical configuration with 1.5 µm diameter polystyrene spheres and found a "minimum feature size of 88 nm beyond diffraction limit at 365 nm working wavelength" [39].

Based on the given examples, the nanoscale FDTD method has several advantages for application on nano-optics. Simulation relating to the nano-object can be possible with help of the nanoscale FDTD method. The technique can also be applied as a useful tool for developing of new nano-optics apparatus. There is no doubt that the application of the nanoscale FDTD method will lead to several new breakthroughs in nano-optics.

The role of the FDTD method can also be seen in oncology. This might be more difficult to imagine than the already mentioned ones in microscopy, radiology, and optics. Indeed, some parts of oncology deal with electromagnetic waves. First, diagnosis of cancer is based on imaging. The use of x-rays and other electromagnetic waves, especially for MRI, is the direct relationship. Second, the treatment of cancer can be based on radiation and electromagnetic waves. The use of many radioactive materials in the treatment can be seen. The good examples can be seen in uterine cervix cancer treatment. Also, indirect use of radiation, especially for x-ray, to stop the growth of tumor cells is a basis in cancer therapy (called radiotherapy). Radio-oncology is the specific area of oncology that deals with radiation and electromagnetic waves in oncology. Much new research in nano-oncology is in progress. There is no doubt that parts of oncology are related to electromagnetic waves as already mentioned; hence, the role of the FDTD method can be imagined. To help the reader better understand the present application, there are many interesting reports on FDTD method applications in nano-oncology. The best referencing pioneer report is by Taflove et al. [40]. In this report, Taflove et al. studied breast cancer diagnosis based on ultrawideband radar techniques [40]. Hagness et al. used the FDTD method to model cancer determination in this study [40]. For the other interesting reports, please see the following list:

- Burfeindt et al. reported on the use of microwave beamforming for treatment of pediatric brain cancer [41]. In this work, FDTD simulations were used for evaluation of the effectiveness of the technique [41]. Burfeindt et al. concluded that "microwave beamforming has the potential to create localized heating zones in the head models for focus locations that are not surrounded by large amounts of cerebrospinal fluid" [41].
- Wang and Gong reported on left-hand metamaterial (LHM) lens applicator for microwave hyperthermia of breast cancer [42]. Wang and Gong used two-dimensional FDTD method for assessment and

found that "hyperthermia with the proposed four-lens applicator of moderate LHM losses could be effective in achieving desired power deposition in a heterogeneous breast model" [42].

- Zastrow et al. reported on noninvasive patient-specific microwave hyperthermia treatment of breast cancer [43]. In this work, Zastrow et al. used FDTD simulations for evaluating the focusing and selective heating efficacy in breast phantoms [43]. Zastrow et al. concluded that "beamforming is a robust method of non-invasively focusing microwave energy at a tumor site in breasts of varying volume and breast tissue density" [43].
- Yacoob and Hassan reported on using the FDTD method to analyze a noninvasive hyperthermia system for brain tumors [44].

## 7.3 FDTD Method for Nanomedicine

Electromagnetic waves are all around us. The light or photon is the most basic wave that everyone knows. Light makes vision possible. As a wave, the wavelength is usually defined at the nano-level. Hence, any phenomena relating to light and other electromagnetic waves are actual topics in nanoscience. Both naked-eye visible or nonvisible waves can be explained in nanoscience. There is no doubt that the phenomenon in nanoscale cannot be simply assessed by simple technique. Naked-eye assessment seems to be not possible. *In vivo* or *in vitro* studies become very difficult. Therefore, in the past, the study of electromagnetic waves in biomedical science was an extremely hard topic. Hence, finding a new technique has been the hope for many centuries. Several limitations of the classical approaches have stimulated scientists to find new channels. The blooming of the new science, nanoscience, might be the answer. Indeed, the application of the nano-technique for study of electromagnetic waves is possible. Several tools, including the nanophotonics and nanoplasmonics tools, are available. Based on those new tools, the solution of complicated electromagnetic wave–related problems can be simply derived. For sure, it is faster than using the classical approaches. The answering of the queries can be for the basic structural or functional aspects (Table 7.1). Advanced nanoscience is useful for this purpose. With the use of a new concept, the computational approach, the problem becomes simple. The *in silico* technique can be the new acceptable solution for many questions [3,4]. The nanoscaled phenomenon can be managed by the informatics technology. A specific branch of nanoscience, namely nanoinformatics, plays an important role at this point. At present, many simulations and manipulations via computational approaches demonstrate the usefulness of nanoinformatics [45].

**TABLE 7.1**

Application of the Computational Technique for Solving of Nanomedicine Problem

| Applications | Details |
| --- | --- |
| 1. Structural approach | This focuses on solving the structural problem. It is meant to help one visualize or clarify the component of the nanostructure of interest. Many new computational-based tools can help predict composition and structure of nanomolecules. Also, those new tools can help construct the figures representing the interested nano-objects. |
| 2. Functional approach | This focuses on learning the process. Step-by-step process or pathway of the nano-phenomenon is specifically focused on in the functional approach. Many new computational-based tools can help conclude, clarify, and predict path of the focused nano-phenomenon. |

Focusing on the present concepts, the computational modeling and designing of interested nano-objects is feasible. With the use of many new computation tools, the analysis can be done. This can close the gap between physics and biology and result in an actual integrated bridging science. Such hybrid science is actually useful. The use of nanomedicine manipulation also falls within this scope. Through the nanomedicine approach, clarification or prediction can be performed as already noted. The FDTD method is a mathematically based technique that can be applied in nanomedicine. As a mathematical technique, the linkage to computational technique can be simply done. Since computers can help manipulate and manage the complex mathematical problem, integration of the computational approach to facilitate use of the FDTD method to solve the nanomedicine questions can be possible. More details as well as interesting examples of important reports can be seen in the next section.

### 7.3.1 A Summarization on Important Reports on Application of the FDTD Method in Nanomedicine

As earlier noted, the computational FDTD method has many advantages in nanomedicine. Here, the author will summarize important reports on the application of the FDTD method in nanomedicine (Table 7.2) [46–50].

1. Application for clarification—A good example of application of computational FDTD method is the usage in modeling and simulation relating to the nanomaterials. Clarification of new

**TABLE 7.2**

Some Important Reports on the Computational FDTD Method in Nanomedicine

| Authors | Details |
|---|---|
| Ray et al. [46] | Ray et al. reported their assessment on metal-enhanced fluorescence of the tryptophan analogue N-acetyl-L-tryptophanamide (NATA) on silver nanostructured surfaces [46]. In this study, Ray et al. used the FDTD method to model a tryptophan-wavelength dipole near a spherical silver particle [46]. |
| Bonmassar et al. [47] | Bonmassar et al. proposed new electrode set based on polymer thick film organic substrate, an organic absorbable, flexible, and stretchable electrode grid for intracranial use [47]. FDTD method simulations were used for assessment of the new electrode in this work [47]. |
| H'dhili et al. [48] | H'dhili et al. used two-dimensional numerical calculations based on the FDTD method to assess local optical field enhancement detected at the end of an apertureless SNOM tip [48]. |
| Wust et al. [49] | Wust et al. used the FDTD method for assessing scanning E-field sensor device for online measurements in annular phased-array systems [49]. Wust et al. concluded that the new system was useful for patient-specific hyperthermia planning [49]. |
| Hallaj et al.[30] | Hallaj et al. used the FDTD method to model the propagation of finite-amplitude sound in a homogeneous thermoviscous fluid [30]. This technique is useful for determining biological tissue in response to a pulse of focused ultrasound [30]. |

nanomaterial structure with help of computational FDTD method is possible. Several published reports have been previously shown in this chapter.

2. Application for prediction—A good example of application of computational FDTD method is the new nanomaterial design. Prediction of action of nanoparticles under difference conditions with help of computational FDTD method simulating is also feasible.

As a summary, it can be seen that computational FDTD method can be applicable for both diagnosis and therapy in nanomedicine.

- In diagnosis (nanodiagnosis)—Computational FDTD method in nanodiagnosis is the main research area at present. The aim is usually to answer the questions on microscopy and optics as previously discussed.

- In treatment (nanotherapy)—The use of computational FDTD method in nanotherapy is possible. The most important application is the use in cancer therapy in clinical oncology as earlier mentioned.

## 7.4 FDTD Method for Manipulating Biomedical Work

Based on computational technology, biomedical data can be manipulated at a low cost within a very short time. To manipulate these biomedical data, the concepts of informatics have to be applied. Basically, the computational tools are needed. A selection of good and appropriate tool is the basis. This is the first step and becomes the key determinant for success. For practitioners, learning the new tools and training to use them is very important. It is necessary that practitioners understand and correctly use the new computational tools for managing the problems. Because the tools have been continuously developed and implemented, continuous education to get updated on this topic is the basic requirement. To get updated, reading the specific books, journals, and websites is suggested. Here, the author will briefly summarize some important available FDTD method tools. These advanced tools can be useful for managing nanomedicine works.

### 7.4.1 Usefulness of FDTD Method in Nanomedicine

Basically, the FDTD method directly deals with electromagnetic waves. It is an actual nanoscale story. Using the computational FDTD method to solve a problem or to simulate a case is possible. This can be very useful to biomedical research questions relating to electromagnetic wave.

Several programs can help with the computational FDTD method. MATLAB is a good basic example. The ways that MATLAB can be used for the FDTD method problems will be further discussed.

1. Creating of graphical model—MATLAB can help with the generation of a graphical model. For example, this is a case of using MATLAB to create a model of electromagnetic pattern at different angles and radii from a central nanovessel. The writing of MATLAB code in MABLAB Command Window is shown below:

```
>> theta = 0:0.2:4*pi;
>> rho = theta/(4*pi);
>> polar(theta, rho)
>>
```

2. Solving the problem of time function equation—Differentiation and integration are common mathematical manipulations in the FDTD method. Solving these mathematical problems with MATLAB can be fine. For example, this is a case of using MATLAB for solving

the time function of left nanolaser magnitude after absorption by focused breast cancer cell and further performing Fourier transform. An example of the specific code is shown here:

```
>> N = 10;% Number of absorption
>> T0 = 0.5;% Period
>> n = 2*N;
>> t = linspace (0, T0, n+1);
>> t (end) = [];
>> f = sawtooth (t, T0);
>> Fn = fft(f);
>> Fn = [conj (Fn (N + 1)) Fn (N+2 : end) Fn (1:N+1)];
>> Fn = Fn/n;
>> a0 = Fn (N + 1);
>> an = 2*real (Fn (N+2 : end));
>> bn = -2*imag (Fn (N+2 : end));
>>% Fourier series coefficients for this sawtooth waveform
>> idx = -N : N ;
>> stem (idx, abs (Fn))
>> xlabel ('Absorption index (n)')
>> title ('Absorbed spectrum line')
>>
```

### 7.4.2 Important Nanoscale FDTD Method Tools for Biomedical Work

There are several computational FDTD method tools for management of biomedical data. Those important computation tools are hereby listed.

1. FDTD Solutions Knowledge Base—FDTD Solutions Knowledge Base is a specific database focusing on the information for the FDTD method.

2. FDTD Solutions—FDTD Solutions is designed for electromagnetic wave problem solving for the design, analysis, and optimization of nanophotonic devices, processes, and materials.

3. FDTD.org—FDTD.org is the database on the literature on the FDTD method.

4. XFDTD—XFDTD is a novel three-dimensional FDTD software program for electromagnetics. XFDTD applications include antenna impedance and radiation, scattering, microstrip, specific absorption rate, electromagnetic compatibility, and shielding. It is a commercial program and is accessible via http://www.remcominc.com/. This tool can be applied in medical imaging. The good example is the report by Ho on using XFDTD for studying the safety of metallic implants in magnetic resonance imaging [51].

5. ToyFDTD.html—ToyFDTD.html is a series of codes demonstrating implementation of three-dimensional FDTD in simulation. It is

accessible via http://www.toyfdtd.org/ or http://www.borg.umn.edu/toyfdtd/.

6. MEEP—MEEP is a free software program for FDTD method manipulation. It can be useful for simulation. It is accessible via http://ab-initio.mit.edu/wiki/index.php/Meep.

7. FIDELITY—FIDELITY is a software program for simulation. It makes use of the FDTD method for modeling microwave circuits, components, antennae, and other high-speed and high-frequency circuitry.

8. WOLFSIM—WOLFSIM is a computational program for FDTD electromagnetic simulator. Either one- or two-dimensional structure can be modeled under condition of incident radiation.

## 7.5 Common Applications of FDTD Method in General Nanomedicine Practice

1. Situation 1: Using the computational FDTD method to determine laser scattering pattern in flow cytometry—This is an example of using the computational FDTD method in nanomedicine. Flow cytometry is an important tool in biomedical study. This makes use of laser, a kind of electromagnetic wave, for radiating to focus studied objects, and the interpretation of the results is done by specific detector to collect the scattering pattern. It can be useful for determining and tracking of cells. This kind of determination is widely used at present in many diseases including infectious and oncological disorders. The FDTD method can be successfully used in tracking of scattering. Simulation of light-cell interaction is the main usefulness in case of medical flow cytometry technology [52]. The good example is the recent report by Yu et al. [53]. In that work, an automated image processing software based on the computational FDTD method for short-time-Fourier-transform algorithm was developed for help analysis.

2. Situation 2: Using the computational FDTD method for studying radio-resistant cancer—This is another example of using the computational FDTD method in oncology. Basically, the use of radiation to attack cancerous cells is the focus in radiotherapy. The response of the cells is expected by de-sizing and degradation. However, the resistance can be seen in some cases and this is the problem of radio-resistant cancer. The studying of the dispersion of radiation in tissue is the important step to realize this problem. The use of the computational FDTD method can be good for such simulation study. A good

example can be seen in the report by Nagaoka and Watanabe [54]. Indeed, to test any new radiotherapy tool, the phantom model study as well as the further computational FDTD method to correspond the exact body's geometry, beam, and exposure time is the novel concept to increase the efficacy of the wave-based therapy (including hyperthermia therapy, radiotherapy, microwave therapy, etc.) for cancer [55].

## 7.5.1 Examples of Nanomedicine Research Based on FDTD Method Application

### 7.5.1.1 Example 1: A Study to Model the Structure of Hemoglobin Anantharaj

Basically, hemoglobin is an important molecule within the human body. It is a red pigment within human red blood cells (erythrocytes). It is the particle that supplies the red color to blood. It is useful for blood circulation to supply good oxygenation to the tissue. Focusing on the hemoglobin structure, it is a king of metalloprotein within erythrocytes. The corresponding parts of hemoglobin include iron molecule and globin molecules. Hemoglobin defects are important conditions in medicine. Thalassemia and hemoglobinopathy are the two most problematic types. To know the pathogenesis of each abnormal hemoglobin defect, clarification on its abnormal structure is required since this can give further data for finding new diagnostic and therapeutic approaches.

Hemoglobin is an extremely small molecule that cannot be seen by the naked eye. It is considered to be in nano-level. In basic laboratory, the hemoglobin is investigated as amount in biochemistry, not structure. So complex techniques such as x-ray diffraction are required for hemoglobin study. Since the x-ray is a kind of electromagnetic wave, there is no doubt that the computational FDTD method technique can be applied. Focusing on investigation of abnormal hemoglobin, classical x-ray diffraction study needs the exact sample of hemoglobin. This presents difficulty since some hemoglobin defects are extremely rare. Hence, the applied technique for modeling of the hemoglobin structure is very useful. A good previous referencing work is on modeling hemoglobin at optical frequency with help of the computational FDTD method technique [56].

Here the author shows a model of structure of a rare hemoglobin defect, hemoglobin Anantharaj. The computational structural analysis of this hemoglobinopathy is done. Pathophysiologically, alpha 11(A9)Lys $\to$ Glu is the main genetic basis for hemoglobin Anantharaj [57,58]. This hemoglobin has no abnormal elongation but a single point alteration. Such abnormal alteration within the alpha globin of the hemoglobin Anantharaj molecule is believed to be the etiological factor for the abnormal presentation. Although the primary structure of hemoglobin Anantharaj is well known, the specific knowledge on its secondary and tertiary structure is not well studied

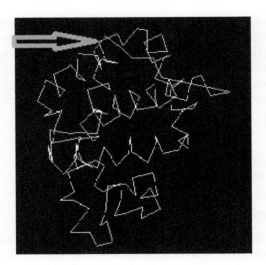

**FIGURE 7.1**
Calculated secondary structures of hemoglobin Anantharaj.

and mentioned. In this work, a standard bioinformatic approach is used for assessing the secondary and tertiary structures based on the available information on amino acid sequence of this hemoglobin. A standard computer-based study was used for protein structure modeling in this work. Starting from database searching to get the sequence of interest, further simulating for the abnormal primary amino acid sequence by direct assignment is done. Then the secondary structure was predicted using NNPREDICT server. Next the tertiary structure modeling by CPHmodels 2.0 Server prediction was done. (This protocol is the standard published protocol to evaluate the abnormal hemoglobin [59].) The predicted results for secondary and tertiary structures of the focused hemoglobin defect are shown in Figure 7.1. Based on the prediction, the abnormal deletion of helix occurs at the abnormal coding site, and this is concomitant with the underlying abnormal primary structure within the hemoglobin Anantharaj. This abnormality is believed to be the underlying cause of the detected abnormality. The loss of helical position might affect the iron-carrying status that results in the iron deficit that can be seen in cases with hemoglobin Anantharaj [57,58].

### 7.5.1.2 Example 2: A Study to Analyze the Effect of Photon Stimulation of the Nanomembrane of Retina in a Case of Laser Surgery

A photon is considered a packet of light. It is usually mentioned when one talks about the twilight characteristics of light particles. In modern physics, the photon is classified as an elementary particle. Based on quantum theory, the quantum of light and other electromagnetic waves can be the photon. The importance is the energy that the photon carries. Basically, the effect of photon stimulation is an important consideration in human tissue. Laser is a kind of electromagnetic wave that might cause damage at the cellular level. Laser exposure might lead to coagulation if high dosage is absorbed. However, the medical laser is still in use and there are several medical applications such as in dermatology and ophthalmology [60]. In ophthalmology, the most important part for every human eye is the retina. Retina is a kind of biomembrane embedded in the deepest part of the eyeball. Physiologically, the retina consists of a single layer of retinal cells. The retina tissue is highly sensitive to light and electromagnetic waves or it is sensitizable by photon. The stimulating reaction at the retina is considered a photoelectric phenomenon. The optics of the eye passes the receptor at the retina and the caught light signal will be further processed and passed via optic nerve to the brain for further central processing and interpretation. Comparing to the camera, the retina acts as the film. As already mentioned, light activates the receptor at the retina, and this is the basic process responsible for human vision. The friendly electromagnetic wave for the eye is the visible light. However, in real life, several electromagnetic rays come to the human eye and they can further reach the retina and react with the retinal membrane. At present, it is confirmed that a variable-wavelength laser has the capability to reach the retina and cause damage [61]. Ophthalmologists who use laser surgery need to take this fact into consideration. The laser beam has to be well focused on the desired lesion, and close observation of the side effects after laser surgery is indicated.

Based on the advanced nanotechnology, the modeling of the change of retina membrane corresponding to the laser stimulation is feasible. The stimulation is an actual electromagnetic phenomenon that occurs at a period of time. Hence, the FDTD method can play a role at this point. Here, the change of pattern in case with laser exposure is presented. For modeling, the simple computational finite analysis approach is done via a standard computer program [62]. As a primary assumption, the membrane is assumed to be very small and flat, and then it is assigned to be a simple two-dimensional structure with uniform thickness. The result can be seen in Figure 7.2. An irregular change on the retinal nanomembrane in case of laser exposure can be seen. This finding corresponds to the recent report on the fluctuation of local temperature depending on diameter of exposure that is verified by the FDTD method approach [62]. Indeed, it is widely accepted in medicine that the eye can be endangered if exposed to nonvisible light for a considerable period of time.

### 7.5.1.3 Example 3: A Study to Clarify the Pattern of Microwave Absorption on the Hyperthermia Treatment for Benign Prostatic Hyperplasia

The prostate gland is a specific organ of human beings. Prostate gland is within the genitourinary system. Its location is at the neck of the urinary bladder. Generally, the prostate gland is a soft tissue organ surrounding the

**A. At normal physiological stage**

|   | 1 | 1 | 1 |   |
|---|---|---|---|---|
| 1 | 1 | 1 | 1 | 1 |
| 1 | 1 | 1 | 1 | 1 |
| 1 | 1 | 1 | 1 | 1 |
|   | 1 | 1 | 1 |   |

**B. With dengue fever (assuming at 38 degree Celcius)**

|   | 1.0245 | 1.0245 | 1.0245 |   |
|---|--------|--------|--------|---|
| 1 | 1.0184 | 1.0122 | 1.0184 | 1 |
| 1 | 1.0046 | 1.0052 | 1.0046 | 1 |
| 1 | 1.0028 | 1.0032 | 1.0028 | 1 |
|   | 1 | 1 | 1 |   |

**FIGURE 7.2**
Model of photon flux change in retinal membrane due to dengue fever.

prostatic male urethra. The main function of the prostate gland is secretion of prostatic fluid, which is needed for completeness of spermatic fluid. The prostate disorder is a common finding in aging males and this has become an important public health threat. Millions of male elderly develop prostate problems, either benign prostatic hyperplasia or prostate cancer. Screening test by rectal examination is recommended; however, the high prevalence of prostate cancer can be observed around the world. Prostate cancer has become one of the most common malignancies among aged males. Screening for this cancer has been implemented for years; however, it is still not successful to control this cancer. To manage the prostate tumor, cutting of the tumor is the basic strategy. The surgical technique for prostate surgery has been continuously developed. Classically, the open surgical management is used. However, newer methods such as transurethral surgery are implemented. Apart from the basic classical surgery, there are also some interesting new therapeutic approaches. The use of nanotechnology for microwave-induced hyperthermia is an advanced concept [63]. An important application is presently on benign prostatic hyperplasia treatment.

Indeed, microwave is a kind of electromagnetic wave that is well known. The microwave can generate heat and this is widely applied in the present day [64]. The application of microwave in medicine is interesting. At first, the microwave was described for its hazardous effect [65–67]. The cellular injuries due to heat can be expected. Nevertheless, medical scientists later recognized the usefulness of heat from microwave. The application for surgery is the exact usefulness of microwave [68,69]. There is no doubt that heat can destroy the cell. Hence, if the heat is controlled for direction and applied to unwanted cells such as cancerous cells, a therapeutic result can be expected [68,69]. This technique is applied in many kinds of surgery including gastrointestinal [70–72] and urological [73–75] surgeries. The surgical techniques that are based on microwave application are considered minimally invasive and can be useful to increase surgical tolerability of the patients.

Focusing on prostate tumor, the use of microwave surgery is confirmed for its usefulness [73–75]. At first, the technique is applied for the benign prostatic hypertrophy [75–77]. "Decreased risks for retrograde ejaculation, treatment for strictures, hematuria, blood transfusions, and the transurethral resection syndrome" was reported for microwave surgery for management of benign prostatic hypertrophy [76]. It is confirmed that the microwave approach is superior and less invasive [78]. For application in prostate cancer, microwave surgery can be applied as a local cancer therapy [79,80]. It is approved for effectiveness and it is also confirmed for usefulness in case of recurrent prostate cancer [81]. For treatment, it has to start from modeling of the focused area for surgical manipulation. This is an actual deal on medical electromagnetic wave; hence, the use of the computational FDTD technique can be imagined. Here, the author gives an example of using computational modeling for prediction of area getting the appropriate microwave on the period of therapeutic process. For simulating, the emGine Environment, a new computational simulation tool, is used. Briefly, emGine Environment is a tool for manipulating electromagnetic field simulation in three dimensions and also allows solving Maxwell's equations in time-domain fashion. In this case, the focused area is a 2 cm nodule and the beam is focused onto the central part at the period of 5 minutes. The simulation based on this condition produces the model showing the appropriate absorbed area based on angle gridding figure as shown in Figure 7.3. Indeed, this kind of simulation is simple and can be applied for any other cancer treatment. The good examples are breast cancer, cervical cancer, and skin cancer. The basic concept is to simulate the microwave generated from the therapeutic probe [82]. It is presently advised that the planning for the best beaming can be the most important step that determines the success of any tumor management. The data from simulation via computational FDTD technique can also be useful in renewing and redesigning new surgical ablation probes and catheters [82,83].

### 7.5.1.4 Example 4: A Study on Difference Laser Wavelength and Loss on Management of Wart

Laser is a well-known technology at present. It is presently used in several applications [84–91]. Briefly, laser is a device that deals with light; hence, it is an actual application that deals with electromagnetic waves [84–91]. The process of light manipulation is the basic principle of any laser device [84–91]. The device firstly emits the electromagnetic radiation, light, through a specific process called optical amplification to get the desired laser [92,93]. During modification, the stimulation of photon emission is done [84–91]. Basically, a laser device has to consist of a gain medium, which is needed for energy supply [92,93]. Also, a laser device requires the optical feedback

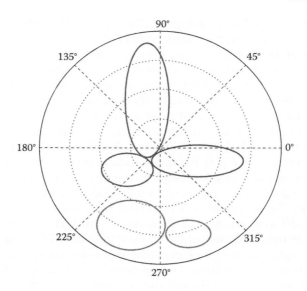

**FIGURE 7.3**
Figure showing the appropriate microwave absorption area on benign prostatic hyperplasia therapy (dark gray line—highly appropriate; light gray line—highly inappropriate).

portion [92,93]. The gain medium is a specific material allowing light amplification. The emission stimulation is used as already mentioned. The stimulating process is done on a specific wavelength within the gain medium aiming at amplification to increase the power [92, 93]. Based on the already described process, there is no doubt that laser devices can provide energetic light that can be used to attack.

As already mentioned, laser has energy. Hence, the energy of laser can be useful. There are many ways to make use of the laser energy [94–98]. Some destructive applications include its use in weapons while the constructive applications are in engineering and medicine. The application of laser in medicine has been continuously developed. In medicine, laser can be applied in both diagnosis and therapy [99,100]. Considering the laser tools in medicine, several kinds are available at present [101]. The good examples of medical laser tools include $CO_2$ lasers [102–105], diode lasers [106–108], dye lasers [109–111], excimer lasers [112–114], fiber lasers [115–117], gas lasers [118], and free-electron lasers [119–123]. As already noted, these laser tools can be applied for both diagnostic and therapeutic purposes. Focusing on diagnostic purposes, there are many laser-based analyzers. The good examples include those used in mammography [124,125], clinical microscopy [126,127], flow cytometry [128,129], and optical coherence tomography [130–132]. It can be seen that the laser-based diagnostic tools can be useful for many medical branches including oncology [133–135], ophthalmology [136–138], and hematology [139–142]. Focusing on thera-peutic purposes, there are also many laser-based therapeutic tools. The use of medical laser can be seen in angioplasty [143–146], lithotripsy [147–151], ocular laser *in situ* keratomileusis (LASIK) [152–155], and photocoagulation [156–158]. The medical therapy that can make use of laser tool can be either general surgery [159–162], cancer surgery [163–166], or plastic-cosmetic sur-gery [167–170]. The examples of those surgeries including baby face prepa-ration [170], scar and tattoo removal [171–174], and cancerous mass removal [163–166] (such as prostatectomy [175–177]).

In the present day, the nanolaser surgery is widely used in dermatol-ogy. This can be for either general or cosmetic purposes. Basically, laser can be used for management of many skin problems such as acne [178–180], melasma [181–183], and wart [184–186]. At first, laser was usually used for tattoo removal. However, with improvement on techniques, the laser can be used for other more complex skin lesions such as vascular problems (e.g., port-wine stains [187] and spider nevus [188]) and pigment problems (e.g., freckles [189] and melisma [181–183]). A wart is a skin tumor caused by a viral infection. Pathologically, the wart is a kind of skin lesion. It is described as a small rough growth with cauliflower shape. It can be seen elsewhere on the human body but it is commonly seen on the foot. At present, the wart is considered a benign neoplasm. The root cause is a viral infection. Based on the present available evidence, it is confirmed that human papil-loma virus (HPV) is the causative pathogen of wart [190–192]. Of interest, HPV virus is widely studied at present. It is accepted that HPV virus is con-sidered an oncogenic pathogen. It can cause cellular abnormality and result in abnormal growth of cells. The neoplasm is the final result. Many cancers, including uterine cervix cancer and penile cancer, are confirmed for their relationship to HPV infection [190–192]. Virologically, HPV virus consists of many types. More than 10 types of HPV can be seen. Focusing on wart, it can be caused by many kinds of HPV types. Hence, there are many variet-ies of wart as well. Because wart is an infectious disease, one can acquire the infection from another person. Skin contact is the main mode of disease transmission. The virus is indeed a highly contagious agent and the trans-mission occurs easily if one contacts the infected skin lesion. Wart is usually problematic and long-lasting in human beings. The well-known kinds of wart in medicine include common wart [193–195], flat wart [196–198], genital wart [199–201], mosaic wart [202–204], periungual wart [205–207], and plan-tar wart [208–210]. Common wart [193–195] or verruca vulagaris is a raised wart with roughened surface. Common wart usually occurs on the hand or foot. Flat wart [196–198] or verruca plana is a small, smooth, flattened wart. It can be seen in groups and is common on the face, neck, and hand. Genital wart [199–201] or venereal wart is the specific kind of wart seen on genitalia. It can be called condyloma acuminatum or verruca acuminata. This kind of wart is a sexually transmitted disease. It is usually problematic in medical practice because it is a risk factor for further development of cancer. Mosaic wart [202–204] is usually seen in groups. It is common on the hand or foot. Periungual wart [205–207] is the specific ward on the nail. It is usually cauliform. Plantar wart [208–210] or verruca plantaris is a wart on the foot. It can be sometimes painful. It is common at pressure points on the soles. Any kind of already mentioned wart is problematic and pathological and requires proper medical management. Wart is considered a condition that needs surgical management because of its relationship with skin cancer if left untreated. Therefore, wart is usually managed early by a general physician in clinical practice.

Warts are usually managed by removal surgery [211,212]. Based on advancement in medicine, the new mode of surgery shifts to the use of laser surgery [212]. The nanolaser is presently used for management of the wart. It is proposed that laser is the best ablative approach [212]. With use of laser, three-fourths of the cases achieve complete removal of wart [212]. However, an important concern is the side effects, including bleeding and pain. Those side effects are believed to relate to the scattering out or loss of the laser [213]. Hence, the measurement of loss is an important factor in determining the success of nanolaser treatment. The loss is usually measured at the therapeutic site and was found to vary by wavelength and frequency (time function); the loss can be presented as "concentration = $k$ × wavelength × frequency." This is an actual phenomenon relating to electromagnetic waves with time function consideration. Hence, the application of the computational FDTD method can be possible. An example of a relationship between change and finalized loss is presented in Table 7.3. In this model, the variation of wavelength is the experimental parameter, the observed loss is the observed parameter, and time is the fixed controlled parameter. Of interest, the model shows that the best wavelength is between 580 and 590 nm, which is concordant with the present dermatological suggestion [212]. Finally, in addition to measurement of loss, the computational FDTD technique can also be useful for adjustment of the microscopic viewing of the dermatological lesion [213]. The good example can be seen in a recent report by Simon and Dimarzio [214].

**TABLE 7.3**

Laser Wavelength and Insertional Loss

| Wavelength (nm) | Loss (Joules/cm²) |
|---|---|
| 610 | 10.4 |
| 600 | 9.8 |
| 590 | 6.8 |
| 580 | 6.9 |
| 570 | 10.6 |

## 7.6 Conclusion

Computational technology is widely applied in several sciences at present. With the use of computational approach, solutions to many difficult questions can be derived. Computational technology application in nanomedicine provides good advantage in this specific field. With use of the computational approach, several manipulations can be done. This can be helpful for answering complicated questions in nanomedical science. Of several computational approaches in nanomedicine, the use of the computational technique to support the study of nanoscale electromagnetic waves is very interesting. The application of the FDTD method can serve this purpose. The computational FDTD method can help medical scientists answer the complex questions relating to several phenomena. Approaches on both structural and functional issues are available. *In silico* clarification and prediction of nano-electromagnetic wave relating phenomena can be done with the help of the FDTD method. In addition, there are many available computational tools for management of biomedical data in nanomedicine that are based on the FDTD method at present. Those new FDTD method–based computational tools can be very useful for many studies in nanomedicine. General practitioners have to recognize and influence use of those tools.

## References

1. Gehlenborg N, O'Donoghue SI, Baliga NS, Goesmann A, Hibbs MA, Kitano H, Kohlbacher O, Neuweger H, Schneider R, Tenenbaum D, Gavin AC. Visualization of omics data for systems biology. *Nat Methods*. 2010 Mar;7(3 Suppl):S56–68.
2. Haarala R, Porkka K. The odd omes and omics. *Duodecim*. 2002;118(11):1193–5.
3. Haddish-Berhane N, Rickus JL, Haghighi K. The role of multiscale computational approaches for rational design of conventional and nanoparticle oral drug delivery systems. *Int J Nanomedicine*. 2007;2(3):315–31.
4. Saliner AG, Poater A, Worth AP. Toward in silico approaches for investigating the activity of nanoparticles in therapeutic development. *IDrugs*. 2008 Oct;11(10):728–32.
5. Lapthorn C, Pullen F, Chowdhry BZ. Ion mobility spectrometry-mass spectrometry (IMS-MS) of small molecules: Separating and assigning structures to ions. *Mass Spectrom Rev*. 2012 Aug 31. doi: 10.1002/mas.21349. [Epub ahead of print]
6. Zhong Y, Hyung SJ, Ruotolo BT. Ion mobility-mass spectrometry for structural proteomics. *Expert Rev Proteomics*. 2012;9(1):47–58.
7. Baumbach JI, Eiceman GA. Ion mobility spectrometry: arriving onsite and moving beyond a low profile. *Appl Spectrosc*. 1999 Sep;53(9):338A–355A.

8. Burlingame AL, Baillie TA, Derrick PJ. Mass spectrometry. *Anal Chem.* 1986 Apr;58(5):165R–211R.
9. Kiser RW, Sullivan RE. Mass spectrometry. *Anal Chem.* 1968 Apr;40(5):273R+.
10. Biemann K. Mass spectrometry. *Annu Rev Biochem.* 1963;32:755–80.
11. Xu G, Gehring PM, Ghosh VJ, Shirane G. High-q-resolution neutron scattering technique using triple-axis spectrometers. *Acta Crystallogr A.* 2004 Nov;60(Pt 6):598–603.
12. Mitchell PW, Paul DM. Low-temperature spin-wave excitations in nickel, by neutron triple-axis spectroscopy. *Phys Rev B Condens Matter.* 1985 Sep 1;32(5):3272–3278.
13. Nieh MP, Yamani Z, Kucerka N, Katsaras J, Burgess D, Breton H. Adapting a triple-axis spectrometer for small angle neutron scattering measurements. *Rev Sci Instrum.* 2008 Sep;79(9):095102.
14. DeKalb EL, Kniseley RN, Fassel VA. Optical emission spectroscopy as an analytical tool. *Ann N Y Acad Sci.* 1966 Jan 20;137(1):235–61.
15. Evers DJ, Hendriks B, Lucassen G, Ruers T. Optical spectroscopy: current advances and future applications in cancer diagnostics and therapy. *Future Oncol.* 2012 Mar;8(3):307–20.
16. Bensalah K, Fleureau J, Rolland D, Rioux-Leclercq N, Senhadji L, Lavastre O, Guillé F, Patard JJ, de Crevoisier R. Optical spectroscopy: a new approach to assess urological tumors. *Prog Urol.* 2010 Jul;20(7):477–82.
17. Barradas NP, Vieira A. Artificial neural network algorithm for analysis of Rutherford backscattering data. *Phys Rev E Stat Phys Plasmas Fluids Relat Interdiscip Topics.* 2000 Oct;62(4 Pt B):5818–29.
18. Schubert R, Dittmar A. Surface analysis of Bio-Sinter ceramic specimens using Rutherford backscattering spectrometry following subcutaneous implantation. *Z Exp Chir Transplant Kunstliche Organe.* 1987;20(1):3–13.
19. Krieger UK, Huthwelker T, Daniel C, Weers U, Peter T, Lanford WA. Rutherford backscattering to study the near-surface region of volatile liquids and solids. *Science.* 2002 Feb 8;295(5557):1048–50.
20. Jeynes C, Barradas NP, Szilágyi E. Accurate determination of quantity of material in thin films by Rutherford backscattering spectrometry. *Anal Chem.* 2012 Jul 17;84(14):6061–9.
21. Mann S. Life as a nanoscale phenomenon. *Angew Chem Int Ed Engl.* 2008; 47(29):5306–20.
22. Whatmore RW. Nanotechnology—what is it? Should we be worried? *Occup Med (Lond).* 2006 Aug;56(5):295–9.
23. O'Brien TP, Bult CJ, Cremer C, Grunze M, Knowles BB, Langowski J, McNally J, Pederson T, Politz JC, Pombo A, Schmahl G, Spatz JP, van Driel R. Genome function and nuclear architecture: from gene expression to nanoscience. *Genome Res.* 2003 Jun;13(6A):1029–41.
24. Wickline SA, Lanza GM. Molecular imaging, targeted therapeutics, and nanoscience. *J Cell Biochem Suppl.* 2002;39:90–7.
25. Behari J. Principles of nanoscience: an overview. *Indian J Exp Biol.* 2010 Oct; 48(10):1008–19.
26. Colliex C. From electron energy-loss spectroscopy to multi-dimensional and multi-signal electron microscopy. *J Electron Microsc (Tokyo).* 2011;60 Suppl 1:S161–71.

27. Meckenstock R. Invited Review Article: Microwave spectroscopy based on scanning thermal microscopy: resolution in the nanometer range. *Rev Sci Instrum.* 2008 Apr;79(4):041101.

28. Chung T, Lim Y, Lee IM, Lee SY, Choi J, Roh S, Kim KY, Lee B. A compact light concentrator by the use of plasmonic faced folded nano-rods. *Opt Express.* 2011 Oct 10;19(21):20751–60.

29. Kawakami Y, Kanai A, Kaneta A, Funato M, Kikuchi A, Kishino K. Micromirror arrays to assess luminescent nano-objects. *Rev Sci Instrum.* 2011 May;82(5):053905.

30. Foteinopoulou S, Vigneron JP, Vandenbem C. Optical near-field excitations on plasmonic nanoparticle-based structures. *Opt Express.* 2007 Apr 2;15(7):4253–67.

31. Liu C, Huang Z, Lu F, Zheng W, Hutmacher DW, Sheppard C. Near-field effects on coherent anti-Stokes Raman scattering microscopy imaging. *Opt Express.* 2007 Apr 2;15(7):4118–31.

32. Lu Y, Zhou C, Wang S, Wang B. Polarization-dependent Talbot effect. *J Opt Soc Am A Opt Image Sci Vis.* 2006 Sep;23(9):2154–60.

33. H'dhili F, Bachelot R, Rumyantseva A, Lerondel G, Royer P. Nano-patterning photosensitive polymers using local field enhancement at the end of apertureless SNOM tips. *J Microsc.* 2003 Mar;209(Pt 3):214–22.

34. Chung T, Lim Y, Lee IM, Lee SY, Choi J, Roh S, Kim KY, Lee B. A compact light concentrator by the use of plasmonic faced folded nano-rods. *Opt Express.* 2011 Oct 10;19(21):20751–60.

35. Su R, Tang D, Ding W, Chen L, Zhou Z. Efficient transmission of crossing dielectric slot waveguides. *Opt Express.* 2011 Feb 28;19(5):4756–61.

36. Zhao L, Li Y, Qi J, Xu J, Sun Q. Light focusing by the unique dielectric nano-waveguide array. *Opt Express.* 2009 Sep 14;17(19):17136–43.

37. Ahmadi A, Ghadarghadr S, Mosallaei H. An optical reflectarray nanoantenna: the concept and design. *Opt Express.* 2010 Jan 4;18(1):123–33.

38. Kawakami Y, Kanai A, Kaneta A, Funato M, Kikuchi A, Kishino K. Micromirror arrays to assess luminescent nano-objects. *Rev Sci Instrum.* 2011 May;82(5):053905.

39. Li S, Du C, Dong X, Shi L, Luo X, Wei X, Zhang Y. Superlens nano-patterning technology based on the distributed polystyrene spheres. *Opt Express.* 2008 Sep 15;16(19):14397–403.

40. Taflove HA, Bridges JE. Two-dimensional FDTD analysis of a pulsed microwave confocal system for breast cancer detection: Fixed-focus and antenna-array sensors. *IEEE Trans Biomed Eng.* 1988; 45 (12): 1470–1479.

41. Burfeindt MJ, Zastrow E, Hagness SC, Van Veen BD, Medow JE. Microwave beamforming for non-invasive patient-specific hyperthermia treatment of pediatric brain cancer. *Phys Med Biol.* 2011 May 7;56(9):2743–54.

42. Wang G, Gong Y. Metamaterial lens applicator for microwave hyperthermia of breast cancer. *Int J Hyperthermia.* 2009;25(6):434–45.

43. Zastrow E, Hagness SC, Van Veen BD. 3D computational study of non-invasive patient-specific microwave hyperthermia treatment of breast cancer. *Phys Med Biol.* 2010 Jul 7;55(13):3611–29.

44. Yacoob SM, Hassan NS. FDTD analysis of a noninvasive hyperthermia system for brain tumors. *Biomed Eng Online.* 2012 Aug 14;11(1):47. [Epub ahead of print]

45. De La Iglesia D, Chiesa S, Kern J, Maojo V, Martin-Sanchez F, Potamias G, Moustakis V, Mitchell JA. Nanoinformatics: new challenges for biomedical informatics at the nanolevel. *Stud Health Technol Inform.* 2009;150:987–91.

46. Ray K, Chowdhury MH, Szmacinski H, Lakowicz JR. Metal-enhanced intrinsic fluorescence of proteins on silver nanostructured surfaces towards label-free detection. *J Phys Chem C Nanomater Interfaces.* 2008;112(46):17957–17963.

47. Bonmassar G, Fujimoto K, Golby AJ. PTFOS: Flexible and absorbable intracranial electrodes for magnetic resonance imaging. *PLoS One.* 2012;7(9):e41187.

48. H'dhili F, Bachelot R, Rumyantseva A, Lerondel G, Royer P. Nano-patterning photosensitive polymers using local field enhancement at the end of apertureless SNOM tips. *J Microsc.* 2003 Mar;209(Pt 3):214–22.

49. Wust P, Berger J, Fähling H, Nadobny J, Gellermann J, Tilly W, Rau B, Petermann K, Felix R. Scanning E-field sensor device for online measurements in annular phased-array systems. *Int J Radiat Oncol Biol Phys.* 1999 Mar 1;43(4):927–37.

50. Hallaj IM, Cleveland RO. FDTD simulation of finite-amplitude pressure and temperature fields for biomedical ultrasound. *J Acoust Soc Am.* 1999 May;105(5):L7–12.

51. Ho HS. Safety of metallic implants in magnetic resonance imaging. *J Magn Reson Imaging.* 2001 Oct;14(4):472–7.

52. Tanev S, Sun W, Pond J, Tuchin VV, Zharov VP. Flow cytometry with gold nanoparticles and their clusters as scattering contrast agents: FDTD simulation of light-cell interaction. *J Biophotonics.* 2009 Sep;2(8-9):505–20.

53. Yu S, Zhang J, Moran MS, Lu JQ, Feng Y, Hu XH. A novel method of diffraction imaging flow cytometry for sizing microspheres. *Optics Express.* 2012; 20 (20): 22245–51.

54. Nagaoka T, Watanabe S. Multi-GPU accelerated three-dimensional FDTD method for electromagnetic simulation. *Conf Proc IEEE Eng Med Biol Soc.* 2011;2011:401–4.

55. Zastrow E, Hagness SC, Van Veen BD, Medow JE. Time-multiplexed beamforming for noninvasive microwave hyperthermia treatment. *IEEE Trans Biomed Eng.* 2011 Jun;58(6):1574–84.

56. Heh DY, Tan EL. Modeling hemoglobin at optical frequency using the unconditionally stable fundamental ADI-FDTD method. *Biomed Opt Express.* 2011 Apr 12;2(5):1169–83.

57. Pootrakul S, Kematorn B, Na-Nakorn S, Suanpan S. A new haemoglobin variant: haemoglobin Anantharaj alpha 11 (A9) lysine replaced by glutamic acid. *Biochim Biophys Acta.* 1975 Sep 9;405(1):161–6.

58. Svasti J, Surarit R, Srisomsap C, Pravatmuang P, Wasi P, Fucharoen S, Blouquit Y, Galacteros F, Rosa J. Identification of Hb Anantharaj [alpha 11(A9)Lys->Glu] as Hb J-Wenchang-Wuming [alpha 11(A9)Lys->Gln]. *Hemoglobin.* 1993 Oct;17(5):453–5.

59. Wiwanitkit V. Secondary and tertiary structure aberration of alpha globin chain in haemoglobin Q-India disorder. *Indian J Pathol Microbiol.* 2006 Oct;49(4):491–4.

60. Leffell DJ, Thompson JT. Lasers in dermatology and ophthalmology. *Dermatol Clin.* 1992 Oct;10(4):687–700.

61. Bessette FM, Nguyen LC. Laser light: its nature and its action on the eye. *CMAJ.* 1989 Dec 1;141(11):1141–8.

62. Roder PB, Pauzauskie PJ, Davis EJ. Nanowire heating by optical electromagnetic irradiation. *Langmuir.* 2012 Oct 12. [Epub ahead of print]

63. Muschter R. Local treatment of prostate cancer using thermal-ablative energy. *Urologe A.* 2009 Jul;48(7):729–39.

64. Davydov BI. Electromagnetic radiofrequency radiation (microwaves): principles and criteria of standardization, threshold dose levels. *Kosm Biol Aviakosm Med.* 1985 May-Jun;19(3):8–21.

65. Servantie B. Damage criteria for determining microwave exposure. *Health Phys.* 1989 May;56(5):781–6.

66. Wilkening GM, Sutton CH. Health effects of nonionizing radiation. *Med Clin North Am.* 1990 Mar;74(2):489–507.

67. McRee DI. Potential microwave injuries in clinical medicine. *Annu Rev Med.* 1976;27:109–15.

68. Jaulerry C, Bataini JP, Brunin F, Gaboriaud G. Present clinical status of hyperthermia associated with radiotherapy (author's transl). *Bull Cancer.* 1981;68(3):261–7.

69. Kawahara H, Odagiri S, Fujita H, Nagano T, Murakami M. Hyperthermia in cancer therapy. *J UOEH.* 1984 Sep 1;6(3):307–15.

70. Seki T, Inoue K. Microwave coagulation therapy of hepatocellular carcinoma. *Nihon Naika Gakkai Zasshi.* 1995 Dec 10;84(12):2024–7.

71. Yanaga K. Current status of hepatic resection for hepatocellular carcinoma. *J Gastroenterol.* 2004 Oct;39(10):919–26.

72. Khatri VP, McGahan J. Non-resection approaches for colorectal liver metastases. *Surg Clin North Am.* 2004 Apr;84(2):587–606.

73. Froeling FM, de la Rosette JJ, Debruyne FM. Minimally invasive urology. *Ned Tijdschr Geneeskd.* 1994 Aug 27;138(35):1756–60.

74. Devonec M, Tomera K, Perrin P. Review: transurethral microwave thermotherapy in benign prostatic hyperplasia. *J Endourol.* 1993 Jun;7(3):255–9.

75. De La Rosette JJ. Microwave treatment of BPH: still an option? *Arch Esp Urol.* 1994 Nov;47(9):889–94.

76. Hoffman RM, Monga M, Elliott SP, Macdonald R, Langsjoen J, Tacklind J, Wilt TJ. Microwave thermotherapy for benign prostatic hyperplasia. *Cochrane Database Syst Rev.* 2012 Sep 12;9:CD004135. doi: 10.1002/14651858.CD004135.pub3.

77. Barry Delongchamps N, Robert G, Descazeaud A, Cornu JN, Azzouzi AR, Haillot O, Devonec M, Fourmarier M, Ballereau C, Lukacs B, Dumonceau O, Saussine C, de la Taille A; Comité des troubles mictionnels de l'homme de l'Association française d'urologie. Surgical management of benign prostatic hyperplasia by thermotherapy and other emerging techniques: A review of the literature by the LUTS committee of the French Urological Association. *Prog Urol.* 2012 Feb;22(2):87–92.

78. Reich O, Seitz M, Gratzke C, Schlenker B, Walther S, Stief C. Benign prostatic hyperplasia (BPH): Surgical therapy options. *Urologe A.* 2010 Jan;49(1):113–26.

79. Muschter R. Local treatment of prostate cancer using thermal-ablative energy. *Urologe A.* 2009 Jul;48(7):729–39.

80. Ahmed S, Lindsey B, Davies J. Emerging minimally invasive techniques for treating localized prostate cancer. *BJU Int.* 2005 Dec;96(9):1230–4.

81. Sherar MD, Trachtenberg J, Davidson SR, Gertner MR. Interstitial microwave thermal therapy and its application to the treatment of recurrent prostate cancer. *Int J Hyperthermia.* 2004 Nov;20(7):757–68.

82. Brannan JD. Electromagnetic measurement and modeling techniques for microwave ablation probes. *Conf Proc IEEE Eng Med Biol Soc.* 2009;2009:3076–8.

83. Nevels RD, Arndt GD, Raffoul GW, Carl JR, Pacifico A. Microwave catheter design. *IEEE Trans Biomed Eng.* 1998 Jul;45(7):885–90.

84. Polanyi TG. Laser physics. *Otolaryngol Clin North Am.* 1983 Nov;16(4):753–74.

85. Nolan LJ. Laser physics and safety. *Clin Podiatr Med Surg.* 1987 Oct;4(4):777–86.
86. Polanyi TG. Physics of the surgical laser. *Int Adv Surg Oncol.* 1978;1:205–15.
87. De Felice E. Shedding light: laser physics and mechanism of action. *Phlebology.* 2010 Feb;25(1):11–28.
88. Nolan LJ. Laser physics and safety. *Clin Podiatr Med Surg.* 1987 Oct;4(4):777–86.
89. Ashiboff R. Introduction to lasers. *Semin Dermatol.* 1994 Mar;13(1):48–59.
90. Monaco WA, Barker FM II. Laser hazards and safety. *Optom Clin.* 1995;4(4):1–15.
91. Gregory RO. Laser physics and physiology. *Clin Plast Surg.* 1998 Jan;25(1):89–93.
92. Huether SE. How lasers work. *AORN J.* 1983 Aug;38(2):207–15.
93. Doiron DR, Profio AE. Laser instrumentation and safety. *Clin Chest Med.* 1985 Jun;6(2):209–17.
94. Tajima T. Laser acceleration and its future. *Proc Jpn Acad Ser B Phys Biol Sci.* 2010;86(3):147–57.
95. Skanes AC, Klein GJ, Krahn AD, Yee R. Advances in energy delivery. *Coron Artery Dis.* 2003 Feb;14(1):15–23.
96. Takac S, Stojanović S. Characteristics of laser light. *Med Pregl.* 1999 Jan–Feb;52(1-2):29–34.
97. Goebel KR. Fundamentals of laser science. *Acta Neurochir Suppl.* 1994;61:20–33.
98. Hillenkamp F. Laser radiation tissue interaction. *Health Phys.* 1989 May;56(5): 613–6.
99. Gibson KF, Kernohan WG. Lasers in medicine—a review. *J Med Eng Technol.* 1993 Mar–Apr;17(2):51–7.
100. Viherkoski E. Lasers in medicine. *Ann Chir Gynaecol.* 1990;79(4):176–81.
101. Takac S, Stojanović S, Muhi B. Types of medical lasers. *Med Pregl.* 1998 Mar–Apr;51(3-4):146–50.
102. Gueissaz F, Ruffieux C. The $CO_2$ laser in dermatology. *Rev Med Suisse Romande.* 1989 Jun;109(6):457–63
103. Landthaler M, Haina D, Hohenleutner U, Seipp W, Waidelich W, Braun-Falco O. The $CO_2$ laser in dermatotherapy—use and indications. *Hautarzt.* 1988 Apr;39(4):198–204
104. Rinaldi F, Cecca E. The use of $CO_2$ laser in ambulatory dermatologic surgery. *Hautarzt. G Ital Dermatol Venereol.* 1990 Sep;125(9):XLIII–XLVI.
105. Schick RO, Schick MP. $CO_2$ laser surgery in veterinary dermatology. *Clin Dermatol.* 1994 Oct–Dec;12(4):587–9.
106. Moriarty AP. Diode lasers in ophthalmology. *Int Ophthalmol.* 1993–1994; 17(6): 297–304
107. Balles MW, Puliafito CA. Semiconductor diode lasers: a new laser light source in ophthalmology. *Int Ophthalmol Clin.* 1990 Spring;30(2):77–83.
108. Surgical diode lasers. *Health Devices.* 1998 Mar;27(3):84–92.
109. Beck EM, Vaughan ED Jr., Sosa RE. The pulsed dye laser in the treatment of ureteral calculi. *Semin Urol.* 1989 Feb;7(1):25–9.
110. Grasso M, Bagley D, Sullivan K. Pulsed dye laser lithotripsy—currently applied to urologic and biliary calculi. *J Clin Laser Med Surg.* 1991 Oct;9(5):355–9.
111. Borovoy MA, Borovoy M, Elson LM, Sage M. Flash lamp pulsed dye laser (585 nm). Treatment of resistant verrucae. *J Am Podiatr Med Assoc.* 1996 Nov;86(11):547–50.
112. Seitz B, Langenbucher A, Naumann GO. Perspectives of excimer laser-assisted keratoplasty. *Ophthalmologe.* 2011 Sep;108(9):817–24.
113. Frentzen M, Koort HJ, Thiensiri I. Excimer lasers in dentistry: future possibilities with advanced technology. *Quintessence Int.* 1992 Feb;23(2):117–33.

114. Seiler T. The excimer laser. An instrument for corneal surgery. *Ophthalmologe.* 1992 Apr;89(2):128–33.
115. Tünnermann A, Schreiber T, Limpert J. Fiber lasers and amplifiers: an ultrafast performance evolution. *Appl Opt.* 2010 Sep 1;49(25):F71–8.
116. Jackson SD, Lauto A. Diode-pumped fiber lasers: a new clinical tool? *Lasers Surg Med.* 2002;30(3):184–90.
117. Wang Z, Chocat N. Fiber-optic technologies in laser-based therapeutics: threads for a cure. *Curr Pharm Biotechnol.* 2010 Jun;11(4):384–97.
118. Bloom AL. Gas lasers. *Appl Opt.* 1966 Oct 1;5(10):1500–14.
119. Ullrich J, Rudenko A, Moshammer R. Free-electron lasers: new avenues in molecular physics and photochemistry. *Annu Rev Phys Chem.* 2012;63:635–60.
120. Schlichting I, Miao J. Emerging opportunities in structural biology with X-ray free-electron lasers. *Curr Opin Struct Biol.* 2012 Oct;22(5):613–26.
121. Neutze R, Moffat K. Time-resolved structural studies at synchrotrons and X-ray free electron lasers: opportunities and challenges. *Curr Opin Struct Biol.* 2012 Oct;22(5):651–9.
122. Edwards GS, Allen SJ, Haglund RF, Nemanich RJ, Redlich B, Simon JD, Yang WC. Applications of free-electron lasers in the biological and material sciences. *Photochem Photobiol.* 2005 Jul–Aug;81(4):711–35.
123. O'Shea PG, Freund HP. Free-electron lasers. Status and applications. *Science.* 2001 Jun 8;292(5523):1853–8.
124. Bílková A, Janík V, Svoboda B. Computed tomography laser mammography. *Cas Lek Cesk.* 2010;149(2):61–5.
125. Richter DM. Computed tomographic laser mammography, a practical review. *Nihon Hoshasen Gijutsu Gakkai Zasshi.* 2003 Jun;59(6):687–93.
126. Seidenari S, Arginelli F, Bassoli S, Cautela J, French PM, Guanti M, Guardoli D, König K, Talbot C, Dunsby C. Multiphoton laser microscopy and fluorescence lifetime imaging for the evaluation of the skin. *Dermatol Res Pract.* 2012; 2012: 810749.
127. Földes-Papp Z, Demel U, Tilz GP. Laser scanning confocal fluorescence microscopy: an overview. *Int Immunopharmacol.* 2003 Dec;3(13–14):1715–29.
128. Tkaczyk ER, Tkaczyk AH. Multiphoton flow cytometry strategies and applications. *Cytometry A.* 2011 Oct;79(10):775–88.
129. Telford WG. Lasers in flow cytometry. *Methods Cell Biol.* 2011;102:375–409.
130. Shapiro HM. Lasers for flow cytometry. *Curr Protoc Cytom.* 2004 Feb;Chapter 1:Unit 1.9.
131. Castro Lima V, Rodrigues EB, Nunes RP, Sallum JF, Farah ME, Meyer CH. Simultaneous confocal scanning laser ophthalmoscopy combined with high-resolution spectral-domain optical coherence tomography: a review. *J Ophthalmol.* 2011;2011:743670.
132. Tegetmeyer H. Application of optical coherence tomography in paediatric neuroophthalmology. *Klin Monbl Augenheilkd.* 2011 Oct;228(10):868–73.
133. Lowes LE, Goodale D, Keeney M, Allan AL. Image cytometry analysis of circulating tumor cells. *Methods Cell Biol.* 2011;102:261–90.
134. Eichert S, Möhrle M, Breuninger H, Röcken M, Garbe C, Bauer J. Diagnosis of cutaneous tumors with in vivo confocal laser scanning microscopy. *J Dtsch Dermatol Ges.* 2010 Jun;8(6):400–10.
135. Gemoll T, Roblick UJ, Habermann JK. MALDI mass spectrometry imaging in oncology (Review). *Mol Med Report.* 2011 Nov–Dec;4(6):1045–51.
136. Davidson SI. Ophthalmology. *Practitioner.* 1976 Oct;217(1300 SPEC NO):596–601.

137. Weale RA. Physics and ophthalmology. *Phys Med Biol.* 1979 May;24(3):489–504.
138. Zhang W, Kogure A, Yamamoto K, Hori S. Use of the laser speckle flowgraphy in posterio rfundus circulation research. *Chin Med J (Engl).* 2011 Dec;124(24): 4339–44.
139. Geneviève F, Godon A, Marteau-Tessier A, Zandecki M. Automated hematology analysers and spurious counts. Part 2. Leukocyte count and differential. *Ann Biol Clin (Paris).* 2012 Mar–Apr;70(2):141–54.
140. Dalal BI. Clinical applications of molecular haematology: flow cytometry in leukaemias and myelodysplastic syndromes. *J Assoc Physicians India.* 2007 Aug;55: 571–3.
141. Kickler TS. Clinical analyzers. Advances in automated cell counting. *Anal Chem.* 1999 Jun 15;71(12):363R–365R.
142. Knüchel R. Analysis by flow cytometry and cell sorting. Report of current status and perspectives for pathology. *Pathologe.* 1994 Apr;15(2):85–95.
143. Petrosian IuS, Kipshidze NN, Putilin SA. Laser angioplasty: current status, problems and prospects. *Kardiologiia.* 1986 Aug;26(8):114–8.
144. Müller FS, Norden C, Dähne F. Percutaneous transluminal laser angioplasty. *Z Arztl Fortbild (Jena).* 1990;84(6):261–6.
145. Forrester JS. Laser angioplasty. Now and in the future. *Circulation.* 1988 Sep;78(3):777–9.
146. Steg PG, Ménasché P. Utilization of laser arterial angioplasty. *Ann Vasc Surg.* 1989 Jan;3(1):86–94.
147. Hofmann R, Hartung R. Laser lithotripsy of ureteral calculi. *Urol Res.* 1990;18 Suppl 1:S49–55.
148. Sosa RE. Laser lithotripsy: an update. *Semin Urol.* 1991 Aug;9(3):203–5.
149. Zerbib M, Steg A, Belas M, Flam T, Debre B. Laser lithotripsy of ureteral calculi: initial experience with a new pulsed dye laser. *J Lithotr Stone Dis.* 1990 Jan;2(1):39–41.
150. Beck EM, Vaughan ED Jr, Sosa RE. The pulsed dye laser in the treatment of ureteral calculi. *Semin Urol.* 1989 Feb;7(1):25–9.
151. Grasso M, Bagley D, Sullivan K. Pulsed dye laser lithotripsy—currently applied to urologic and biliary calculi. *J Clin Laser Med Surg.* 1991 Oct;9(5):355–9.
152. Wilson SE. LASIK: management of common complications. Laser in situ keratomileusis. *Cornea.* 1998 Sep;17(5):459–67.
153. Farah SG, Azar DT, Gurdal C, Wong J. Laser in situ keratomileusis: literature review of a developing technique. *J Cataract Refract Surg.* 1998 Jul;24(7): 989–1006.
154. Buratto L, Ferrari M. Indications, techniques, results, limits, and complications of laser in situ keratomileusis. *Curr Opin Ophthalmol.* 1997 Aug;8(4):59–66.
155. Carr JD, Stulting RD, Thompson KP, Waring GO III. Laser in situ keratomileusis: surgical technique. *Ophthalmol Clin North Am.* 2001 Jun;14(2):285–94, vii.
156. Trutneva KV, Makarskaia NV. Photocoagulation in diabetic retinopathy (survey of the literature). *Vestn Oftalmol.* 1973;2:82–5.
157. Coscas G, Chaine G. Treatment of diabetic retinopathy with laser photocoagulation (author's transl). *Diabete Metab.* 1979 Sep;5(3):247–59.
158. Protell RL, Silverstein FE, Auth DC. Laser photocoagulation for gastrointestinal bleeding. *Clin Gastroenterol.* 1978 Sep;7(3):765–74.
159. Dixon JA. Current laser applications in general surgery. *Ann Surg.* 1988 Apr;207(4):355–72.

160. Skobelkin OK, Brechov JI, Baschilov WP, Parchomenko JG, Davydov BN, Bogatov WW, Smoljaninov MW, Danilin NA, Schapovalov AM, Kalinnikov WW. Laser in clinical surgery (author's transl). *Zentralbl Chir.* 1982;107(4): 209–13.
161. Joffe SN, Schröder T. Lasers in general surgery. *Adv Surg.* 1987;20:125–54.
162. Daly CJ. The techniques and uses of lasers in general surgery. *Curr Opin Gen Surg.* 1993:3–7.
163. Goldman L. Laser treatment of cancer. *Prog Clin Cancer.* 1967;3:205–20.
164. Hayata Y, Kato H. Laser and cancer therapy. *Gan To Kagaku Ryoho.* 1983 Jun;10(6):1387–94.
165. Aronoff BL. Laser surgery in the treatment of cancer. *Compr Ther.* 1988 May;14(5):64–8.
166. Löfgren L. Cancer treatment can be "tailor-made" with laser surgery and photochemistry. *Lakartidningen.* 1990 Sep 5;87(36):2754, 2757–60.
167. Gregory RO. Applications of lasers in plastic surgery. *J Fla Med Assoc.* 1989 Jul;76(7):595–8.
168. Malm M, Lundeberg T. Laser technology in plastic surgery. *Scand J Plast Reconstr Surg Hand Surg.* 1992;26(1):3–11.
169. Hobby LW. Lasers in plastic surgery: an expanding frontier. *J Med Assoc Ga.* 1991 Nov;80(11):587–93.
170. Toregard BM. Laser applications in cosmetic surgery. *Ann Chir Gynaecol.* 1990;79(4):208–15.
171. Tammaro A, Fatuzzo G, Narcisi A, Abruzzese C, Caperchi C, Gamba A, Parisella FR, Persechino S. Laser removal of tattoos. *Int J Immunopathol Pharmacol.* 2012 Apr–Jun;25(2):537–9.
172. Kent KM, Graber EM. Laser tattoo removal: a review. *Dermatol Surg.* 2012 Jan;38(1):1–13.
173. Choudhary S, Elsaie ML, Leiva A, Nouri K. Lasers for tattoo removal: a review. *Lasers Med Sci.* 2010 Sep;25(5):619–27.
174. Pfirrmann G, Karsai S, Roos S, Hammes S, Raulin C. Tattoo removal—state of the art. *J Dtsch Dermatol Ges.* 2007 Oct;5(10):889–97.
175. Anson K, Seenivasagam K, Miller R, Watson G. The role of lasers in urology. *Br J Urol.* 1994 Mar;73(3):225–30.
176. Williams JH, Chilton CP. New therapeutic options in prostatic enlargement. *Br J Hosp Med.* 1994 May 4–17;51(9):477–81.
177. Smith JA Jr. Laser treatment of the urethra and prostate. *Semin Urol.* 1991 Aug;9(3):180–4.
178. Baker TM. Dermabrasion. As a complement to aesthetic surgery. *Clin Plast Surg.* 1998 Jan;25(1):81–8.
179. Baker TM. Laser resurfacing for facial acne scars. *Cochrane Database Syst Rev.* 2001;(1):CD001866.
180. Hirsch RJ, Lewis AB. Treatment of acne scarring. *Semin Cutan Med Surg.* 2001 Sep;20(3):190–8.
181. Arora P, Sarkar R, Garg VK, Arya L. Lasers for treatment of melasma and post-inflammatory hyperpigmentation. *J Cutan Aesthet Surg.* 2012 Apr;5(2):93–103.
182. Ball Arefiev KL, Hantash BM. Advances in the treatment of melasma: a review of the recent literature. *Dermatol Surg.* 2012 Jul;38(7 Pt 1):971–84.
183. Situm M, Kolić M, Bolanca Z, Ljubicić I, Misanović B. Melasma—updated treatments. *Coll Antropol.* 2011 Sep;35 Suppl 2:315–8.

184. Ockenfels HM, Hammes S. Laser treatment of warts. *Hautarzt.* 2008 Feb;59(2): 116–23.
185. Park HS, Kim JW, Jang SJ, Choi JC. Pulsed dye laser therapy for pediatric warts. *Pediatr Dermatol.* 2007 Mar–Apr;24(2):177–81.
186. Benson RC Jr. Laser use in open surgery and external lesions. *Urol Clin North Am.* 1986 Aug;13(3):421–34.
187. Faurschou A, Olesen AB, Leonardi-Bee J, Haedersdal M. Lasers or light sources for treating port-wine stains. *Cochrane Database Syst Rev.* 2011 Nov 9;(11):CD007152.
188. Tan OT, Gilchrest BA. Laser therapy for selected cutaneous vascular lesions in the pediatric population: a review. *Pediatrics.* 1988 Oct;82(4):652–62.
189. Shah S, Alster TS. Laser treatment of dark skin: an updated review. *Am J Clin Dermatol.* 2010 Dec 1;11(6):389–97.
190. Orth G, Jablonska S, Breitburd F, Favre M, Croissant O. The human papillomaviruses. *Bull Cancer.* 1978;65(2):151–64.
191. Rivera Y, Vázquez Botet M, Sánchez J. Recent concepts in human papillomavirus. *P R Health Sci J.* 1988 Dec;7(3):233–43.
192. Grussendorf-Conen EI. Infections with human pathogenetic papillomaviruses. *Z Hautkr.* 1989 Aug 15;64(8):627–8, 633.
193. Lewis MR, Lutterbeck EF. Verucae: a clinical review. *J Am Podiatry Assoc.* 1967 Jun;57(6):263–8.
194. Pass F. Warts. Biology and current therapy. *Minn Med.* 1974 Oct;57(10):844–7, 852.
195. Coskey RJ. What's new about warts. *Cutis.* 1976 Oct;18(4):527–31.
196. Egawa K. Verruca plana juvenilis. *Ryoikibetsu Shokogun Shirizu.* 1999;(25 Pt 3): 259–61.
197. Pavithra S, Mallya H, Pai GS. Extensive presentation of verruca plana in a healthy individual. *Indian J Dermatol.* 2011 May;56(3):324–5.
198. Bouquot JE. Oral and maxillofacial pathology case of the month. Verruca plana (flat wart; Verruca plana juvenilis). *Tex Dent J.* 2005 Jun;122(6):579, 584–6.
199. Yanofsky VR, Patel RV, Goldenberg G. Genital warts: a comprehensive review. *J Clin Aesthet Dermatol.* 2012 Jun;5(6):25–36.
200. Dunne EF, Friedman A, Datta SD, Markowitz LE, Workowsk: KA. Updates on human papillomavirus and genital wants and counseling messages from the 2012 Sexually Transmitted Diseases Treatment Guidelines. *Clin Infect Dis.* 2011 Dec; 53 Suppl 3: S143–52.
201. Stanley MA. Genital human papilloma virus infections: current and prospective therapies. *J Gen Virol.* 2012 Apr;93(Pt 4):681–91.
202. Angello LJ, Massik P. Mosaic verrucae: a case report. *J Natl Assoc Chirop.* 1957 Sep;47(9):465.
203. Herbst S. Mosaic wart. A case report. *J Am Podiatry Assoc.* 1966 Jul;56(7):331.
204. Heimlich HA. Mosaic verrucae. *J Natl Assoc Chirop.* 1952 Oct;42(10):27.
205. Moghaddas N. Periungual verrucae diagnosis and treatment. *Clin Podiatr Med Surg.* 2004 Oct;21(4):651–61, viii.
206. Moghaddas N. Periungual verrucae.Diagnosis and treatment. *Clin Podiatr Med Surg.* 1995 Apr;12(2):189–99.
207. Tosti A, Piraccini BM. Warts of the nail unit: surgical and nonsurgical approaches. *Dermatol Surg.* 2001 Mar;27(3):235–9.
208. Lichon V, Khachemoune A. Plantar warts: a focus on treatment modalities. *Dermatol Nurs.* 2007 Aug;19(4):372–5.

209. Watkins P. Identifying and treating plantar warts. *Nurs Stand*. 2006 Jun 28–Jul 4;20(42):50–4.
210. Landsman MJ, Mancuso JE, Abramow SP. Diagnosis, pathophysiology, and treatment of plantar verruca. *Clin Podiatr Med Surg*. 1996 Jan;13(1):55–71.
211. Mulhem E, Pinelis S. Treatment of nongenital cutaneous warts. *Am Fam Physician*. 2011 Aug 1;84(3):288–93.
212. Ockenfels HM, Hammes S. Laser treatment of warts. *Hautarzt*. 2008 Feb;59(2):116–23.
213. Raulin C, Kimmig W, Werner S. Laser therapy in dermatology and esthetic medicine. Side effects, complications and treatment errors. *Hautarzt*. 2000 Jul;51(7):463–73.
214. Simon B, Dimarzio CA. Simulation of a theta line-scanning confocal microscope. *J Biomed Opt*. 2007 Nov–Dec;12(6):064020.

# Appendix A: Material and Physical Constants

## CONTENTS

## A.1 Common Material Constants

**TABLE A.1**

Approximate Conductivity at 20°C

|  | Material | Conductivity (S/m) |
|---|---|---|
| 1. | Conductors | |
|  | Silver | $6.3 \times 10^7$ |
|  | Copper (standard annealed) | $5.8 \times 10^7$ |
|  | Gold | $4.5 \times 10^7$ |
|  | Aluminum | $3.5 \times 10^7$ |
|  | Tungsten | $1.8 \times 10^7$ |
|  | Zinc | $1.7 \times 10^7$ |
|  | Brass | $1.1 \times 10^7$ |
|  | Iron (pure) | $10^7$ |
|  | Lead | $5 \times 10^7$ |
|  | Mercury | $10^6$ |
|  | Carbon | $3 \times 10^7$ |
|  | Water (sea) | 4.8 |
| 2. | Semiconductors | |
|  | Germanium (pure) | 2.2 |
|  | Silicon (pure) | $4.4 \times 10^{-4}$ |
| 3. | Insulators | |
|  | Water (distilled) | $10^{-4}$ |
|  | Earth (dry) | $10^{-5}$ |
|  | Bakelite | $10^{-10}$ |
|  | Paper | $10^{-11}$ |
|  | Glass | $10^{-12}$ |
|  | Porcelain | $10^{-15}$ |

*(continued)*

**TABLE A.1 (CONTINUED)**

Approximate Conductivity at 20°C

| Material | Conductivity (S/m) |
|----------|--------------------|
| Mica | $10^{-15}$ |
| Paraffin | $10^{-15}$ |
| Rubber (hard) | $10^{-15}$ |
| Quartz (fused) | $10^{-17}$ |
| Wax | $10^{-17}$ |

**TABLE A.2**

Approximate Dielectric Constant and Dielectric Strength

| Material | Dielectric Constant (or Relative Permittivity) (Dimensionless) | Strength, $E$ (V/m) |
|----------|------------------------|---------------------|
| Barium titanate | 1200 | $7.5 \times 10^6$ |
| Water (sea) | 80 | — |
| Water (distilled) | 8.1 | — |
| Nylon | 8 | — |
| Paper | 7 | $12 \times 10^6$ |
| Glass | 5–10 | $35 \times 10^6$ |
| Mica | 6 | $70 \times 10^6$ |
| Porcelain | 6 | — |
| Bakelite | 5 | $20 \times 10^6$ |
| Quartz (fused) | 5 | $30 \times 10^6$ |
| Rubber (hard) | 3.1 | $25 \times 10^6$ |
| Wood | 2.5–8.0 | — |
| Polystyrene | 2.55 | — |
| Polypropylene | 2.25 | — |
| Paraffin | 2.2 | $30 \times 10^6$ |
| Petroleum oil | 2.1 | $12 \times 10^6$ |
| Air (1 atm) | 1 | $3 \times 10^6$ |

**TABLE A.3**

Relative Permeability

| | Material | Relative Permeability, $\mu_r$ |
|----|----------|-------------------------------|
| 1. | Diamagnetic | |
| | Bismuth | 0.999833 |
| | Mercury | 0.999968 |
| | Silver | 0.9999736 |

(*continued*)

**TABLE A.3 (CONTINUED)**

Relative Permeability

| | Material | Relative Permeability, $\mu_r$ |
|---|---|---|
| | Lead | 0.9999831 |
| | Copper | 0.9999906 |
| | Water | 0.9999912 |
| | Hydrogen (STP) | $\simeq 1.0$ |
| 2. | Paramagnetic | |
| | Oxygen (STP) | 0.999998 |
| | Air | 1.00000037 |
| | Aluminum | 1.000021 |
| | Tungsten | 1.00008 |
| | Platinum | 1.0003 |
| | Manganese | 1.001 |
| 3. | Ferromagnetic | |
| | Cobalt | 250 |
| | Nickel | 600 |
| | Soft iron | 5000 |
| | Silicon iron | 7000 |

**TABLE A.4**

Approximate Conductivity for Biological Tissue

| Material | Conductivity (S/m) | Frequency |
|---|---|---|
| Blood | 0.7 | 0 (DC) |
| Bone | 0.01 | 0 (DC) |
| Brain | 0.1 | $10^2$–$10^6$ Hz |
| Breast fat | 0.2–1 | 0.4–5 GHz |
| Breast tumor | 0.7–3 | 0.4–5 GHz |
| Fat | 0.1–0.3 | 0.4–5 GHz |
| | 0.03 | $10^2$–$10^6$ Hz |
| Muscle | 0.4 | $10^2$–$10^6$ Hz |
| Skin | 0.001 | 1 kHz |
| | 0.1 | 1 MHz |

**TABLE A.5**

Approximate Dielectric Constant for Biological Tissue

| Material | Dielectric Constant (Relative Permittivity) | Frequency |
|---|---|---|
| Blood | $10^5$ | 1 kHz |
| Bone | 3000–10,000 | 0 (DC) |
| Brain | $10^7$ | 100 Hz |
| | $10^3$ | 1 MHz |
| Breast fat | 5–50 | 0.4–5 GHz |
| Breast tumor | 47–67 | 0.4–5 GHz |
| Fat | 5 | 0.4–5 GHz |
| | $10^6$ | 100 Hz |
| | 10 | 1 MHz |
| Muscle | $10^6$ | 1 kHz |
| | $10^3$ | 1 MHz |
| Skin | $10^6$ | 1 kHz |
| | $10^3$ | 1 MHz |

## A.2 Physical Constants

| Quantity | Best Experimental Value | Approximate Value for Problem Work |
|---|---|---|
| Avogadro's number (/kg mol) | $6.0228 \times 10^{26}$ | $6 \times 10^{26}$ |
| Boltzmann constant (J/k) | $1.38047 \times 10^{-23}$ | $1.38 \times 10^{-23}$ |
| Electron charge (C) | $-1.6022 \times 10^{-19}$ | $-1.6 \times 10^{-19}$ |
| Electron mass (kg) | $9.1066 \times 10^{-31}$ | $9.1 \times 10^{-31}$ |
| Permittivity of free space (F/m) | $8.854 \times 10^{-12}$ | $\dfrac{10^{-9}}{36\pi}$ |
| Permeability of free space (H/m) | $4\pi \times 10^{-7}$ | $12.6 \times 10^{-7}$ |
| Intrinsic impedance of free space ($\Omega$) | 376.6 | $120\pi$ |
| Speed of light in free space or vacuum (m/s) | $2.9979 \times 10^8$ | $3 \times 10^8$ |
| Proton mass (kg) | $1.67248 \times 10^{-27}$ | $1.67 \times 10^{-27}$ |
| Neutron mass (kg) | $1.6749 \times 10^{-27}$ | $1.67 \times 10^{-27}$ |
| Planck's constant (J s) | $6.6261 \times 10^{-34}$ | $6.62 \times 10^{-34}$ |
| Acceleration due to gravity (m/s$^2$) | 9.8066 | 9.8 |
| Universal constant of gravitation (m$^2$/kg s$^2$) | $6.658 \times 10^{-11}$ | $6.66 \times 10^{-11}$ |
| Electron volt (J) | $1.6030 \times 10^{-19}$ | $1.6 \times 10^{-19}$ |
| Gas constant (J/mol K) | 8.3145 | 8.3 |

# Appendix B: Photon Equations, Index of Refraction, Electromagnetic Spectrum, and Wavelength of Commercial Laser

## CONTENTS

## B.1 Photon Energy, Frequency, Wavelength

| | |
|---|---|
| Photon energy (J) | Planck's constant × frequency |
| Photon energy (eV) | $\dfrac{\text{Planck's constant} \times \text{frequency}}{\text{Electron charge}}$ |
| Photon energy (cm$^{-1}$) | $\dfrac{\text{Frequency}}{\text{Speed of light in vacuum}}$ |
| Photon frequency (Hz) | $\dfrac{1\ \text{(cycle)}}{\text{Period (s)}}$ |
| Photon wavelength (μm) | $\dfrac{\text{Speed of light in free space}}{\text{Frequency}}$ |

## B.2 Index of Refraction for Common Substances

| Substance | Index of Refraction |
|---|---|
| Air | 1.000293 |
| Diamond | 2.24 |
| Ethyl alcohol | 1.36 |
| Fluorite | 1.43 |
| Fused quartz | 1.46 |
| Crown glass | 1.52 |
| Flint glass | 1.66 |
| Glycerin | 1.47 |
| Ice | 1.31 |

*(continued)*

| Substance | Index of Refraction |
|-----------|---------------------|
| Polystyrene | 1.49 |
| Rock salt | 1.54 |
| Water | 1.33 |

## B.3 Electromagnetic Spectrum

**TABLE B.1**

Approximate Common Optical
Wavelength Ranges of Light

| Color | Wavelength |
|-------|------------|
| Ultraviolet region | 10–380 nm |
| Visible region: | 380–750 nm |
| Violet | 380–450 nm |
| Blue | 450–495 nm |
| Green | 495–570 nm |
| Yellow | 570–590 nm |
| Orange | 590–620 nm |
| Red | 620–750 nm |
| Infrared | 750 nm$^{-1}$ mm |

## B.4 Wavelengths of Commercially Available Lasers

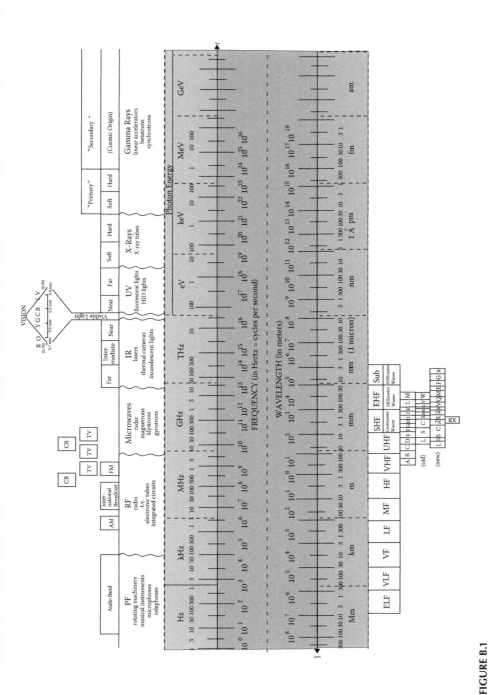

**FIGURE B.1**

Simplified chart of the electromagnetic spectrum. (From J. C. Whitaker, *The Electronics Handbook*. Boca Raton, FL: CRC Press, 1996. ISBN 0-8493-8345-5).

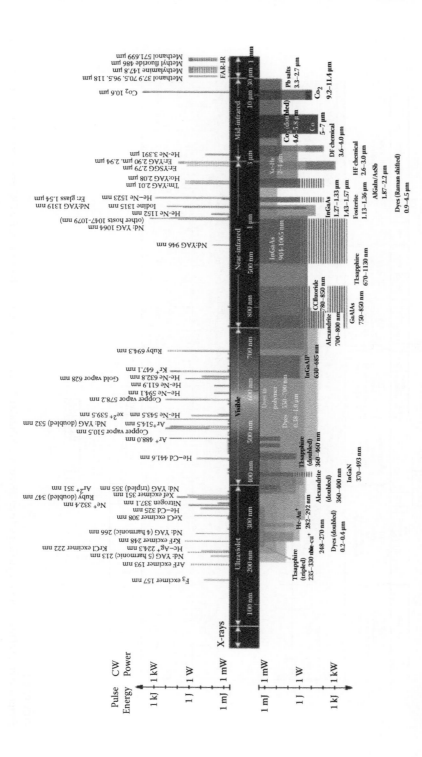

**FIGURE B.2**
Wavelengths of commercially available lasers. (From M. J. Weber, *Handbook of Laser Wavelengths*. Boca Raton, FL: CRC Press, 1999, ISBN 0-8493-3508-6.) (See color figure.)

## References

1. M. J. Weber, *Handbook of Laser Wavelengths*. Boca Raton, FL: CRC Press, 1999. ISBN 0-8493-3508-6.
2. J. C. Whitaker, *The Electronics Handbook*. Boca Raton, FL: CRC Press, 1996. ISBN 0-8493-8345-5.

# Appendix C: Symbols and Formulas

**CONTENTS**

## C.1 Greek Alphabet

| Uppercase | Lowercase | Name |
|---|---|---|
| A | $\alpha$ | Alpha |
| B | $\beta$ | Beta |
| Γ | $\gamma$ | Gamma |
| Δ | $\delta$ | Delta |
| E | $\varepsilon$ | Epsilon |
| Z | $\zeta$ | Zeta |
| H | $\eta$ | Eta |
| Θ | $\theta, \vartheta$ | Theta |
| I | $\iota$ | Iota |
| K | $\kappa$ | Kappa |
| Λ | $\lambda$ | Lambda |
| M | $\mu$ | Mu |
| N | $\nu$ | Nu |
| Ξ | $\xi$ | Xi |
| O | $o$ | Omicron |
| Π | $\pi$ | Pi |
| P | $\rho$ | Rho |
| Σ | $\sigma$ | Sigma |
| Y | $\upsilon$ | Upsilon |
| Φ | $\varphi, \phi$ | Phi |
| X | $\chi$ | Chi |
| Ψ | $\psi$ | Psi |
| Ω | $\omega$ | Omega |

## C.2 International System of Units (SI) Prefixes

| Power | Prefix | Symbol | Power | Prefix | Symbol |
|---|---|---|---|---|---|
| $10^{-35}$ | Stringo | — | $10^{0}$ | — | — |
| $10^{-24}$ | Yocto | y | $10^{1}$ | Deka | da |
| $10^{-21}$ | Zepto | z | $10^{2}$ | Hecto | h |
| $10^{-18}$ | Atto | $A$ | $10^{3}$ | Kilo | k |
| $10^{-15}$ | Femto | f | $10^{6}$ | Mega | M |
| $10^{-12}$ | Pico | p | $10^{9}$ | Giga | G |
| $10^{-9}$ | Nano | n | $10^{12}$ | Tera | T |
| $10^{-6}$ | Micro | $\mu$ | $10^{15}$ | Peta | P |
| $10^{-3}$ | Milli | m | $10^{18}$ | Exa | E |
| $10^{-2}$ | Centi | c | $10^{21}$ | Zetta | Z |
| $10^{-1}$ | Deci | d | $10^{24}$ | Yotta | Y |

## C.3 Trigonometric Identities

$$\cot\theta = \frac{1}{\tan\theta}, \quad \sec\theta = \frac{1}{\cos\theta}, \quad \csc\theta = \frac{1}{\sin\theta}$$

$$\tan\theta = \frac{\sin\theta}{\cos\theta}, \cot\theta = \frac{\cos\theta}{\sin\theta}$$

$$\sin^2\theta + \cos^2\theta = 1, \tan^2\theta + 1 = \sec^2\theta, \cot^2\theta + 1 = \csc^2\theta$$

$$\sin(-\theta) = -\sin\theta, = \cos(-\theta) = \cos\theta, \tan(-\theta) = -\tan\theta$$

$$\csc(-\theta) = -\csc\theta, \sec(-\theta) = \sec\theta, \cot(-\theta) = -\cot\theta$$

$$\cos(\theta_1 \pm \theta_2) = \cos\theta_1 \cos\theta_2 \pm \sin\theta_1 \sin\theta_2$$

$$\sin(\theta_1 \pm \theta_2) = \sin\theta_1 \sin\theta_2 \pm \cos\theta_1 \cos\theta_2$$

$$\tan(\theta_1 \pm \theta_2) = \frac{\tan\theta_1 \pm \tan\theta_2}{1 \mp \tan\theta_1 \tan\theta_2}$$

$$\cos\theta_1 \cos\theta_2 = \frac{1}{2}[\cos(\theta_1 + \theta_2) + \cos(\theta_1 - \theta_2)]$$

$$\sin\theta_1 \sin\theta_2 = \frac{1}{2}[\cos(\theta_1 - \theta_2) + \cos(\theta_1 + \theta_2)]$$

$$\sin\theta_1 \cos\theta_2 = \frac{1}{2}[\sin(\theta_1 + \theta_2) + \sin(\theta_1 - \theta_2)]$$

$$\cos\theta_1 \sin\theta_2 = \frac{1}{2}[\sin(\theta_1 + \theta_2) + \sin(\theta_1 - \theta_2)]$$

$$\sin\theta_1 + \sin\theta_2 = 2\sin\left(\frac{\theta_1 + \theta_2}{2}\right)\cos\left(\frac{\theta_1 - \theta_2}{2}\right)$$

$$\sin\theta_1 - \sin\theta_2 = 2\cos\left(\frac{\theta_1 + \theta_2}{2}\right)\sin\left(\frac{\theta_1 - \theta_2}{2}\right)$$

$$\cos\theta_1 + \cos\theta_2 = 2\cos\left(\frac{\theta_1 + \theta_2}{2}\right)\cos\left(\frac{\theta_1 - \theta_2}{2}\right)$$

$$\cos\theta_1 - \cos\theta_2 = 2\sin\left(\frac{\theta_1 + \theta_2}{2}\right)\sin\left(\frac{\theta_1 - \theta_2}{2}\right)$$

$$a \sin \theta - b \cos \theta = \sqrt{a^2 + b^2} \, \cos(\theta + \phi), \text{ where } \phi = \tan^{-1}\left(\frac{b}{a}\right)$$

$$a \sin \theta - b \cos \theta = \sqrt{a^2 + b^2} \, \sin(\theta + \phi), \text{ where } \phi = \tan^{-1}\left(\frac{b}{a}\right)$$

$$\cos(90° - \theta) \sin \theta, \ \sin(90° - \theta) = \csc \theta, \ \tan(90° - \theta) = \cot \theta$$

$$\cot(90° - \theta) \tan \theta, \ \sec(90° - \theta) = \csc \theta, \ \csc(90° - \theta) = \sec \theta$$

$$\cos(\theta \pm 90°) = \mp \sin \theta \ \sin(\theta \pm 90°) = \pm \sin \theta, \ \tan(\theta \pm 90°)$$

$$= -\cot \theta$$

$$\cos(\theta \pm 180°) = -\cos \theta \ \sin(\theta \pm 180°) = -\sin \theta, \ \tan(\theta \pm 180°)$$

$$= \tan \theta$$

$$\cos 2\theta = \cos^2 \theta - \sin^2 \theta, \ \cos 2\theta = 1 - 2 \sin^2 \theta, \ \cos 2\theta$$

$$= 2 \cos^2 \theta - 1$$

$$\sin 2\theta = 2 \sin \theta \cos \theta, \ \tan 2\theta = \frac{2 \tan \theta}{1 - \tan^2 \theta}$$

$$\cos 3\theta = 4 \cos^3 \theta - 3 \sin \theta$$

$$\sin 3\theta = 3 \sin \theta - 4 \sin^3 \theta$$

$$\sin \frac{\theta}{2} = \pm \sqrt{\frac{1 - \cos \theta}{2}}, \ \cos \frac{\theta}{2} \pm \sqrt{\frac{1 + \cos \theta}{2}},$$

$$\sin \theta = \frac{e^{j\theta} - e^{-j\theta}}{2j}, \quad \cos \theta = \frac{e^{j\theta} + e^{-j\theta}}{2} \ (j = \sqrt{-1}),$$

$$\tan \theta = \frac{e^{j\theta} + e^{-j\theta}}{j(e^{j\theta} + e^{-j\theta})}$$

$$e^{\pm j\theta} = \cos \theta \pm j \sin \theta \ \text{(Euler's identity)}$$

$$1 \text{ rad} = 57.296°$$

$$\pi = 3.1416$$

## C.4 Hyperbolic Functions

$$\cosh x = \frac{e^x + e^{-x}}{2}, \sinh x = \frac{e^x - e^{-x}}{2}, \tanh x = \frac{\sinh x}{\cosh x}$$

$$\cosh x = \frac{1}{\tanh x}, \operatorname{sech} x = \frac{1}{\cosh x}, \cosh x = \frac{1}{\sinh x}$$

$$\sin jx = j \sinh x, \cos jx = \cosh x$$

$$\sinh jx = j \sin x, \cosh jx = \cos x$$

$$\sin(x \pm jy) = \sin x \cosh y \pm j \cos x \sinh y$$

$$\cos(x \pm jy) = \cos x \cosh y \pm j \sin x \sinh y$$

$$\sinh(x \pm y) = \sinh x \cosh y \pm \cosh x \sinh y$$

$$\cosh(x \pm y) = \cosh x \cosh y \pm \sinh x \sinh y$$

$$\sinh(x \pm jy) = \sinh x \cos y \pm j \cosh x \sin y$$

$$\cosh(x \pm jy) = \cosh x \cos y \pm j \sinh x \sin y$$

$$\tanh(x \pm jy) = \frac{\sinh 2x}{\cosh 2x + \cos 2y} \pm j \frac{\sin 2y}{\cos 2x + \cos 2y}$$

$$\cosh^2 x - \sinh^2 x = 1$$

$$\operatorname{sech}^2 x + \tanh^2 x = 1$$

## C.5 Complex Variables

A complex number can be written as

$$z = x = jy = r\angle\theta = re^{j\theta} = r(\cos\theta + j\sin\theta),$$

where

$$x = \operatorname{Re} z = r \cos\theta$$

$$y = \operatorname{Im} z = r \sin\theta$$

$$r = |z| = \sqrt{x^2 + y^2}, \theta = \tan^{-1}\left(\frac{y}{x}\right)$$

$$j = \sqrt{-1}, \frac{1}{j} = -j, j^2 = -1$$

The complex conjugate of

$$z = z^* = x - jy = r\angle -\theta = re^{-j\theta} = r(\cos\theta - j\sin\theta)(e^{j\theta})^n = e^{jn\theta} = \cos n\theta + j\sin n\theta$$

(de Moivre's theorem).
If $z_1 = x_1 + jy_1$ and $z_2 = x_2 + jy_2$, then only if $x_1 = x_2$ and $y_1 = y_2$

$$z_1 \pm z_2 = (x_1 + x_2) \pm j(y_1 + y_2)$$

$$z_1 z_2 = (x_1 x_2 - y_1 y_2) + j(x_1 y_2 + x_2 y_1) = r_1 r_2 e^{j(\theta_1 + \theta_2)} = r_1 r_2 \angle \theta_1 + \theta_2$$

$$\frac{z_1}{z_2} = \frac{(x_1 + jy_1)}{(x_2 + jy_2)} \cdot \frac{(x_2 + jy_2)}{(x_2 + jy_2)} = \frac{x_1 x_2 + y_1 y_2}{x_2^2 + y_2^2} = \frac{r_1}{r_2} e^{j(\theta_1 + \theta_2)} = \frac{r_1}{r_2} \angle \theta_1 - \theta_2$$

$\ln(re^{j\theta}) = \ln r + \ln e^{j\theta} = \ln r + j\theta + j2m\pi$ ($m$ = integer)

$$\sqrt{z} = \sqrt{x + jy} = \sqrt{r}(e^{j\theta/2})\sqrt{r}\angle\theta/2$$

$$z^n = (x + jy)^n = r^n e^{jn\theta} = r^n \angle n\theta \quad (n = \text{integer})$$

$$z^{1/n} = (x + jy)^{1/n} = r^{1/n} e^{j\theta/n} = r^{1/n} \angle \theta/2 + 2\pi m / n, \quad (m = 0, 1, 2, \ldots, n-1)$$

## C.6 Table of Derivatives

| | |
|---|---|
| $y = c$ (constant) | $\dfrac{dy}{dx} = 0$ |
| $cx^n$ ($n$ any constant) | $cnx^{n-1}$ |
| $e^{ax}$ | $ae^{ax}$ |
| $ax (a > 0)$ | $a^x \ln a$ |
| $\ln x$ ($x > 0$) | $\dfrac{1}{x}$ |
| $\dfrac{c}{x^a}$ | $\dfrac{-ca}{x^{a+1}}$ |
| $\log_a x$ | $\dfrac{\log_a e}{x}$ |
| $\sin ax$ | $a \cos ax$ |

| | |
|---|---|
| cos $ax$ | $-a \sin ax$ |
| tan $ax$ | $-a\sec^2 ax = \dfrac{a}{\cos^2 ax}$ |
| cot $ax$ | $-a\csc^2 ax = \dfrac{-a}{\sin^2 ax}$ |
| sec $ax$ | $\dfrac{a \sin ax}{\cos^2 ax}$ |
| csc $ax$ | $\dfrac{-a\cos ax}{\sin^2 ax}$ |
| arcsin $ax = \sin^{-1} ax$ | $\dfrac{a}{\sqrt{1-a^2x^2}}$ |
| arccos $ax = \cos^{-1} ax$ | $\dfrac{-a}{\sqrt{1-a^2x^2}}$ |
| arctan $ax = \tan^{-1} ax$ | $\dfrac{a}{1+a^2x^2}$ |
| arccot $ax = \cot^{-1} ax$ | $\dfrac{-a}{1+a^2x^2}$ |
| sinh $ax$ | $a \cosh ax$ |
| cosh $ax$ | $a \sinh ax$ |
| tanh $ax$ | $\dfrac{a}{\cosh^2 ax}$ |
| $\sinh^{-1} ax$ | $\dfrac{a}{\sqrt{1-a^2x^2}}$ |
| $\cosh^{-1} ax$ | $\dfrac{a}{\sqrt{a^2x^2-1}}$ |
| $\tanh^{-1} ax$ | $\dfrac{a}{1-a^2x^2}$ |
| $u(x) + \upsilon(x)$ | $\dfrac{du}{dx} + \dfrac{du}{dx}$ |
| $u(x)\,\upsilon(x)$ | $u\dfrac{du}{dx} + \upsilon\dfrac{du}{dx}$ |
| $\dfrac{u(x)}{\upsilon(x)}$ | $\dfrac{1}{\upsilon^2}\left(\upsilon\dfrac{du}{dx} - u\dfrac{d\upsilon}{dx}\right)$ |
| $\dfrac{1}{\upsilon(x)}$ | $\dfrac{-1}{\upsilon^2}\dfrac{d\upsilon}{dx}$ |
| $y(\upsilon(x))$ | $\dfrac{dy}{d\upsilon}\dfrac{d\upsilon}{dx}$ |
| $y(\upsilon(u(x)))$ | $\dfrac{dy}{d\upsilon}\dfrac{d\upsilon}{dx}\dfrac{du}{dx}$ |

## C.7 Table of Integrals

$$\int a\,dx = ax + c \ (c \text{ is an arbitrary constant})$$

$$\int x\,dy = xy - \int y\,dx$$

$$\int x^n\,dx = \frac{x^{n+1}}{n+1} + c, (n \neq -1)$$

$$\int \frac{1}{x}\,dx = \ln - |x| + c$$

$$\int e^{ax}\,dx = \frac{e^{ax}}{a} + c$$

$$\int a^x\,dx = \frac{a^x}{\ln a} + c \quad \text{for} \quad (a > 0)$$

$$\int \ln x\,dx = x\ln x - x + c \quad \text{for} \quad (x > 0)$$

$$\int \sin ax\,dx = \frac{-\cos ax}{a} + c$$

$$\int \cos ax\,dx = \frac{\sin ax}{a} + c$$

$$\int \tan ax\,dx = \frac{-\ln|\cos ax|}{a} + c$$

$$\int \cot ax\,dx = \frac{-\ln|\cos ax|}{a} + c$$

$$\int \sec ax\,dx = \frac{-\ln\left(\dfrac{1-\sin ax}{1+\sin ax}\right)}{2a} + c$$

$$\int \csc ax\,dx = \frac{-\ln\left(1-\cos ax/1+\cos ax\right)}{2a} + c$$

$$\int \frac{1}{x^2 + a^2}\,dx = \frac{\tan^{-1}(x/a)}{a} + c$$

$$\int \frac{1}{x^2 - a^2}\,dx = \frac{\ln(x - a/x + a)}{2a} + c \quad \text{or} \quad \frac{\tanh^{-1}(x/a)}{a} + c$$

$$\int \frac{1}{a^2 - x^2}\,dx = \frac{\ln(x + a/x - a)}{2a} + c$$

$$\int \frac{1}{\sqrt{a^2 - x^2}}\,dx = \sin^{-1}(x/a) + c$$

$$\int \frac{1}{\sqrt{a^2 - x^2}}\,dx = \frac{\sinh^{-1}(x/a)}{a} + c \quad \text{or} \quad \ln\left(x + \sqrt{x^2 + a^2}\right) + c$$

$$\int \frac{1}{\sqrt{x^2 - a^2}}\,dx = \ln\left(x + \sqrt{x^2 + a^2}\right) + c$$

$$\int \frac{1}{x\sqrt{x^2 - a^2}}\,dx = \frac{\sec^{-1}(x/a)}{a} + c$$

$$\int xe^{ax}\,dx = \frac{(ax - 1)e^{ax}}{a^2} + c$$

$$\int x \cos ax\,dx = \frac{\cos ax + ax \sin ax}{a^2} + c$$

$$\int x \sin ax\,dx = \frac{\sin ax + ax \cos ax}{a^2} + c$$

$$\int x \ln x\,dx = \frac{x^2}{2} \ln x - \frac{x^2}{4} + c$$

$$\int xe^{ax}\,dx = \frac{e^{ax}(ax - 1)}{a^2} + c$$

$$\int e^{ax} \cos bx\,dx = \frac{e^{ax}(a \cos bx + b \sin bx)}{a^2 + b^2} + c$$

$$\int e^{ax} \sin bx\,dx = \frac{e^{ax}(-b \cos bx + a \sin bx)}{a^2 + b^2} + c$$

$$\int \sin^2 x\,dx = \frac{x}{2} - \frac{\sin 2x}{4} + c$$

$$\int \cos^2 x\,dx = \frac{x}{2} - \frac{\sin 2x}{4} + c$$

$$\int \tan^2 x\, dx = \tan x - x + c$$

$$\int \cot^2 x\, dx = \cot x - x + c$$

$$\int \sec^2 x\, dx = \tan x + c$$

$$\int \csc^2 x\, dx = -\cot x + c$$

$$\int \sec x \tan x\, dx = \sec x + c$$

$$\int \csc x \cot x\, dx = -\csc x + c$$

## C.8 Table of Probability Distributions

| 1. Discrete Distribution | Probability | Expectation (Mean) $\mu$ | Variance $\sigma^2$ |
|---|---|---|---|
| Binomial $B(n,p)$ | $\binom{n}{r} p^r (1-p)^{n-r} = \dfrac{n!\, p^r q^{n-r}}{r!(n-r)!}$, <br> $r = 0, 1, \ldots, n$ | $np$ | $np(1-p)$ |
| Geometric $G(p)$ | $(1-p)^{r-1} p$ | $\dfrac{1}{p}$ | $\dfrac{1-p}{p^2}$ |
| Poisson $p(\lambda)$ | $\dfrac{\lambda^n e^{-\lambda}}{n!}$ | $\lambda$ | $\Lambda$ |
| Pascal (negative binomial $NB(r,p)$) | $\binom{x-1}{r-1} p^r (1-p)^{x-r}$, <br> $x = r, r+1, \ldots$ | $\dfrac{r}{p}$ | $\dfrac{r(1-p)}{p^2}$ |
| Hypergeometric $H(N,n,p)$ | $\dfrac{\binom{Np}{r}\binom{N-Np}{n-r}}{\binom{N}{n}}$ | $np$ | $np(1-p)\dfrac{N-n}{N-1}$ |

| 2. Continuous Distribution | Density $f(x)$ | Expectation (Mean) $\mu$ | Variance $\sigma^2$ |
|---|---|---|---|
| Exponential $E(\lambda)$ | $\begin{cases} \lambda e^{-\lambda x}, & x \geq 0 \\ 0 & x \geq 0 \end{cases}$ | $\dfrac{1}{\lambda}$ | $\dfrac{1}{\lambda^2}$ |
| Uniform $U(a,b)$ | $\begin{cases} \dfrac{1}{b-a}, & a < x < b \\ 0, & \text{elsewhere} \end{cases}$ | $\dfrac{a+b}{2}$ | $\dfrac{(b-a)^2}{12}$ |
| Standardized normal $N(0,1)$ | $\varphi(x) = \dfrac{e^{-x^2/2}}{\sqrt{2\pi}}$ | $0$ | $1$ |
| General normal | $\dfrac{1}{\sigma} \varphi\left(\dfrac{x-\mu}{\sigma}\right)$ | $\mu$ | $\sigma^2$ |
| Gamma $\Gamma(n,\lambda)$ | $\dfrac{\lambda^n}{\Gamma(n)} x^{n-1} e^{-\lambda x}$ | $\dfrac{n}{\lambda}$ | $\dfrac{n}{\lambda^2}$ |
| Beta $\beta(p,q)$ | $a_{p,q} x^{p-1}(1-x)^{q-1}, 0 \leq x \geq 1$ $a_{p,q} = \dfrac{\Gamma(p+q)}{\Gamma(p)\Gamma(q)}, p > 0, q > 0$ | $\dfrac{p}{p+q}$ | $\dfrac{p}{(p+q)^2(p+q+1)}$ |
| Weibull $W(\lambda,\beta)$ | $\lambda^\beta \beta x^{\beta-1} e^{-(\lambda x)}, x \geq 0$ $F(x) = 1 - e^{-(\lambda x)\beta}$ | $\dfrac{1}{\lambda} \Gamma\left(1+\dfrac{1}{\beta}\right)$ | $\dfrac{1}{\lambda^2}(A-B)$ $A = \Gamma\left(1+\dfrac{2}{\beta}\right)$ $B = \Gamma^2\left(1+\dfrac{1}{\beta}\right)$ |
| Rayleigh $R(\sigma)$ | $\dfrac{x}{\sigma^2} e^{-x^2/2\sigma^2}, x \geq 0$ | $\sigma\sqrt{\dfrac{\pi}{2}}$ | $2\sigma^2\left(1-\dfrac{\pi}{4}\right)$ |

## C.9  Summations (Series)

### C.9.1  Finite element of terms

$$\sum_{n=0}^{N} a^n = \frac{1-a^{N+1}}{1-a}; \sum_{n=0}^{N} na^n = a\left(\frac{1-(N+1)a^N + Na^{N+1}}{(1-a)^2}\right)$$

$$\sum_{n=0}^{N} n = \frac{N(N+1)}{2}; \sum_{n=0}^{N} n^2 = \frac{N(N+1)(2N+1)}{6}$$

$$\sum_{n=0}^{N} n(n+1) = \frac{N(N+1)(N+2)}{3};$$

$$(a+b)^N = \sum_{n=0}^{N} NC_n a^{N-n} b^n, \quad \text{where} \quad NC_n = NC_{N-n} = \frac{NP_n}{n!} = \frac{N!}{(N-n)!n!}$$

### C.9.2  Infinite element of terms

$$\sum_{n=0}^{\infty} x^n = \frac{1}{1-x}, (|x|<1); \quad \sum_{n=0}^{\infty} nx^n = \frac{1}{(1-x)^2}, (|x|<1)$$

$$\sum_{n=0}^{\infty} n^k x^n = \lim_{a\to 0}(-1)^k \frac{\partial^k}{\partial a^k}\left(\frac{x}{x-e^{-a}}\right), (|x|<1); \sum_{n=0}^{\infty} \frac{(-1)^n}{2n+1} = 1 - \frac{1}{3} + \frac{1}{5} - \frac{1}{7} + \ldots = \frac{1}{4}\pi$$

$$\sum_{n=0}^{\infty} \frac{1}{n^2} = 1 + \frac{1}{2^2} + \frac{1}{3^2} + \frac{1}{4^2} + \cdots = \frac{1}{6}\pi^2$$

$$e^x = \sum_{n=0}^{\infty} \frac{x^n}{n!} = 1 + \frac{1}{1!}x + \frac{1}{2!}x^2 + \frac{1}{3!}x^3 + \ldots$$

$$a^x = \sum_{n=0}^{\infty} \frac{(\ln a)^n x^n}{n!} = 1 + \frac{(\ln a)x}{1!} + \frac{(\ln a)^2 x^2}{2!} + \frac{(\ln a)^3 x^3}{3!} + \cdots$$

$$\ln(1 \pm x) = \sum_{n=1}^{\infty} \frac{(\pm 1)^n x^x}{n} = \pm x - \frac{x^2}{2} \pm \frac{x^3}{3} - \cdots, \quad (|x| < 1)$$

$$\sin x = \sum_{n=1}^{\infty} \frac{(-1)^n x^{2x+1}}{(2n+1)!} = x - \frac{x^3}{3!} \pm \frac{x^5}{5!} - \frac{x^7}{7!} \cdots$$

$$\cos x = \sum_{n=0}^{\infty} \frac{(-1)^n x^{2n}}{(2n)!} = 1 - \frac{x^2}{2!} \pm \frac{x^4}{4!} - \frac{x^6}{6!} + \cdots$$

$$\tan x = x + \frac{x^3}{3} + \frac{2x^5}{15} + \cdots, \quad (|x| < 1)$$

$$\tan^{-1} x = \sum_{n=0}^{\infty} \frac{(-1)^n x^{2n+1}}{(2n+1)} = x - \frac{x^3}{3} + \frac{x^5}{5} - \frac{x^7}{7} + \cdots, \quad (|x| < 1)$$

## C.10 Logarithmic Identities

$$\log_e a = \ln a \,(\text{natural logarithm})$$

$$\log_{10} a = \ln a \,(\text{natural logarithm})$$

$$\log ab = \log a + \log b$$

$$\log \frac{a}{b} = \log a - \log b$$

$$\log a^n = n \log a$$

## C.11  Exponential Identities

$$e^x = 1 + x + \frac{x^2}{2!} + \frac{x^3}{3!} + \frac{x^4}{4!} + ..., \quad \text{where} \quad e \simeq 2.7182$$

$$e^x e^y = e^{x+y}$$

$$(e^x)^n = e^{nx}$$

$$\ln e^x = x$$

## C.12  Approximations for Small Quantities

If $|a| \ll 1$, then

$$\ln(1+a) \simeq a$$

$$e^a \simeq 1 + a$$

$$\sin a \simeq a$$

$$\cos a \simeq 1$$

$$\tan a \simeq a$$

$$(1+a)^n \simeq 1 \pm na$$

## C.13  Matrix Notation and Operations

### C.13.1  Matrices

A *matrix* is a rectangular array of elements arranged in rows and columns. The array is commonly enclosed in brackets. Let a matrix $A$ (expressed in boldface as **A** or in bracket as [$A$]) have $m$ rows and $n$ columns; then the matrix can be expressed by

$$
\mathbf{A} = [A] = \begin{bmatrix}
a_{11} & a_{12} & . & . & . & a_{1j} & . & . & . & a_{1n} \\
a_{21} & a_{22} & . & . & . & a_{2j} & . & . & . & a_{2n} \\
. & . & . & . & . & . & . & . & . & . \\
. & . & . & . & . & . & . & . & . & . \\
a_{i1} & a_{i2} & . & . & . & a_{ij} & . & . & . & a_{in} \\
. & . & . & . & . & . & . & . & . & . \\
. & . & . & . & . & . & . & . & . & . \\
. & . & . & . & . & . & . & . & . & . \\
a_{m1} & a_{m2} & . & . & . & a_{mj} & . & . & . & a_{mn}
\end{bmatrix} \dots
$$

where the element $aij$ has two subscripts, of which the first denotes to the row $i$th and the second denotes to the column $j$th which the element locates in the matrix. A matrix with $m$ rows and $n$ columns, $[A]$, is defined as a matrix of order or size $m \times n$ ($m$ by $n$), or an $m \times n$ matrix. A vector is a matrix that consists of only one row or one column.

### C.13.1.1 Location of an Element in a Matrix

$$
Let\ A = \begin{bmatrix}
a_{11} & a_{12} & a_{13} & a_{14} \\
a_{21} & a_{22} & a_{23} & a_{24} \\
a_{31} & a_{32} & a_{33} & a_{34} \\
a_{41} & a_{42} & a_{43} & a_{44}
\end{bmatrix}\ \text{is matrix with size } 4 \times 4
$$

where
$a_{11}$ is the element a at row 1 and column 1.
$a_{12}$ is the element a at row 1 and column 2.
$a_{32}$ is the element a at row 3 and column 2.

### C.13.2 Special Common Types of Matrices

a. If $m \neq n$, then the matrix $[A]$ is called *rectangular matrix.*
b. If $m = n$, then the matrix $[A]$ is called *square matrix of order n.*
c. If $m = 1\ and\ n > 1$, then the matrix $[A]$ is called *row matrix or row vector.*
d. If $m > 1\ and\ n = 1$, then the matrix $[A]$ is called *column matrix or column vector.*
e. If $m = 1\ and\ n = 1$, then the matrix $[A]$ is called *a scalar.*

f. A *real matrix* is a matrix whose elements are all real.

g. A *complex matrix* is a matrix whose elements may be complex.

h. A *null matrix* is a matrix whose elements are all zero.

i. An *identity* (or *unit*) *matrix*, [*I*] or **I**, is a square matrix whose elements are equal to zero except those located on its *main diagonal* elements, which are unity (or one). *Main diagonal* elements have equal row and column subscripts. The main diagonal runs from the upper left corner to the lower right corner. If the elements of an identity matrix are denoted as $e_{ij}$, then

$$e_{ij} = \begin{cases} 1 & i = j \\ 0 & i \neq j \end{cases}$$

j. A *diagonal matrix* is a square matrix which has zero elements everywhere except on its main diagonal. That is, for diagonal matrix, $a_{ij} = 0$ when $i \neq j$ and not all $a_{ii}$ are zero.

k. A *symmetric matrix* is a square matrix whose elements satisfy the condition $a_{ij} = a_{ji}$ for $i \neq j$.

l. An *antisymmetric* (or *skew symmetric*) *matrix* is a square matrix whose elements $a_{ij} = -a_{ji}$ for $i \neq j$, and $a_{ii} = 0$.

m. A *triangular matrix* is a square matrix whose all elements on one side of the diagonal are zero. There are two types of triangular matrices: first, an upper triangular **U** whose elements below the diagonal are zero, and second, a lower triangular **L**, whose elements above the diagonal are all zero.

n. A *partitioned* (or *block*) *matrix* is a matrix that is divided by horizontal and vertical lines into smaller matrices called *submatrices* or *blocks*.

### C.13.3 Matrix Operations

#### C.13.3.1 Transpose of a Matrix

The *transpose* of a matrix $\mathbf{A} = [a_{ij}]$ is donated as $\mathbf{A}^T = [a_{ji}]$ and is obtained by interchanging the rows and columns in matrix **A**. Thus, if a matrix **A** is of order $m \times n$, then $\mathbf{A}^T$ will be of order $n \times m$.

#### C.13.3.2 Addition and Subtraction

Addition and subtraction can only be performed for matrices of the same size. The addition is accomplished by adding corresponding elements of each matrix. For addition, $\mathbf{C} = \mathbf{A} + \mathbf{B}$ implies that $c_{ij} = a_{ij} + b_{ij}$.

Now, the subtraction is accomplished by subtracting corresponding elements of each matrix. For subtraction, $C = A - B$ implies that $c_{ij} = a_{ij} - b_{ij}$ where $c_{ij}, a_{ij},$ and $b_{ij}$ are typical elements of the $C$, $A$, and $B$ matrices, respectively.

Both $A$ and $B$ matrices are in the same size $m \times n$. The resulting matrix $C$ is also of size $m \times n$.

Matrix addition and subtraction are associative:

$$A + B + C = (A + B) + C = A + (B + C)$$

$$A + B - C = (A + B) - C = A + B(B - C)$$

Matrix addition and subtraction are commutative:

$$A + B = B + A$$

$$A - B = -B + A$$

### C.13.3.3 Multiplication by Scalar

A matrix is multiplied by a scalar by multiplying each element of the matrix by the scalar. The multiplication of a matrix $A$ by a scalar $c$ is defined as

$$cA = [ca_{ij}]$$

The scalar multiplication is commutative.

### C.13.3.4 Matrix Multiplication

The product of two matrices is $C = AB$ if and only if the number of columns in $A$ is equal to the number of rows in $B$. The product of matrix $A$ of size $m \times n$ and matrix $B$ of size $n \times r$ results in matrix $C$ of size $m \times r$. Then, $c_{ij} = \sum_{k=1}^{n} a_{ik} b_{kj}$. That is, the $(ij)$th component of matrix $C$ is obtained by taking the dot product

$$c_{ij} = (i\text{th row of } A) \cdot (j\text{th column of } B)$$

Matrix multiplication is associative:

$$ABC = (AB)C = A (BC)$$

Matrix multiplication is distributive:

$$A (B + C) = AB + AC$$

Matrix multiplication is not commutative:

$$AB \neq BA$$

### C.13.3.5  Transpose of Matrix Multiplication

The transpose of matrix multiplication is usually denoted $(AB)^T$ and is defined as

$$(AB)^T = B^T A^T$$

### C.13.3.6  Inverse of Square Matrix

The inverse of a matrix $A$ is denoted by $A^{-1}$. The inverse matrix satisfies

$$AA^{-1} = A^{-1}A = I$$

A matrix that possesses an inverse is called *nonsingular matrix* (or *invertible matrix*). A matrix without an inverse is called a *singular matrix*.

### C.13.3.7  Differentiation of a Matrix

The differentiation of a matrix is differentiation of every element of the matrix separately. To emphasize, if the elements of the matrix $A$ are a function of $t$, then

$$\frac{dA}{dt} = \left[ \frac{da_{ij}}{dt} \right]$$

### C.13.3.8  Integration of a Matrix

The integration of a matrix is integration of every element of the matrix separately. To emphasize, if the elements of the matrix $A$ are a function of $t$, then

$$\int A \, dt = \left[ \int a_{ij} \, dt \right]$$

### C.13.3.9  Equality of Matrices

Two matrices are equal if they have the same size and their corresponding elements are equal.

### C.13.4 Determinant of a Matrix

The determinant of a square matrix $\mathbf{A}$ is a scalar number denoted by $|\mathbf{A}|$ or det $\mathbf{A}$.

The value of a second-order determinant is calculated from

$$\det \begin{bmatrix} a_{11} & a_{12} \\ a_{21} & a_{22} \end{bmatrix} = \begin{vmatrix} a_{11} & a_{12} \\ a_{21} & a_{22} \end{vmatrix} = a_{11}a_{22} - a_{12}a_{21}$$

By using the sign rule of each term, the determinant is determined by the first row in the diagram $\begin{vmatrix} + & - & + \\ - & + & - \\ + & - & + \end{vmatrix}$.

The value of a third-order determinant is calculated in form

$$\det \begin{bmatrix} a_{11} & a_{12} & a_{13} \\ a_{21} & a_{22} & a_{23} \\ a_{31} & a_{32} & a_{33} \end{bmatrix} = \begin{vmatrix} a_{11} & a_{12} & a_{13} \\ a_{21} & a_{22} & a_{23} \\ a_{31} & a_{32} & a_{33} \end{vmatrix}$$

$$= a_{11} \begin{vmatrix} a_{22} & a_{23} \\ a_{32} & a_{33} \end{vmatrix} - a_{12} \begin{vmatrix} a_{21} & a_{23} \\ a_{31} & a_{33} \end{vmatrix} + a_{13} \begin{vmatrix} a_{21} & a_{22} \\ a_{31} & a_{32} \end{vmatrix}$$

## C.14 Vectors

1. Vector derivative

a. Cartesian coordinates

| | |
|---|---|
| Coordinates | $(x,y,z)$ |
| Vector | $A = A_x\, a_x + A_y\, a_y + A_z\, a_z$ |
| Gradient | $\nabla A = \dfrac{\partial A}{\partial x} a_x + \dfrac{\partial A}{\partial y} a_y + \dfrac{\partial A}{\partial z} a_z$ |
| Divergence | $\nabla \cdot A = \dfrac{\partial A_x}{\partial_x} + \dfrac{\partial A_y}{\partial_y} + \dfrac{\partial A_z}{\partial_z}$ |

Curl

$$\nabla \times A = \begin{vmatrix} a_x & a_y & a_z \\ \dfrac{\partial}{\partial_x} & \dfrac{\partial}{\partial_y} & \dfrac{\partial}{\partial_z} \\ A_x & A_y & A_z \end{vmatrix}$$

$$= \left( \frac{\partial A_z}{\partial_y} - \frac{\partial A_y}{\partial_z} \right) a_x + \left( \frac{\partial A_x}{\partial_z} - \frac{\partial A_z}{\partial_x} \right) a_y + \left( \frac{\partial A_y}{\partial_x} - \frac{\partial A_x}{\partial_y} \right) a_z$$

Laplacian

$$\nabla^2 A = \frac{\partial^2 A}{\partial x^2} + \frac{\partial^2 A}{\partial y^2} + \frac{\partial^2 A}{\partial z^2}$$

## b. Cylindrical coordinates

Coordinates     $(\rho, \phi, z)$

Vector     $A = A_\rho a_\rho + A_\phi a_\phi + A_z a_z$

Gradient     $\nabla A = \dfrac{\partial A}{\partial \rho} a_\rho + \dfrac{1}{\rho} \dfrac{\partial A}{\partial \phi} a_\phi + \dfrac{\partial A}{\partial z} a_z$

Divergence     $\nabla \cdot A = \dfrac{1}{\rho} \dfrac{\partial}{\partial \rho} (\rho A_\rho) + \dfrac{\partial A_\phi}{\partial \phi} + \dfrac{\partial A_z}{\partial z}$

Curl

$$\nabla \times A = \frac{1}{\rho} \begin{vmatrix} a_\rho & \rho a_\phi & a_z \\ \dfrac{\partial}{\partial \rho} & \dfrac{\partial}{\partial \phi} & \dfrac{\partial}{\partial z} \\ A_\rho & \rho A_\phi & A_z \end{vmatrix}$$

$$= \left( \frac{1}{\rho} \frac{\partial A_z}{\partial \phi} - \frac{\partial A_\phi}{\partial z} \right) a_\rho + \left( \frac{\partial A_\rho}{\partial z} - \frac{\partial A_z}{\partial \rho} \right) a_\phi + \frac{1}{\rho} \left( \frac{\partial}{\partial x} (\rho A_\phi) - \frac{\partial A_\rho}{\partial \rho} \right) a_z$$

Laplacian     $\nabla^2 A = \dfrac{1}{\rho} \dfrac{\partial}{\partial \rho} \left( \rho \dfrac{\partial A}{\partial \rho} \right) + \dfrac{1}{\rho^2} \dfrac{\partial^2 A}{\partial \phi^2} + \dfrac{\partial^2 A}{\partial z^2}$

## c. Spherical coordinates

Coordinates     $(r, \theta, \phi)$

Vector     $A = A_r a_r + A_\theta a_\theta + A_\phi a_\phi$

Gradient     $\nabla A = \dfrac{\partial A}{\partial r} a_r + \dfrac{1}{r} \dfrac{\partial A}{\partial \theta} a_\theta + \dfrac{1}{r \sin \theta} \dfrac{\partial A}{\partial \phi} a_\phi$

Divergence     $\nabla \cdot A = \dfrac{1}{r^2} \dfrac{\partial}{\partial r} \left( r^2 A_r \right) + \dfrac{1}{r \sin \theta} \dfrac{\partial}{\partial \theta} (A_\theta \sin \theta) + \dfrac{1}{r \sin \theta} \dfrac{\partial A_\phi}{\partial \phi}$

$$\nabla \times A = \frac{1}{r^2 \sin\theta} \begin{vmatrix} a_r & ra_\theta & (r\sin\theta)a_\phi \\ \dfrac{\partial}{\partial r} & \dfrac{\partial}{\partial \theta} & \dfrac{\partial}{\partial \phi} \\ A_r & rA_\theta & (r\sin\theta)A_\phi \end{vmatrix}$$

Curl

$$= \frac{1}{r\sin\theta}\left(\frac{\partial}{\partial\theta}(A_\phi \sin\theta) - \frac{\partial A_\theta}{\partial\phi}\right)a_r +$$

$$\frac{1}{r}\left(\frac{1}{\sin\theta}\frac{\partial A_r}{\partial\phi} - \frac{\partial}{\partial r}(rA_\phi)\right)a_\theta + \frac{1}{r}\left(\frac{\partial}{\partial r}(rA_\theta) - \frac{\partial A_r}{\partial\theta}\right)a_\phi$$

Laplacian $\quad \nabla^2 A = \dfrac{1}{r^2}\dfrac{\partial}{\partial r}\left(r^2 \dfrac{\partial A}{\partial r}\right) + \dfrac{1}{r^2 \sin\theta}\dfrac{\partial}{\partial\theta}\left(\sin\theta \dfrac{\partial A}{\partial\theta}\right) + \dfrac{1}{r^2 \sin\theta}\dfrac{\partial^2 A}{\partial\phi^2}$

2. Vector identity

   a. Triple products

   $$\mathbf{A}\ (\mathbf{B} \times \mathbf{C}) = \mathbf{B}\ (\mathbf{C} \times \mathbf{A}) = \mathbf{C} \cdot (\mathbf{A} \times \mathbf{B})$$

   $$\mathbf{A}\ (\mathbf{B} \times \mathbf{C}) = \mathbf{B}(\mathbf{A} \cdot \mathbf{C}) - \mathbf{C}(\mathbf{A} \cdot \mathbf{B})$$

   b. Product rules

   $$\nabla(fg) = f(\nabla g) + g(\nabla f)$$

   $$\nabla(\mathbf{A} \cdot \mathbf{B}) = \mathbf{A} \times (\nabla \times \mathbf{B}) + \mathbf{B} \times (\nabla \times \mathbf{A}) + (\mathbf{A} \cdot \nabla)\mathbf{B} + (\mathbf{B} \times \nabla)\mathbf{A}$$

   $$\nabla \cdot (f\!A) = f(\nabla \cdot \mathbf{A}) + \mathbf{A} \cdot (\nabla f)$$

   $$\nabla(\mathbf{A} \times \mathbf{B}) = \mathbf{B} \cdot (\nabla \times \mathbf{B}) - \mathbf{A} \cdot (\nabla \times \mathbf{B})$$

   $$\nabla \times (f\mathbf{A}) = f(\nabla \times \mathbf{A}) - \mathbf{A} \times (\nabla f) = \nabla \times (f\!A) = f(\nabla \times \mathbf{A}) + (\nabla f) \times \mathbf{A}$$

   $$\nabla \times (\mathbf{A} \times \mathbf{B}) = (\mathbf{B} \cdot \nabla)\mathbf{A} - (\mathbf{A} \cdot \nabla)\mathbf{B} + \mathbf{A}(\nabla \cdot \mathbf{B}) - (\nabla \cdot \mathbf{A})$$

   c. Second derivative

   $$\nabla \cdot (\Delta \times \mathbf{A}) = 0$$

   $$\nabla \times (\nabla f) = 0$$

   $$\nabla \cdot (\nabla f) = \nabla^2 f$$

   $$\nabla \times (\Delta \times \mathbf{A}) = \nabla \cdot (\Delta \cdot \mathbf{A}) - \nabla^2 \mathbf{A}$$

d.   Addition, division, and power rules

$$\nabla(f + g) = \nabla f + \nabla g$$

$$\nabla(\mathbf{A} + \mathbf{B}) = \nabla \cdot \mathbf{A} + \nabla \cdot \mathbf{B}$$

$$\nabla \times (\mathbf{A} \times \mathbf{B}) = \nabla \times \mathbf{A} + \nabla \times \mathbf{B}$$

$$\nabla\left(\frac{f}{g}\right) = \frac{g(\nabla f) - f(\nabla g)}{g^2}$$

$$\nabla f^n = nf^{n-1}\nabla f \ (n = \text{integer})$$

3. Fundamental theorems
   a.   Gradient theorem

$$\int_a^b (\nabla f) \cdot dl = f(b) - f(a)$$

   b.   Divergence theorem

$$\int_{volume} (\nabla \cdot \mathbf{A}) dv = \oint_{surface} \mathbf{A} \cdot ds$$

   c.   Curl (Stokes) theorem

$$\int_{surface} (\nabla \times \mathbf{A}) \cdot ds = \oint_{line} \mathbf{A} \cdot dl$$

   d.   $$\oint_{line} f\, dl = - \oint_{surface} \nabla f \times ds$$

   e.   $$\oint_{surface} f\, ds = - \oint_{volume} \nabla f\, dv$$

   f.   $$\oint_{surface} \mathbf{A} \times ds = - \int_{volume} \nabla \times \mathbf{A}\, dv$$

# Index